AF344428

FROM THE
BIG BANG
TO THE
HUMAN PREDICAMENT

Outline of an Ultimate Evolutionary Synthesis

By Nikolai Eberhardt

Pentland Press, Inc.
England•USA•Scotland

PUBLISHED BY PENTLAND PRESS, INC.
5122 Bur Oak Circle, Raleigh, North Carolina 27612
United States of America
919-782-0281

ISBN 1-57197-101-7
Library of Congress Catalog Card Number 97-075983

Eberhardt, Nikolai.
From the Big Ban to the human predicament : outline
of an ultimate evolutionary synthesis / Nikolai Eberhardt. –
1st ed.
p. cm.
Includes bibliographical references and index.
Preassigned LCCN: 97-75983
ISBN: 1-57197-101-7

1. Nervous system–Evolution. 2. Evolution (Biology)
3. Comparative neurobiology. 4. Neurophysiology.
5. Neuropsychology. I. Title.

QP356.E34 1998   156
                                 QBI98-788

To my children and granchildren as a guide to survival
in the interesting times ahead,

and to all who are on the quest of building a better future
for humankind on this planet.

TABLE OF CONTENTS

PART II: BEYOND HUMAN NATURE

Acknowledgements

I have to thank the late Donald T. Campbell for a great deal of intellectual challenge, especially through the faculty seminar on cultural evolution that he conducted in the fall of 1990, and during the seminar that we both led on the contents of this book while it was in the process of being written. He, the doctorate students, and the many undergraduates who later took my courses all made a great contribution by giving me the opportunity to try out on them the bulk of my ideas for the first time, to receive critique, and occasionally to encounter well-deserved bafflement. Robert Heilbroner was so kind to read the manuscript, to encourage me, and to provide some very constructive critique. My thanks also go to Francis Crick, who wrote some brief comments on an early version, and to Paul Brutsche (of the C. G. Jung Institute in Zurich), who talked with me about Jung and disagreed.

I also have to thank the many colleagues and friends with whom I had the opportunity to discuss the matters that have become this book. Not all of them have agreed with my views, but all have contributed to its progress. Among them are Kathy Alpaugh, Lois Banta-Eberhardt, Nicholas Balabkins, Mark Bickhard, Silvio Eberhardt, John Karakash, and Arturs and Dzintra Kalnins.

My thanks also go out to all who were so kind as to share their drawings or photographs with me, thus significantly improving this work: Irven and Nancy DeVore, Frans DeWaal, Jane Goodall, David Hubel and Torsten Wiesel, Tuevo Kohonen, Ricardo Martinez Murillo (of the Cajal Institute), Ken Miller and Michael Stryker, and John E. Yellen.

However, nobody but myself is responsible for what this book has become.

Introduction

This book is about fulfilling a dream that is at least three thousand years old. It mainly has been dreamt by thinkers within the succession of civilizations called "Western," but by no means was it absent in other cultures. It is the dream to understand and to explain things as they really are, and not to be satisfied with a fabric of invented stories passed on to soothe the anxieties and discomforts of a life with no known purpose.

This dream, however, is quite different from today's Holy Grail of science: grand unification, a mathematical construct designed to prove that all phenomena can be reduced to one "mother of all equations." This may be challenging to some. But it is like the challenge of a game of chess. It does not touch on what counts in the reality of a mostly troubled human existence. Besides, very likely, the mysteries of the universe are beyond the reach of our naturally evolved mind. They cannot be solved in a real sense because our brain is not supernatural. It clicks away mechanically like the rest of that, admittedly, marvelous enterprise of nature.

But the mysteries of the human condition are an entirely different matter. What would count here are answers to questions like these: Why is human history such a violent, bloody mess? What can we do to make life more pleasant? Why is there religion and religious fanaticism? On a more abstract, but equally important, level are other questions: How can we know? What is truth? Does free will exist? To consider these, cosmology will be found to have only marginal value. Here, the mechanics of evolution and of the brain are the obvious points of departure, and they are leading us to surprising discoveries about ourselves and our predicament.

In the current intellectual climate, such questions make people throw up their arms in bafflement. The questions generally are left to philosophers, who talk much but do not answer them either. Let me call such issues "existential questions." Real answers, if any, can only come when all relevant results of the natural sciences are recognized as building blocks of a unified theory of ourselves, of our relation to the world, and of our society. But such unified theory, today, is in a dismal state. Despite the phenomenal progress in the hard natural sciences during this past century, there is no coherent view available that takes full advantage of that huge body of knowledge. Instead, one finds philosophical haggling between adherents of various "schools of thought," by specialists whose horizon, by necessity, has remained limited, because to penetrate the depth of any advanced field today is already in itself the work of a lifetime. As a result, we know a lot, but understand little.

In psychology, not long ago, the behaviorists still fought with the ethologists. The former claimed the human mind to be a "clean slate" at birth; the latter, on the contrary, studied innate (instinctual) behavior. Then sociobiology wanted to supersede both, but many of its proponents seem to have fallen for a simplistic genetic determinism. The former behaviorists have become "cognitive scientists," often with a vacuous fundamental belief—that the mind, somehow, (of all things!) resembles a computer program.

In physics, the great success of quantum theory and relativity has led to a kind of epistemological sleep (with the exception of Gerhard Vollmer, whose work, however, has largely remained unrecognized). It's as if the rigors of physical research made obsolete any questions about the nature of physical insight and its limits. This lack of epistemological contemplation has opened the gates to "postmodernist" critique, which has put itself completely outside of the rigorous intellectual tradition of the past, drawing instead on the subjectivity of existential philosophy and declaring science merely a "social construct." I think this antirational fashion is a direct consequence of the lack of a unifying synthesis.

The purpose of this book is to show that such a unifying synthesis is possible and that the existential questions, today, for the first time may be given real answers from the natural sciences. They quite naturally emerge from evolutionary biology, animal behavior (ethology), ape research, neuroscience of the brain, and self-adapting artificial neural networks. In this synthesis, a unifying conceptual framework may be seen in evolutionary epistemology.

Epistemology (or the theory of how we know) traditionally has been a branch of philosophy, and thus had to remain speculative at best. But now, following the ideas of Konrad Lorenz and Donald T. Campbell, it has become possible to give it an empirical foundation by studying the insightful adaptation in animals throughout the ascending phyla. In this framework, evolution can be seen as a process of knowledge-gathering, of which we are the ultimate and most advanced heirs. That is why searching for pure knowledge—or shall I say "self-orientation"—is innate to the human species. This search was powerful enough that Socrates chose to die for it, Galileo was put on trial, and others let themselves be burnt at the stake. This quest, so dominant in Western civilization in our days, after two and a half thousand years of toil, is finally nearing fulfillment[i]. If that is so, we will have understood our predicament, adapted to the realities of our existence and to the realities of our small planet, and hence added much to our survival fitness. Then there will be a future for civilized humanity on Earth. Otherwise, there will be none.

So far, however, uncertainty reigns supreme in human matters, and new insights are sorely needed. Our technology-based, market-driven civilization has revealed some deep-rooted problems. Cultural decline is observed worldwide, and the dangers of overpopulation and ecosystem collapse seem to be imminent. Natural science has changed its purpose. Most researchers I know complain that they do not work on what would really interest them, but rather on where the money is. Today's science is not driven by the quest for knowledge, but by the "invisible hand" of a technology-driven market, whose values are not necessarily

---

[i] Recently E.O. Wilson, the father of sociobiology, has expressed similar views in his book: *Consilience: The Unity of Knowledge* (Alfred A. Knopf, 1998.) This was not available as of this writing.

human. More and more, our fateful course begins to look like a Faustian bargain. What previously were genuine seekers of truth are made to metamorphose into another type, already cast by the genius of Goethe: Faust's assistant Wagner, the intelligence with a narrow horizon, occupied with trivia that have no bearing to the Promethean zeal of an ascending humanity.

In more detail, what are the main insights that have made the ultimate evolutionary synthesis possible? The first actually comes from the philosophical foundation to it all, which has been much strengthened by the scientific progress of this century. It is best described by the word "physicalism," which states that in understanding the workings of the universe, and of ourselves, there is no inherent need for an agent or force outside of the dynamic interactions experimentally studied in the physical sciences. Nevertheless, in the strict sense of Karl Popper's critique, the above sentence is not a statement of science, but has to remain a statement of a foundational belief that is neither testable nor refutable in principle. As such, it has been attacked by spirit-matter dualists, who have the need to believe in a "prime intelligence," a creator, or whatever mystical entity arranges or intervenes, or has intervened in such a way as to make this miraculous enterprise of cosmos and life work. Two typical representatives of such Platonic dualism have been C. Lloyd Morgan (1852-1936), who believed that life "emerges" from such realms of the eternal, and Henry Bergson (1859-1941), the philosopher of "vitalism."

Life, indeed, is so complex, and at the same time such a beautifully ordered system, that it is difficult to see that it all came about as a result of merely physical dynamics. This intuitive difficulty has been the main pillar of the critique of physicalism. But now the situation has very much changed with the advent of simulations on fast computers. As it turns out, emergence of ordered complexity, such as life, is an inherent characteristic of the dynamic of systems of many interacting parts, also known as "complex systems." Even such central features of life as sexual reproduction, parasitism, and symbiosis are seen to emerge spontaneously from the logic of simple rules. Thus, the main critique of physicalism has been invalidated, even though it (or any other philosophical foundation) can never be proven with certainty. It can only be strengthened or made more convincing. The present evolutionary synthesis intends to further increase the level of confidence in physicalism: If all of life's complexity, including human intelligence and art, even consciousness, can be explained in principle on grounds of physicalism—wouldn't this amount to an almost absolute level of confidence in the approach? My readers, indeed, should expect this to be one of the results of carrying further this synthesis.

Since the reader is invited to participate in an intellectual journey through many continents of knowledge, I include in figure 0.1 a road map, a rough outline of how the synthesis will unfold in its major steps. This should help to visualize what is being developed on the next two or three pages. It is also hoped that later on, while studying this not undemanding text, my reader will occasionally turn

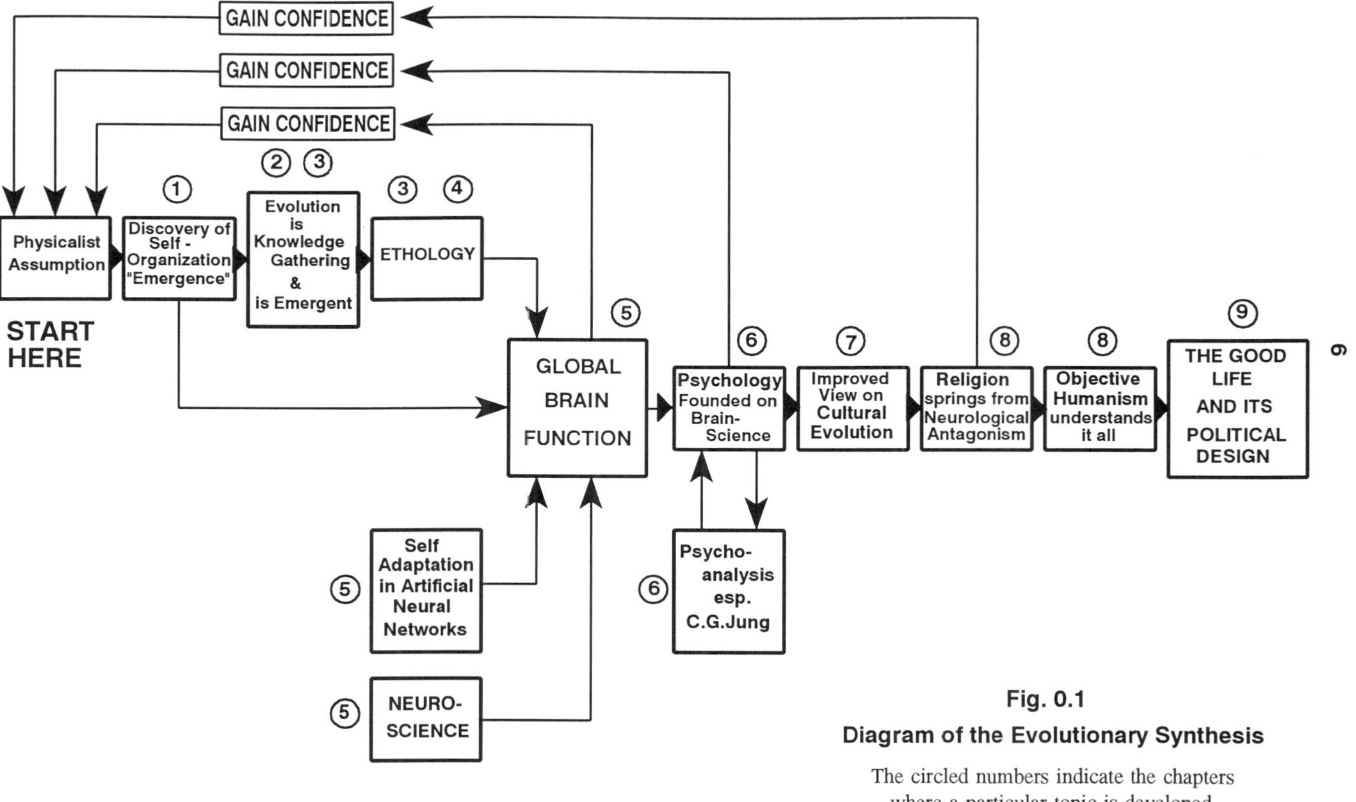

Fig. 0.1

Diagram of the Evolutionary Synthesis

The circled numbers indicate the chapters
where a particular topic is developed.

back and consult this map in order to take bearings as to how any particular later topic will relate to the whole enterprise.

Any synthesis is built from parts and in steps. After an introduction to some recent insights about self-organization that shed new light on the physics of life (chapter 1), I will begin by showing the coherence, or continuity, of all biological evolution up to the human level, especially of such mental characteristics as cognition and will. This mainly will be based on ethology, the study of animal behavior throughout the ascending phyla (chapters 2, 3, and 4). Particularly, the insightful behavior of primates will form an important link between animal and human minds.

The second and central building block of the synthesis is a global understanding of the mechanics of brain function (chapter 5). Here, the results of neuroscience can be nicely sorted out and made coherent with the knowledge provided by ethology. After all, evolution of the brain and evolution of behavior are parallel, and both can be studied in surviving species representing various stages of phylogeny. But the main revelation here comes from another application of fast computers: the amazing properties of artificial neural networks that mimic the real networks of the brain. It turns out that such networks have the ability, among other feats, to spontaneously self-adapt to the task of recognizing and classifying externally provided information, such as the information coming from the sense organs. With this in mind, together with the first building block, careful study of brain function reveals the human being as a mechanically functioning entity, even if an enormously complex one, who has the ability to functionally adapt to the external world in order to survive in it and master it. There also is an innate driving force, or "will" that has been phylogenetically acquired. This primarily resides in some genetically determined neural networks of the brain stem and from here exerts a certain amount of control over the process of adaptation. The cognitive, self-adapting networks primarily reside in the neocortex. They can be considered intelligent, even though in the animal world they are subservient to the brain stem's "will." Therefore our individuality appears to be a twofold result of history: a millions-of-years-old genetic history and a short personal history that, to some extent, shapes us according to the environment and culture we are born into. By the way, functional brain architecture provides the final resolution of the old nature-versus-nurture problem.

Ethology and brain science, together, can lay a better foundation for psychology (chapter 6). This is a third building block and the next step in the evolutionary synthesis. Here, an unexpected revelation occurs: The discoveries fit smoothly into psychoanalytic theory of the Jungian kind, as long as this is stripped of its mystical (dualist) component. This is highly satisfying and revealing of the powers of the human mind. It shows that this major and influential intellectual enterprise of a century has not been in vain, even though some dead ballast needs to be shed.

As for the objective biological foundations of human existence, with the inclusion of insights from ethology and neuroscience, at this point the evolutionary synthesis of human nature will have come full cycle. The physicalist assumption will have been lifted to a high degree of certainty by an overwhelming number of detailed results of experimental natural science that all fit this larger and highly satisfying model. Human nature can be considered understood.

But beyond human nature, the second defining feature of humanity is that it generates culture and cultural tradition, of which religion is an important ingredient. Therefore the second part of this book is devoted to explaining, on the basis of the newly-found understanding of human nature, why this is so; to explaining the origins of cultural universals and religion, and to explaining why human history so far has been such an unfortunate and bloody affair (chapter 7). Ultimately this should lead to the discovery of what the good life may be, and how this may be fostered by better political design and through effectively spreading the new insights, together with their moral empowerment. This very much looks like the fulfillment of the long quest of Western civilization.

In chapter 8, religion will be a major topic. Note that in figure 0.1 even from here an arrow feeds back, adding to confidence in the physicalist assumption. That is because religion is found to originate from an opposition between the ancient world of the innate brain stem "will" and the learned neocortical attachment to the present-day world. No spiritualist assumptions are necessary. This affirms and explains the inner need for religion and defines its purpose. But it also strongly questions the prevailing view and practice of religion today. Nevertheless, I believe that some of the more radical theologians of the day may not find much ground to quarrel with my views.

With this in mind, anyone should be awed by the Promethean dimensions of the evolutionary synthesis. It is risky and is related to the mythologically forbidden fruit from the tree of knowledge. The path of humanity toward objective knowledge and enlightenment through the natural sciences calls for great courage and responsibility, because traditional religion, for many, is still a life-sustaining force. Any revolution that destroys, but does not have a better order to offer, actually would be a step backwards.

However, I myself have deeply experienced, and will try to demonstrate to my readers at the end of chapter 8, that the humanism that results from understanding ourselves objectively, as well as understanding our place as the pinnacle of this mysterious, emerging nature, causes a quasi-religious conversion experience. This induces more ethical conviction and responsibility than any religion was ever able to muster before. Perhaps one should not lament too much about a possible demise of religion as we know it today. With it would disappear religious wars, murderous fanaticism, ugly intolerance, and an irresponsibility founded on the illusion that someone else will take care of things for us. I also think that creative religious minds may very well be able to assimilate yet another strong message from the world of objective knowledge and find new ways to celebrate the uniqueness of the human spirit and its natural place of responsibility—without engaging in self-deception. One striking example of such creativity is the German renegade Catholic theologian Eugen Drewermann, who is hailed as a new Luther by some. Not only do his views seem to be in harmony with what this book will develop from a completely different point of departure; also, in the name of better self-orientation, he has called for a scientific synthesis of the kind presented here (Schönborn 1993). He has even outlined the main ingredients: complexity theory, ethology, neuroscience, depth psychology á la C. G. Jung, and an explanation of human culture based on these. Not unexpectedly, however, Drewermann remains a spirit-matter dualist. Theology still has a long way to go before really assimilating science.

In order to further whet the reader's appetite, I will now give a preview of some of the central insights to be derived. There are two pillars on which all conclusions rest.

The First Pillar

Human consciousness has evolved by Darwinian selection as an assembly machine of incomplete pieces of sensory information about the situation in the external world. The machine constructs a global picture that may be far from correct. Nevertheless, this picture permits responses and actions that are more realistic than without it. This is the survival value of consciousness.

The neuro-psychologist Michael Gazzaniga has called this facility the "central interpreter-storyteller." In this book a different term is introduced—"the emotional puzzle assembler"—because the global picture is like a puzzle assembled from pieces. Some are missing; others are made to fit by force. The assembler is "emotional" because the process is under the control of the basic survival will (in the brain stem) that is not rational, but instinctual, and the stories weaved have to have an emotional content, so that they can drive on to forceful action that insures survival.

This machine works according to principles known and well-studied in self-adapting artificial neural networks. Its working is associated with our inner experience of "being conscious." This experience is not different in principle from being hungry or cold, and as such can never be a subject of physics. Only the underlying physiology can be studied by methods of natural science. It will be substantiated that nothing supernatural need be associated with the existence of consciousness, nor is it necessary to invoke some of the mysteries of modern quantum physics, as some noted physicists presently are doing. They can only add to the prevailing confusion.

The Second Pillar

The most basic functional dynamic in all animal brains that have sprouted a neocortex is based on a "dualism" or "antagonism" between brain stem and neocortex. This is a consequence of the fact that the brain stem was much older and the only organ of central control in early organisms. Here, the responses were almost purely genetically determined and hardwired. They were "instinctual." The neocortex, on the other hand, came into existence as an organ of individual adaptation to the concurrent external world. Thus, the center of the "will" remained in the brain stem; "cognition," with its associated memory and adaptive intelligence, mainly happens in the neocortex.

Even though this is a natural, functional and healthy antagonism in all brains that have developed a neocortex, in modern humans the balance has radically tilted in favor of the neocortex, particularly of the left hemisphere consciousness. This has happened because consciousness has become such an overwhelming success in mastering our physical environment through objective knowledge, tools, and technology. We have become entrapped by our own success and have lost the ability to properly reconnect to our inherited biological needs, hedonic wishes, and instincts of the brain stem. As a result, these remain in a state that is unmediated by reason and thus leads to chaos, violence, and sociocultural decline.

In present-day cultural evolution, this dualism is the most urgent problem of adaptation. It needs to be resolved before any further progress of humanity can begin. It is also the neurological key to an objective psychology of the future, to the understanding of what has driven human history to become that bloody medley of violence and arrogance, and to what drives all religions. Our technology-driven, free-market civilization is strongly reinforcing that dualism. What is often referred to as an "inner void" is caused by it, and very likely the frustration over not being able to bridge it leads to apocalyptic beliefs. The appearance of extreme religious sects is a direct indicator that the void is rapidly becoming unbridgeable in a significant part of the world's population.

Religion and the search for objective truth have one goal in common: to close the neurological dualism, to make the person whole. But while the former tries to restore a pre-schizoid paradiseal mental condition, thus running opposite to the forward thrust of evolution, the latter relies on the new emerging powers of reason to heal through the ultimate objective insight about ourselves. If the former is comparable to folk medicine, then the latter would be the scientific medicine needed for an acute, life-threatening disease.

Building on the two pillars, in the second part of this book, "Beyond Human Nature," it will become possible to further refine and expand the coherent theory, to illustrate its explanatory power, and to introduce the reader to the ethical challenges posed by it. Some of the most urgent and practical conclusions from such a realistic understanding of human nature and culture are these (chapter 9):

1. Ever since the political theories of the enlightenment period, the most fundamental assumption about the nature of citizens forming the social contract was this: each individual is an autonomous, self-responsible agent, possibly answerable to God only, whose will and pursuit of happiness has to be

respected absolutely, as long as it is not harmful to others. This turns out to be badly flawed. We are far from being autonomous. Even though there is a certain genetically fixed inventory, upon whose fulfillment happiness depends, much of our wants are a result of the programming of the neocortex by others and by common beliefs, state doctrines, advertising, technological super-stimuli, and the like. Especially in modern technological "media culture," this influence has become excessive and contrary to the desired state of well-being. "Pursuing happiness," as it was meant by the founding fathers, today can lead to self-destruction. Elaboration of this insight leads to the next points:

2. In the current situation, without a determined and systematic communal pursuit of the cultivation of objective truth, esthetic taste, and an effective hygiene of values, we are condemned to an existence not much above the level of apes. Worse: Apes with guns, atom bombs, and poison gas, who thoughtlessly multiply, destroying the planet.

3. The social contract between government and citizen has to be reengineered, so as to better insure such systematic pursuit, and thus to protect and foster creative cultural development, positive education, and hence more happiness for all.

Currently, in the United States, such statements are political anathema. But some European countries have de facto moved in the indicated direction. They have proven to be more pragmatic than the oldest working democracy, which continues to proudly suffer from much unexamined, deadweight dogmatism. Nevertheless the challenge remains enormous everywhere.

It may well be that the impetus for change will ultimately arise through a new religious or quasi-religious paradigm. What is advocated in this book, however, is nothing less than a second and final stage in the political project of enlightenment that two hundred years ago had ushered in democratic systems of government. But today it is not just "freedom and the pursuit of happiness" that are at stake, but the very survival of humankind. In the eighteenth century, the enlightenment philosophers felt empowered by the success of Newton's physics, even though their knowledge was small compared to what we know today. May the emerging comprehensive world view and self-understanding give new empowerment to my fellow citizens of planet Earth to build a viable future.

Finally, a necessary word of advice to my readers concerning the nature of this book: It wants to give an outline of a system of connections between a vast number of specialties. Since it has to cover a broad spectrum, by necessity a compromise had to be made between breadth and depth. Without defeating its own purpose, this cannot be a textbook covering everything from physics to evolution, ethology and neuroscience. On the other hand, in order to reach the general intelligent reader, some introduction to selected basics in these fields was necessary. Otherwise, this synthesis would lack its building material. Therefore, to scientist readers, some sections or even chapters will appear pedestrian, if not superficial. If that happens, just skip the pages. Yet, others may feel threatened by too much hard-to-understand detail. In that case, please don't give up. It is in the nature of a progressing synthesis to shed light backwards on its own points of departure. By explaining how the parts fit into a larger whole, better insight will unfold about the parts in themselves.

But no true enlightenment can ever be obtained without a good measure of effort.

PART ONE

THE

NATURAL FOUNDATIONS
OF HUMAN EXISTENCE

Chapter One

Cosmology as Self Organization or *On The Physics of Life*

If we want to know about our origin, the origin of the universe and the meaning of life, we should not look to a book of mytho-poetry, but open the book of nature. No poet can ever match the beauty of the true story. Physics, as well as Darwinian biology, has acquired an interesting new twist: the concept of emergence. It states that in physical and biological evolution there is a clearly observable trend toward higher ordered complexity. Classical Darwinian selection remains as only one pole in a continuum of processes, whose other pole is self-organization, or the emergence of order. Yet contrary to earlier views, emergence is not a mystical concept at all. It follows simple logic, like the movements of planets; as of this writing, classical physics is sufficient to understand its mechanics. Yet we probably will never know why there is emergence, just as we do not know why there is gravitation. The ultimate mystery remains. Nevertheless, our state as participants in this enterprise of emergent nature becomes much clearer. We see that we can explain ourselves in terms of physics.

Nature Emerges

As far as anyone can tell, we will never know what existed before the Big Bang. Possibly, even the very idea will need modification. Maybe there have been many small bangs instead, or the universe is "inflationary," much larger then previously imagined, with many bangs sprouting all the time in other places.

But one thing is quite certain: *The universe is a dynamic process.* Things change, organize, and evolve. The stars are not unchanging; it is only our short lifetime that creates this illusion. That is why this process of change has remained undetected for so long. Change was not found first in the heavens, but in the process of life here on Earth. But even that evolution was slow enough to have remained undetected until the nineteenth century. One of the oldest beliefs is that this dynamic process and human life have a "higher" purpose. The Greek word for this is "teleology."[1] Teleological belief generally is dualistic. It implies a preexistent life force or divinity outside nature that has set a goal to the unfolding of nature. There seems to be a need for such beliefs because they recur again and again in many forms. This need is akin to the need to weave a mythological story, "a story to live by," in Joseph Campbell's words. And the stories change and evolve according to the state of intellectual development of the author or the culture he or she is a part of. But they all express one and the same wish: *to convince that all the decisions do not rest with us alone, and that somehow we are a meaningful part of a larger whole.*

In building a comprehensive scientific world view, this part of human needs cannot be neglected. When I propose a neuro-psychological explanation for this need in chapter 7, I will not do so in order to explain it away, but to understand its inner reality. And this entire book, in one sense, is weaving a story to live by that is compatible with the high level of insight available today.

Giving in to these needs, a multiplicity of stories has been composed in the past; they range from the spirits of animism, through ancient gods and goddesses, to the stories of impending divine transfiguration of the world, as in Zoroastrian and Christian belief. On the most developed level of this tradition is Teilhard de Chardin's great vision of spiritual evolution from the material point Alpha to the spiritual endpoint Omega. Not surprisingly, Teilhard was a paleontologist as well as a Catholic priest. Some biologists in the earlier part of this century still subscribed to "vitalism," mostly elaborated by Henry Bergson in his *L'évolution créatrice,* or in C. Lloyd Morgan's *Emergent Evolution.* But even among contemporary physicists we find expressions of this basic mythopoetic need. Cosmologists John D. Barrow and Frank J. Tipler posit an *Anthropic Cosmological Principle,* which has been endorsed by one of the best known theoretical physicists, John A. Wheeler. According to this, there is overwhelming evidence in nature that the universe has been designed in such a manner that life, particularly intelligent life, can evolve in it. Thus, the purpose of the universe would be to bring forth humans. Even though such statements do not have Popperian scientific status, because they are not refutable in principle, they seem to fascinate. Among the many arguments are the following:

1. The anomaly of water. Water is the only liquid that (near the freezing point) expands with decreasing temperature. In this way, thermal convection is blocked. Ice swims on top so that the water underneath can stay liquid and warmer, thus protecting aquatic life.

---

[1] From telos, meaning "an end"—the belief that there is an overall purpose and end, for which all nature had been designed

2. Deuterium versus hydrogen. Only a small change in the initial conditions at the Big Bang would have resulted in the universal prevalence of deuterium (heavy hydrogen) instead of hydrogen. Heavy water, D_2O, cannot be used by higher animals.

3. Evolutionary time scale. Small changes in the initial conditions would make the universe develop much faster, leaving insufficient time for intelligent life to evolve.

Probably, most scientists who have ever achieved a sufficiently deep level of understanding of cosmology had such personal speculations about the last causes and the purpose of life. But these are questions asked in the name of a preconceived religion. In particular, a mythology wants to be asserted that has a prime intelligence who has designed the universe, and hence is taking care of things. All the evidence is interesting, but no procedure has ever been invented to test an anthropic principle, nor does it seem likely that one will ever be found.

Thus, the word "emergence" has a historical bias of mysticism. But presently it is being used frequently among researchers of a completely different kind— computer scientists who simulate complex, self-organizing, nonlinear systems on big computers. Stars, galaxies, and cosmic dust under gravitational forces form such a system. We will see that our body and our brain also are such systems. One of the computer scientists, Daniel Hillis, the inventor of the connection machine, must have been aware of the old usage of the word "emergence" when he wrote:

> Emergence offers a way to believe in physical causality while simultaneously maintaining the impossibility of a reductionist explanation of thought. For these who fear mechanistic explanations of the human mind, our ignorance of how local interactions produce emergent behavior offers a reassuring fog in which to hide the soul. (Hillis 1988)

I have to disagree with Hillis. The purpose of this section and the next three sections is to show that, on the contrary, *the new concept of emergence can go a long way in thinning the fog of mystery.*

The poetic priest does not convince. But the conventional Darwinist does not fully succeed either. Fortunately we do not have to choose between the suspended nothingness of conventional Darwinism and the comfortable fog of spiritualism.

Nature does not know of such human epistemological dilemmas. Beginning at the Big Bang, it goes its own way. Let us try to follow it for a while. At the beginning there are elementary particles. They are simple enough. An electron is described by its negative charge, mass, and spin. A proton has positive charge and, likewise, mass and spin. A neutron has only mass and spin. A photon has only energy and polarization. For the moment there is no need to consider more exotic particles like quarks, which themselves are the building blocks of elementary particles; the physicist will forgive me some irrelevant simplifications. Proton, electron, and neutron are the main building blocks of the kind of matter on which life has emerged. Somewhere in the distant past, all the chemical elements around us were formed from elementary particles. Back then there was no Earth, only high temperature and elementary particles. The particles underwent reactions with each other. For instance, if two protons collide at a temperature of about one million degrees centigrade, they may fuse, give up energy, and become a neutron-proton pair, also known as the nucleus of heavy hydrogen. After this, any of the proton-neutron pairs can assimilate another proton, becoming a nucleus of the light isotope of helium. Two of these nuclei can collide again, resulting in a regular helium nucleus with two protons and two neutrons. Two protons and energy are set free in the process. The released energy can maintain the temperature necessary for further nuclear reactions. But no explosive chain reaction is triggered since the gas expands and cools, thus maintaining an equilibrium. This, among some others, is the predominant process that causes our sun to shine.

The physicist will go on to explore the events in great quantitative detail, speaking of reaction cross sections, stabilities, and lifetimes of particles. And when this quantitative causal explanation predicts that if the total mass of a sun exceeds a certain limit, after a given time it will explode as a supernova, and such explosions can be observed with a telescope, a great intellectual feat is accomplished, indeed.

But I think something is amiss in such standard explanations. Freely following Popper (Popper and Eccles 1977), let us imagine an intelligent observer shortly after the Big Bang, with no suns in existence yet. No heavier elements are present; solid matter does not exist. Nevertheless we will do the impossible and give our observer a laboratory in an air-conditioned spaceship. Let us say that he can collect the elementary particles from outside and cool them down. He will discover the forming of a hydrogen atom and be amazed by the spectral lines it emits. What kind of mysterious order can come out of the red-hot chaos around him? He might discover hydrogen nuclear fusion and develop profound theories about elementary particles, similar to our own. Symmetries, weak and strong forces, quarks, et cetera, might very well be part of his concepts.

Yet can he in principle predict the equilibrium condition in future suns and the instability leading to an explosion? Maybe. Probably, he will not be able to foresee the formation of heavy elements in the explosion, nor all the chemistry that would follow at lower temperatures. Certainly, he will not even dream of a temperate planet like Earth or the fragile and complex hydrogen bonds of protein chemistry and the life this will make possible—not to speak of natural selection and its results, such as the intelligence of bonobo chimpanzees or humans.

This unlikely story illustrates that elementary particles and their interactions in themselves are too simple to be considered the sole determinants of all future worlds. There must be more to the story. Looking at the emergence of ever more complex organization of matter and life in cosmology and evolution, all from the same elementary particles, for lack of a better word, we call this process of organization "emergence."

Another approach to this concept is to speak of dynamic systems and the emergence of unexpected new characteristics when the system is assembled from parts that in themselves do not have such characteristics. The elementary particles, with their relatively simple rules of interaction, form a dynamic system. When the rules of the game of interaction are played out, new characteristics—such as the compounds of chemistry—arise. There is nothing irrational or teleological in this matter. *Simple rules, if played over a period of time, reveal an unexpected and unpredictable content—that is, emergence.*

Emergence is seen almost regularly in computer simulations of complex nonlinear systems. In mathematics, highly ordered patterns can even emerge without nonlinearity. An example is the Sierpinsky triangle, as it is generated by a random process found by Michael Barnsley. Its rules are as follows: Draw three points, forming a triangle (fig. 1.1). Assign the numbers 1 and 2 to the first point, 3 and 4 to the second, and 5 and 6 to the third. Now roll a die. Go to the number that came up. Roll again and go on a straight line halfway to the number that came up now. Make a dot there. Roll again and go from that dot halfway to the next number. And so on. After one hundred thousand dots have been plotted with a computer, the result is displayed in figure 1.1. It is an infinite regression of triangles. There is no end to the triangles within triangles if the game is played long enough and the image is printed accurately enough. *Infinite order through randomness is an aspect of emergence.* Repeat the game, and no two series of the randomly drawn numbers will be alike, but all will produce the same triangle.

Another example is the Mandelbrot set (fig. 1.2). Take a point in the x,y coordinate plane. Choose an initial point x_0, y_0. Now, on the computer, go through the sequence[2]

$$x_{n+1} = x_{n2} - y_{n2} + x_0$$

$$y_{n+1} = 2x_n y_n + y_0,$$

starting with $x_1 = y_1 = 0$. If after many iterations the Nth point stays in reasonable nearness to the origin, the point (x_0, y_0) was part of the set (white middle figure). If the sequence diverges (moves out toward infinity), the point was not part of the set (gray scale region). Borders of the set show infinite detail of the most beautiful patterns in every enlargement. Ultimately, the resolution is limited by computer power. One obtains particularly enchanting pictures if the speed of divergence is represented by color instead of gray scale. This has become a popular sport among computer scientists and programmers. Again, a simple rule unveils a hidden universe. But was this universe really just hidden from

---

[2] Actually, this is the mapping $z_{n+1} = z_n^2 + z_0$ in the complex plane.

Fig. 1.1 The Sierpinsly triangle–result of simple rules, randomly played
[Courtesy of Chris Younger, ATT–Bell Labs]

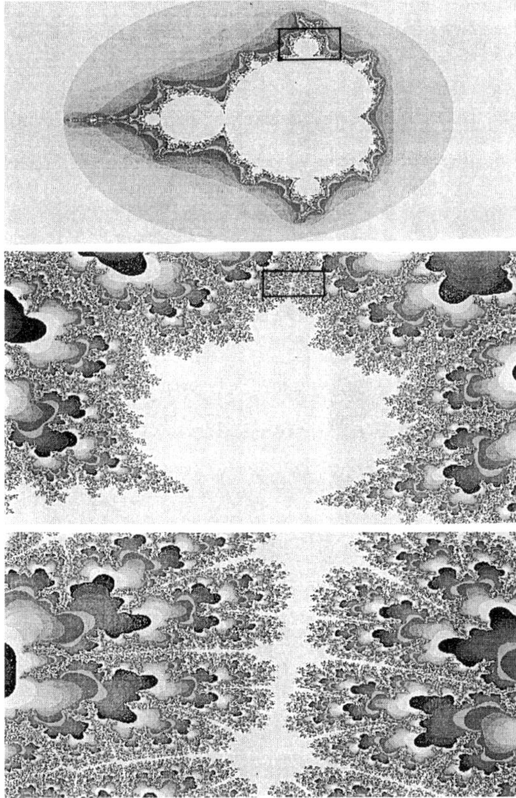

Fig. 1.2 The Mandelbrot set with two successively magnified details
[Courtesy of Joel Elston, Lehigh University]

view? There seems to be a temptation to think so, conforming to platonic idealism: the universal idea somehow preexists in a divine intelligence. However, if the accidents of emergence make the universe form, we end up with an entirely different philosophy: this complex, infinite universe was not in existence before, nor even prefigured in any conceivable way. *It emerged from the rules.* As a matter of fact, it is in principle not possible to predict the infinite detail of the Mandelbrot set without actually playing the game.

Another example is even more striking, particularly when it is seen evolving on the screen of a computer. The "Game of Life," invented by John H. Conway of Cambridge University, is another system of simple rules with unexpected results. Actually, the general idea goes back to John von Neumann, the inventor of the digital computer, and is a special case of a *cellular automaton.* Take a large, checkerboard-type array of fields, or picture cells (abbreviation: "pixels"). A given pixel can be "on" (or "alive") or "off," ("empty," "dead"). Each pixel has eight neighboring pixels (four on the sides and four diagonally). The rules for each pixel cell are the same:

1. If a live pixel has exactly two or three live neighbors, it will survive into the next generation; otherwise, it will die (either from "overpopulation" or "loneliness").
2. Each empty pixel that has exactly three live neighbors will become alive ("give birth").

All pixels are processed simultaneously; this represents a generation. Thus, the distribution of live pixels changes stepwise from one generation (or tick) to the next. Start with a random distribution, or any pattern, and watch the evolution on the screen of a computer.[3] Patterns form and dissolve, and forms emerge that are stable, or move, or oscillate. An entire zoo of forms has been discovered by many enthusiastic players. Figure 1.3 shows a sequence of movement of a form called the "spaceship." Figure 1.4 includes the "glider" in its movement. "Glider guns" like the contraption on the left have been discovered. Its two main forms oscillate up and down, as if mating. Once in each period they collide and emit a glider, which moves diagonally to the bottom right corner. Four of the gliders have been captured in the figure. The simple pattern at the bottom right devours each one on arrival. Therefore it has been called an "eater." Figure 1.5 shows seven successive generations of self-replication. Here, however, the rules are slightly different. Each pixel is considered to have only four neighbors (top, bottom, right and left, known as a "von Neumann neighborhood"). An empty pixel gives birth if it has exactly one neighbor. A live pixel dies if it has two or more neighbors, or none at all. Some theorems have been discovered about this kind of evolution, but this emerged "simple" universe is far from having been completely explored.

Cellular automata of this kind come in many forms. In the "Game of Life," each cell has only two states. A more sophisticated game might have any number of states, akin to energy states, each being represented by a color or a gray scale

---

[3] Jean Michel of Paris, the author of a "Windows"-compatible version of the "Game of Life," has made it publicly available on the Internet.

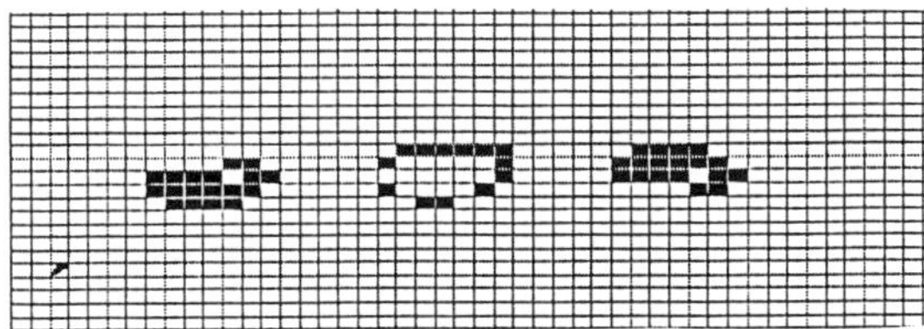

Fig. 1.3
The "spaceship" in its flight.

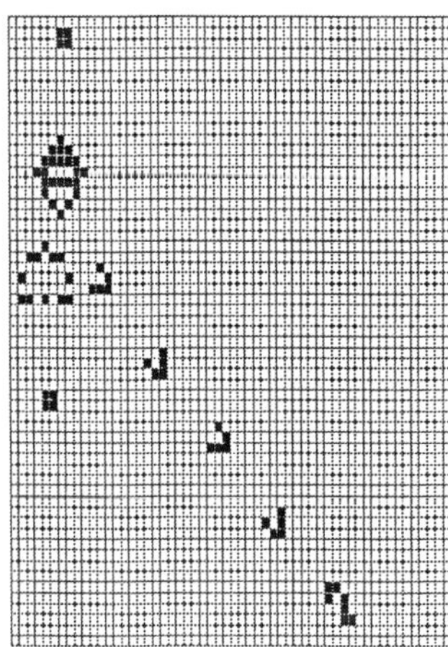

Fig. 1.4
Shape on bottom right devours "gliders," which are
periodically procreated by "glider gun," left.

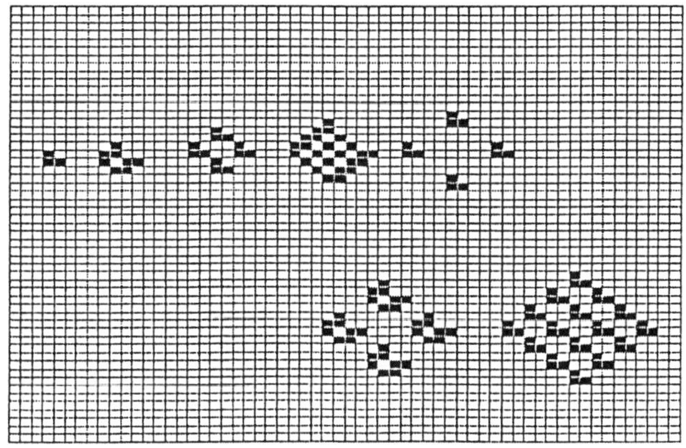

Fig. 1.5
Self replication in the "Game of Life."
[Figs. 1.3 to 1.5 courtesy of Aldo Frigo, Lehigh University]

value. The neighborhood, likewise, can be redefined to include more cells. In a three-dimensional array there would be twenty-six immediate neighbors and ninety-eight neighbors in the next surrounding layer. The rules of interaction can be more complicated, and the emerging patterns more complex. It is not long before they appear miraculous to us. Later in this book, similar processes will pop up at several crucial junctures, every time illuminating the nature of such "miraculous" intricacies of real life.

This dynamic has the same feel as the dynamic of the real universe. Here too, simple particles have energy states and interact with each other according to simple rules, while ordered complexity emerges.

One thing is common to all such games: the "universe" only emerges at a certain threshold of excitation. If the rules are made so that the probability to reach a higher state is too high, after a while there will be a dense, often uniform, saturation pattern everywhere. If, conversely, the probability is too low, all cells will revert to the lowest state (they will die).

Similarly, real life can emerge only in a very narrow range of temperatures—from zero degrees centigrade to roughly forty degrees. In rare cases, primitive life can survive above the boiling point of water, but not much higher.

Cellular automata of this kind have one other feature in common with the real universe and with life: one cannot go back in time; the game is irreversible. A simple check with the "Game of Life," using pencil and paper, will show that a given pattern can be a descendent of many different patterns in the previous generation; a known current state does not tell what the previous state has been. Looking at nature in this way breaks down the borders between cosmology and the evolution of life. This unified enterprise should be referred to as "emergent evolution of the universe," meaning both physical and biological universe.

The preceding three examples of emergence were purely mathematical. In fact, the whole of mathematics itself is an example of emergence of a universe of form from simple rules—the rules of counting. As compared with emergence in the material world of physical evolution, chemical evolution, and the evolution of life, mathematical emergence leads a strange existence—it lacks a substrate until given one by the human mind, in the form of lines in the sand, sketches on paper, or computer memories. We might say mathematical emergence is not primarily natural. Yet it shows emergence all the same. Mathematics, once started, has a life of its own. The old quarrel whether mathematics is constructed or discovered loses its meaning before the concept of emergence. Mathematicians may set rules (axioms); from there mathematics emerges. Seeing emergence at work in such areas, as well as in the evolution of matter and life from elementary particles, underlines the universal presence of this principle. One cannot avoid the conclusion: *Emergence is a fundamental principle in nature*. Particularly in the evolution of mind, as expounded in this writing, the explanatory power of emergence will be considerable.

Yet the concept of emergence neither changes nor contradicts anything known to natural science. On the contrary, it is necessary because neither the Darwinian triad of variation, selection, and retention, nor the causal laws of physics predict any tendency toward increasing complexity. Nevertheless, exactly this happens almost everywhere in physical and biological evolution.

Here, we are witnessing some aspects of a revolutionary development that will not only radically advance the knowledge of the mechanics of biological evolution, but also may shed new light on some of the foundations of physics. Complex nonlinear systems can be generally characterized as consisting of a large number of interacting parts (hence "complex"), such as colliding molecules in a gas, molecules undergoing dynamic chemical reactions in the liquid state, consumers and manufacturers exchanging goods and money, animals competing for ecological space, or interconnected neurons adapting to the reality of the external world. If such interactions follow a nonlinear law, such as the forces between molecules which are not proportional (linear) to their distance, a common characteristic of all the above systems can be found: *It is their ability to self-organize, from initial chaos, to a more ordered process or an ordered pattern.*

A good example of such a dynamic is the free market. The simple desire of every participant to make a profit causes the natural growth of a complex and efficient system of production and distribution of goods. Intervention or "improvement" from outside, such as an artificial plan of a Marxist government, endangers the system. This is not to say that insightful corrections are not possible. But economists still do not understand how to model this dynamic process correctly. Snowflakes are an example of a self-organized and ordered but static pattern. In all such cases order "emerges."

Attempts to understand the mechanics of emergence have been made for quite a while. Stuart A. Kauffman is active at the Santa Fe Institute in New Mexico, which has become the best known center for research in complex systems. Among its founders is Murray Gell-Mann, who set in motion the latest wave of progress in particle physics by positing the existence of quarks. The institute publishes at least one volume of proceedings every year. There are three volumes on "Artificial Life," presenting computer simulations that exhibit emergent properties (Langton 1989, 1992, 1994), (Brooks 1994), (Rietman 1993). For instance, "genetic algorithms" are programs that self-adapt to solve given mathematical tasks. Robots are known whose knowledge grows in similar quasi-biological ways. "Tierra" is a program that causes evolution and competition between simple organisms that are described by sets of "genes." New species form, including parasites. Immunity is seen to develop, as well as a form of sexual reproduction.

In Europe, among others is Hermann Haken at the University of Stuttgart, who has founded an active area of research also investigating self-organization. He calls it "synergetics," or the study of synergies, meaning combined, self-organized actions which emerge from initial chaos (Haken 1978, 1991). He too sees in this a universal phenomenon that can be approached with new unified concepts. Both groups study self-organization in such diverse areas as physics, chemistry, biology, sociology, economy and computation (Haken and Stadler 1990). Another well-known proponent of self-organization is Ilya Prigogine (Nicolis and Prigogine 1977). Lately, a certain degree of disappointment has set in because, still, no general "laws" or equations have been found that govern these exciting phenomena. But if emergence is as fundamental as claimed in this writing, may it be that such conventional concepts as reductionist "laws" are irrelevant here?

As far as biology is concerned, Manfred Eigen has given a superb overview of self-organization at work in molecular biology (Eigen 1992). He gives voice to what seems to be a general consensus. All life is based on one emergent event which, however, could have happened many times in different forms: the assemblage of a molecule that could (a) replicate itself, like some forms of cellular automata do, and more importantly, (b) carry information (or "knowledge") over to subsequent generations. This carrier of knowledge was RNA, the precursor of DNA. *The ability to preserve knowledge across generations represents an emergent level above all previous levels of physical evolution.* A new game, so to speak, is entered. In this game the replicating family of individual molecular clusters, which knows best how to exploit its environment in order to multiply, will always win. Darwinian selection begins. In accordance with point (b), biological evolution will be viewed as a process of knowledge-gathering. This will be the leading theme of chapter 2. Later, the same trend continues in cultural evolution.

Evolution by Natural Selection

When Darwin published *The Origin of Species* in 1856, biologist Thomas Huxley is reported to have said, "How extremely stupid not to have thought of that!" It was a simple idea whose time had come. In nature, the best adapted animals and plants produce more offspring. There is variation between individuals, the cause of which was not yet explained at the time; subsequently, the most efficient breeders and survivors among them have more offspring. There is steady competition for food and space, and there is predation. Thus, the whole living enterprise becomes a dynamic process between change, birth, survival, and death. Today, portions of this process are well-understood and quantified. Over generations, forms, functions, and habits converge toward an equilibrium which is never quite reached. Darwin had worked for twenty years on a collection of observations that would convince his skeptics. But most fellow scientists, especially the younger generation did not need much convincing. Objections from clerical circles were instantly countered by other biologists who jumped on the bandwagon. In 1860 the opposition united forces at a meeting of the British Association for the Advancement of Science, but was soundly defeated by Thomas

Huxley and Joseph Hooker. Thereafter, the Church of England wisely remained silent. The Catholic Church had learned from the disgrace of Galileo's trial. Mainstream Judeo-Christian religion today takes Darwinian evolution as God's way to create.

Yet, despite such a good beginning, it currently does not seem that Darwin's insight has really sunk in as naturally as Newton's physics. There seems to be an inner resistance at work. Many people have been trapped since childhood in various belief systems and feel threatened deep down by the consequences of Darwin's biology. Can our origin remain divine after Darwin? Can there be help and consolation by a personal God, particularly if life is miserable? Can there remain any moral guidelines? And how about death? Apparently, even among many scientists those questions are better left untouched as long as they can be avoided. Existential belief, be it religion or the old brand of humanism, can never be disproved, because it exists in the mind alone and avoids reference to testable reality. It is only possible to present another view that has a stronger power, explains more, and is more helpful. In contrast, Newton's physics is existentially neutral.

For these reasons, physics has had the tendency to be accepted more easily. Yet Darwinian biology is a natural science as well, and is studied with the same methods; actually, it is less difficult, and shares with physics the characteristic that many areas of knowledge represent a quite certain truth. Other areas are somewhat more speculative, and ultimately there remains a mystery, not so much the mystery of the first origin of life, but the mystery of the first origin of the universe and of the mechanistic laws that govern it, including emergence.

Many people without a working experience in natural science tend to discount the whole enterprise if they think they have found a contradiction or two. Even some philosophers of science have a tendency to do so. Such critique is quite appropriate to the logical constructs found in philosophy. They are absolutely correct in mathematics. In natural science, however, a theory rests on a large body of observations and experimental evidence, in detail only known to the specialist. Real contradictions generally lead to an incremental correction of the theory, not to outright refutation. One typical example of this is the orbit of the planet Mercury. It shows deviations from Newton's theory. Nevertheless, Newton's theory of celestial mechanics was generally accepted because it accurately described the orbits of all the other planets and moons. It needed an improvement, though, which was made later by Einstein. No serious physicist before Einstein would have flatly stated that Newton's theory was wrong. But any such theory can always be improved upon. And so it is with Darwinian biology, which presently is being corrected with the concept of self-organization.

Much has happened to Darwinism since Darwin's time. In 1900, Gregor Mendel's rules of inheritance were rediscovered, thus providing one well-defined source of variability. In 1901, H. de Vries published his theory of mutations. After that, genetic biology was dominated by Mendel's rules and de Vries' mutations until 1953, when even the chemical basis of inheritance and the mechanics of mutation were clarified by the discovery of the double helical structure of deoxyribonucleic acid (DNA) by Francis H. Crick and J. D. Watson. Since 1901, the field of genetics has experienced such overwhelming success that at least one important aspect of evolution often became neglected—the constraints imposed by the physicochemical system of the body. This happened predominantly among the more pragmatically-oriented American biologists. Obviously, genes can only construct what the material of construction will permit. And selection takes place on the expressed (constructed) organism—the phenotype. As S. J. Gould and R. C. Lewontin (1979) have pointed out, these constraints are at least as important as the genetic base itself. There is a natural architecture for each building material.

There is an old critique of the neo-Darwinian theory of evolution by random mutation, followed by selection of the fittest. The frequency of mutations, particularly of useful mutations, is found to be too slow. Numerical estimates make it unlikely that all the diversity of life could have evolved in this way in the limited timespan of four billion years.

But life is a complex, dynamic system. If such systems self-organize, then this must be an additional ingredient in evolution. Stuart A. Kauffman (1991, 1993) is one of the proponents of this view. Lately, Murray Gell-Mann (1994) has written a personal philosophy about the role of adaptive complexity in life. The most advanced insights in this matter have been summarized by the Nobel prize-winning biologist Manfred Eigen (1992) in his booklet "Steps Toward Life." It turns out that classical selection from accidental mutants happens only at one end of a continuum of evolutionary processes, and that this pole is approached but never quite reached by higher organisms. Lower organisms evolve nearer to the other pole, that of chaos, where there are large copying errors (mutations) and large genetic variability within a species. In fact, a better word for such populations would be "quasi-species." *The large variability can cause a population to move rapidly toward an environmental niche, as if intentionally directed.* Such rapid and quasi-directed evolution had not been foreseen by classical neo-Darwinian theory. This discovery invalidates the mentioned critique of Darwinian selection being too slow. During most of the evolutionary time, very simple single-cell organisms were undergoing rapid, quasi-directed adaptation. Only after those were perfected could they, again rapidly, assemble into multicellular organisms. The new theory has already been translated into working biotechnology. In Eigen's laboratory, "evolution machines" are rapidly evolving new microorganisms.

Emergent Levels in Darwinian Evolution

In the mathematical examples, the medium of emergence was located in the silicon chips of the computer. What, exactly, is the medium of emergence of multicellular life? Here, we have to make a distinction between slow phylogenetic emergence over a long timespan and short-term ontogenetic emergence.[4] Only the latter can be directly observed in our short life span, at least in advanced life. Here, the medium is the physicochemical system of embryonic and postnatal development. As already mentioned, the internal environment is set up by the possibilities of biochemistry and biophysics. They form the (complex) rules of the game. Only certain reactions, hormonal controls, cell adhesion and recognition mechanisms, and schemes of local induction to specific cell modification and growth are possible since they are constrained by the laws of chemistry. These rules and the available material determine the architecture of the organism. The genetically stored initial conditions are played out in this multidimensional space of permitted biochemical reaction chains. Already, this part of the game promises a vast multitude of complex and highly unexpected results.

But in addition, the initial genetic conditions change slowly over many generations, as the genetic code undergoes mutation and selection. And now, as the organism interacts with the world, the truly unique is selected from the highly unexpected. Thus, biological evolution leads to forms and dynamics which are far above anything imaginable with overly simple mechanistic concepts.

On top of this emergent universe of life there may be defined a superuniverse that continues to emerge after the basic physique of organisms has been established. This is the universe of neurons, neural networks, and sense organs in interaction with given external realities. In this universe, functional knowledge about the world enters into the individual animal from outside. Not much is seen changing in the organism, yet what results is a mind or spirit that from now on animates the animal to more intelligent and more effective use of the given external reality.

This emergent level seems to bear within it a still higher one. Once the mind has reached a certain perfection, it begins to incorporate our own internal world into its perceptions. Mental reflection emerges. When this reflection focuses on the cognitive side of the internal world, thinking emerges. When it's focused on the totality of the vegetative-emotional world as well as the cognitive world, art emerges.

And that is not all. When inner worlds become expressed in writing, art, technology, and tradition, future generations can partake in levels previously reached. Culture emerges like a supra-organism with a lifetime spanning many individuals in many generations.

Konrad Lorenz seems to have been in agreement with all this, but he did not like the word "emergence" because it had already been used by the vitalists with a mystical meaning. But vitalism has become so irrelevant today that one should not feel constrained to give the word a new meaning, as almost everyone has done who has observed this surprising process unfold on the computer. Lorenz preferred "fulguration," which is derived from the Latin word fulgus, meaning the (creative) "spark." This, so to speak, ignites a new emergent level. The concept of

---

[4] Ontogeny is the formation of the individual animal from the fertilized egg. Averaging over individual variations, the result is known as the "phenotype."

"levels of system integration" had been coined a long time ago by L. V. Bertalanffy (1949, 1968) in his "General System Theory."

Loosely related to emergent levels are Karl Popper's three "Worlds" (Popper 1972). They will help us to steer clear of numerous confusions in the course of this writing. According to Popper, all reality can be classified as belonging to one of these "Worlds." World One is the world of physical objects, galaxies, planets, atoms, and oceans. World Two is the mental world of inner experiences. This is equally real. Contrary to his earlier years, the later Popper had the tendency to look at this world in a spiritual context. I take it as built on the physical world, as a higher emergent level. This is one of the defining statements of the "biological naturalist's" view of the mind. World Three consists of all artifacts made by humans, such as books, pyramids, paintings, and tools. The significance of this "World" is that all of its objects, at one time, have been conceived in World Two, thus testifying to its reality. Of course, the artifacts as such are also part of the physical World One, but they have the special status of being meaningful to all human minds (World Two) of all times.

Armed with Popper's concept of the three "Worlds," we can attack the problem of the so-called "psycho-physical dualism." For an example, let me consider music. It goes against common-sense and intuition that Mozart's music could be a result of natural selection from random variants. A thought like this seems, in fact, to render questionable the entire argument for the evolution of complex organisms with novel types of knowledge and inner experiences from primitive forms. How can a single causal chain of events lead from the slimy to the sublime?

But after having seen the "Game of Life" evolve, or the Mandelbrot set form, we can begin to feel that such development, indeed, might not be impossible. We might reexperience the shudder felt by the Pythagoreans vis-à-vis the numbers governing the harmonies on the monochord, if not the universe. Yet a weakness in one link of this chain from "the slimy to the sublime" remains. By now we can imagine self-organized patterns of neural firings in our brain, set in motion by the auditory patterns invented by Mozart. But we cannot see how this translates into the feelings of joy and elation we experience, which the genius obviously had intended us to experience, as he himself must have experienced.

This transition from neural dynamic to inner feelings is the leap from World One to World Two, from the world of physics (or physiology) to the world of psyche. Here, nothing mechanical is to be found; only feelings, images, emotions, compulsions, logical constructs, et cetera exist.

To follow the mechanic of emergent evolution beyond this point is not directly possible. This transition can only be accomplished by an act that may seem to border on faith. But such belief certainly is not without foundation. We can observe neural firing patterns and waveforms in the brain, while the investigated subjects report their inner experiences or move around. We can even electrically stimulate certain patterns of neural firing and see what the subject reports in response, or what action he takes. Chapter 5 will derive an understanding of global brain function from such experiments. *The conclusion can only be that the inner experiences and the firing patterns are aspects of one and the same phenomenon.* In other words, there is no psycho-physical dualism, but a psycho-physical identity. I do not mind if this is called a "psycho-physical parallelism" because, indeed, our experiments show the two different worlds (of Popper) working in parallel. But there is no need whatsoever, nor any evidence, of the psychic world having an existence independent of the firings of neurons. Unfortunately, philosophers generally mean just this when they speak of parallelism.

There is a well-known analogy to this situation in physics. This is the much-discussed wave-particle dualism. It is resolved in the same fashion. Experimental evidence shows that the same phenomenon, depending on the method of observation, will manifest itself as a wave or as a particle. Here, there can be no doubt whatsoever that only one entity is behind these "complementary" appearances.

Emergence as a Scientific Concept

Somehow, the concept of emergence does not seem to jibe with the standard perception of the nature of natural science. That is because science, up to now, was almost purely "reductionist." It reduces the many phenomena to fewer causes, and these further to even fewer subcauses, until only the four fundamental forces are left. And even those are subjected to theories of unification or grand unification. The strength of the reductionist program derives from its being open to mathematical quantification, which can be used to predict exactly the behavior of most physical systems. Our entire high technology rests on the great success of this approach.

Emergence, on the other hand, seems to go in the opposite direction. That is why it took such a long time, and the invention of computers, before the concept could even be formulated. It simply asserts that, under certain conditions, a multiplicity of ordered phenomena will come into existence when the simple rules of interaction between particles are played out on a large scale. As already mentioned, so far, emergence has defied predictive mathematical quantification. No such mathematical approach seems conceivable, at least not with the tools known today. The game always has to be played in order to see what happens. Nothing much can be predicted ahead of time. This was true even in the mathematical games considered. We are at a loss to say why ordered complexity emerges, but we can at least try to measure its progress. It is conceivable that such a measure could be defined from cosmology and from paleontological data.

First of all, ordered complexity should be more clearly defined. This is not trivial. As species evolve, complexity of form is not necessarily increasing. In fact, perfection often lies in simplification. Complexity should be more a measure of functional diversity, or of functional knowledge of the environment. In further evolved life it primarily would measure the degree of adaptive modification that can be accomplished by the nervous system.

But once it is defined, a measure of emergence would show a progressive (exponential?) increase during the timespan from the Big Bang to the appearance of *Homo sapiens*, and beyond that, in the evolution of culture. Some necessary properties of such measure can be stated:

1. *It would have to measure complexity as an average over a certain timespan.* Evolution does not proceed steadily upward; at best it "meanders," and "devolutionary steps" (Campbell 1974b) are well-known, such as the disappearance of eyes in certain species. Or it may be "punctuated," speeding up at times. The extinction of dinosaurs in a catastrophe shows the influence of random events on a large time scale. Mutations represent randomness on a smaller time scale. Thus, the measure of emergent complexity can be a function of evolutionary time only in the sense that it would state probabilities for a certain level of complexity to occur. Such functions are known as "stochastic functions."

2. It should properly account for *"inverse quantity."* As the degree of complexity advances, the number of the complex entities becomes smaller. Most of the matter in the universe is still in the form of lightweight (simple) atoms, if not in the form of elementary particles. Heavier elements are formed later and become scarcer as their weight (complexity) increases. Organic molecules are still rarer, and living entities come still later, and are rarer again. Finally, intelligent life forms are the latest and the rarest.

3. It should account for *inverse fluctuation.* As the number of entities diminishes with complexity, the effects of random events become more pronounced. In that spirit, a potential extinction of Homo sapiens, whether caused by himself or otherwise, would be a random fluctuation in this particular solar system.

4. A mathematically defined measure of the state of emergence of the whole universe would have to be averaged over states on many different planets where life exists. In fact, only if other systems of evolution of life do exist on other planets, would the above measure be meaningful. If we are alone, we might as well consider this Earth an unsystematic and unlikely quirk.

There is a tradition among theoretical biologists to say that life, particularly intelligent life, is just such an unlikely quirk. Stephen J. Gould (1989) has made such a statement. According to this, there is a long chain of unlikely events at the roots of our existence. Every time evolution of life would repeat itself somewhere in the universe, the outcome would be radically different from what happened on Earth, and intelligent life would most likely not appear. In that framework of neo-Darwinian biology we are an unlikely event, indeed. Yet if one recognizes emergence as a fundamental principle, the chain of unlikely events appears to be natural. *If intelligent life is found to exist elsewhere, with Gould's arguments the universal occurrence of emergence would stand verified.*

The Role of Classical Physics

The self-organizing property of the universe presently begins to be recognized even by the most mainstream physicists. But there is another aspect of the current style of thinking about physics in its relation to life, particularly conscious life, that has to be addressed. The present generation of physicists has grown up on the incredible success of quantum theory. It generally has been forgotten that in the first half of this century a break occurred in the nature of physics. Up to the introduction of the quantum concept by Max Planck, all of the understanding could be intuitive. Even though the models built, at times, were complicated and mathematical, ultimately everything could be understood in terms of action and reaction of forces whose workings could be dissected in detail. This is similar to the way that one can visualize the sequence of movements and chemical reactions in an internal combustion engine. This reassured us of our abilities to really understand nature, and ultimately ourselves, in that same framework.

This certainty was shattered by quantum mechanics. Suddenly, things on the atomic scale could no longer be grasped in the same way, but could only be computed with the aid of ad hoc equations which owed their existence only to their ability to quantitatively describe, but not to help in intuitive understanding. Albert Einstein did not like this at all, and for that reason he was never willing to fully embrace the new theory. In a certain way, his reservations foreshadowed today's decline in the social status of physical truth. Now it is often said that modern physics has undermined the "certainties" provided by science, that all knowledge remains relative, and that natural science creates just another myth among many others.

But there is a much better answer to the challenge of modern physics than this epistemological relativism. *Our mind has evolved by natural selection in the medium-size world of everyday experience. Therefore, it cannot be expected to be at home in the microscopic world of atoms.* (Vollmer 1985). (Compare also the "Philosophical Insert 1" on p. 30. This evolutionary argument, while pointing out the limits of our mind's grasp, *also relieves us from the mistaken belief that somehow truth is to be gained by applying our power of intuitive thought to areas for which it was not built.*

Those limits, however, do not necessarily limit the most important truth of all: the truth about ourselves. *What if the human mind can be understood already in terms of classical physics, self-organization and Darwinian evolution?* Then Ockham's razor[5] would require that we leave out the complications of quantum physics, and our scientific world view would remain unbroken. I will try to show that, indeed, we can understand ourselves and our place in the universe on this basic intuitive plane. That quantum mechanics

---

[5] William of Ockham (circa 1290-1349): "What can be explained by the assumption of fewer things, is vainly explained by the assumption of more things" (Runes 1984). Or, raze all unnecessary assumptions.

provides the theoretical foundation of, for instance, biochemistry, does not matter. Chemical reactions also can be studied in the laboratory without any reference to their theoretical foundation in quantum theory.

Some contemporary physicists are unwilling to make this simple evolutionary argument and try to explain the mystery of life, particularly conscious life, with the mysteries of counterintuitive quantum mechanics. Foremost among them is Roger Penrose (1989). But following the evolutionary view, human truth will be found to be much simpler and much more beautiful.

On Freedom of Choice

One of the questions regularly brought up in physicalist discussions of the roots of humanity concerns determinism: Is a physicalist theory of mind deterministic? Stated differently, do we have the freedom of choice?

One could argue that if emergence follows fixed rules, as in Conway's "Game of Life," then the outcome is fully determined by the initial configuration. This game, indeed, is "deterministic." So, if the universe is like the "Game of Life," everything would be predetermined at the beginning. But the study of real physics shows that long and complex chains of cause and effect cease to be deterministic. There is no need to invoke Heisenberg's uncertainty relation, because in classical mechanics there is already the simple effect that infinitesimally small aberrations at the beginning of a causal chain can avalanche into events of gigantic proportions. Workers in the field of computer simulation of large-scale weather dynamics speak of the "butterfly effect": if a butterfly in southern California flaps its wings, causing a microscopic air turbulence, three days later, at least in principle, this can result in a snowstorm in New England. For such reasons, the universe cannot be deterministic[6] in any real sense. *In this aspect, it differs from the "Game of Life."* Therefore, Darwinian evolution cannot be predetermined; it must have a large element of the accidental. This equally applies to the inner "states" of our body and brain. The number of neurons in our brain compares to the number of stars in our galaxy.

The question of freedom of choice can be answered in the same spirit. At first one notes that our thirty thousand to one hundred thousand genes cannot begin to determine a brain in fine detail; it has at least 10^{10} neurons and 10^{13} connections between them. This could not even happen if embryology was deterministic, and the environment was constant. Instead, individual neocortices are programmed by individual experience (see chapter 6). The genetic determinants only establish a general trend in order to satisfy the most fundamental survival needs, such as the appetite for sex, the drive to learn (see the next chapter), and the potential to further adapt the nervous system to the current life situation. The causes for a given individual's particular decision are so many, they are so diverse and not strictly quantitative, and probably not quantifiable, that, as in nondeterministic physics, decisions can be strongly determined neither by the individual's past nor by the genome. When we have

---

6 My use of the word deviates from the classical usage. In the old–abstract philosopical context it was said that when the state of the universe would be completely known at a given time, and would follow classical mechanics, then all the future states would be determined in principle. Thus, the much bespoken "chaotic" behavior in systems that follow classical mechanics has to be called "deterministic chaos," but it is "indeterministic" in any real and practical sense. I am only interested in real and practical things.

convincingly demonstrated that the ego's mechanical side consists of semi-disordered salvoes of neural firings, how can we suspect any sizable mechanical causal determination of any action?

Are we then ruled by "chaotic" or random accidents? Does life equal random noise? Toward the end of the first part of this book, some limited support will be given to an affirmative answer. But there is a much deeper connection—it was said that individual neocortices are programmed by individual experience. Since we are social animals, our conspecifics will figure prominently in that programming. Therefore, suddenly again, we find ourselves in a "Game of Life" situation. Our neighbors will influence us, as we will influence our neighbors in return. The emerging social pattern will be a common "culture" whose norms and beliefs are difficult to escape. Nevertheless, we (our egos[7]) will not feel too constrained. After all, we generally want to fit in socially. This is instinctual and, hence, genetically fixed. *Here is a clear genetic and cultural determination, but it is not experienced as a restriction to freedom of the will.*

Yet we (our conscious egos) find ourselves enmeshed in a web of imponderables. All good drama, from Greek antiquity on, understands the human being as tragic and heroic. Humans are tragic because, even with the best of intentions, we cause suffering to ourselves and to others. We are heroic because we overcome the suffering with great self-sacrifice. *This human tragedy is not caused by the gods, but because the emergent level of human conscious activity is rooted in biological evolution.* This is where the general limits to our freedom of choice become apparent. We can briefly forget about them, but we cannot beat our biological nature for long. If we want to be happy, we can only go with it in an intelligent way, never against it. *That is the true limit to the freedom of our will.*

But our inquiring minds cannot feel satisfied with such general statements. Where, *exactly*, are the limits to our freedom? And what, *exactly*, are the causes of the human tragedy? Those are the most important questions that can be asked today in the name of all humanity. If we can answer them, the new eon of creative progress may, after all, become future reality. In order to find answers, we have to turn to the natural science that deals with the foundations of our existence. Having emerged from the world of animals, we now must turn to biology.

---

[7] Throughout this book, "ego" is used in the Jungian definition as "the center of the field of consciousness."

Chapter Two

Biological Evolution as Knowledge-Gathering or *The Roots of Cognition*

We can say that crystals, in a way, are self-replicating, and so is life. But what makes life a higher and revolutionary emergent level is that knowledge which has led to reproductive success can be stored in the DNA strand and thus passed on to the next generation. Later, the individual organism begins to acquire additional knowledge about the environment. Finally, in multicellular organisms, a nervous system specializes in faster and more efficient knowledge acquisition by the individual. Individuality of any organism is best defined by its particular knowledge. In humans, some of the knowledge will become conscious. Emergence of life is synonymous with the emergence of knowledge. It is amazing that, at the end, our evolved brain can look back at all this, its own origin, and understand it, at least in principle.

The Cell–Principal Machinery of Life

From the red-hot chaos to intelligent life, there are seven clearly separable levels of emergent complexity: quarks, atoms, molecules, condensed matter, organic molecules, living cells, and multicellular organisms. "Condensed matter" physics, as it is known today, includes liquids and solids, including *organic* matter. But due to the variability of the hydrogen bond in organic molecules, these can assemble or "condense" into matter whose complexity is many orders of magnitude above the complexity in inorganic matter. Therefore organic molecules represent a separate emergent level. Living cells are made from organic molecules as elementary particles are made from quarks.

For the moment we will leave out the question of the origin of life and dwell on the living cells as they can be studied today. Up to the present, most of the evolutionary time on this planet has been spent developing the self-replicating cell. The Earth's crust is about 4 billion (4 10^9) years old. The first microfossils of single-cell organisms have been dated at 3.5 billion years, but real multicellular life came into existence only about one-half billion years ago. If time is a measure of difficulty, then mastery of a successful cell chemistry in single-cell organisms already was six-sevenths of the task of building *Homo sapiens*.

For the interested reader in Vignette 1 (p. 284), the basic details of cell biology are reviewed. The degree of organized complexity in the living cell overwhelms us. Specialists still puzzle over many details of this physicochemical machinery. But today there can be no doubt: *machinery it is*.

As viewed from outside, the cell is an automatic chemical factory that cannot only self-replicate; in the human embryo, it can also switch to 254 different modes of operation, when it becomes specialized. Its products are delivered upon receiving external stimuli, and it picks up whatever raw material is needed from the surrounding liquid. Even better—on demand it can change physical shape or deposit solid material, such as bone.

It is quite appropriate to say that it is the knowledge of how to use organic chemistry, particularly protein chemistry, that makes this factory possible. The knowledge is carefully safeguarded in the DNA. The chemical knowledge forms a platform upon which externally visible, knowledgeable behavior is built. Ultimately, this will become the intelligent behavior of higher multicellular animals like ourselves. To appreciate the hugeness of the phenomenon we will have to begin studying the simplest life forms known—the protozoa or first animals.[8]

The *Paramecium caudatum*, or slipper animalcule (fig. 2.1), is a much studied single-cell organism that lives in stagnant fresh water and marshes. In its simple way, without any recognizable nervous system, and without vision, it knows how to succeed in its environment. For instance, when it collides with a solid obstacle it instantly begins to swim backwards for a short distance; then it stops, turns by a random angle, and continues forward again. Repeating this simple sequence, it manages to travel around all obstacles which impede its constant search for food. It is useful to compare this organism with a robot that has been given a simple program. Without that program the robot would run into a wall and continue spinning its wheels in vain until the battery is empty. By installing the program we seem to give it the following simple knowledge: a) You cannot go through a wall, and b) You can continue your search if you back up and change direction.

Of course, I have used anthropomorphic concepts. This robot does not think, nor does the paramecium. Still, both have knowledge about a certain property of obstacles and how to react to them in order to survive. With Konrad Lorenz, I insist on using the word "knowledge," not "information." "Information" describes

---

[8] The virus is actually simpler, but it is a life form that can only exist as a parasite of cellular life.

an empty medium capable of carrying knowledge, not knowledge itself. Moreover, the use of "information" suggests a flow of information toward someone capable of using it. No one like this exists in any organism or animal, or even any human. *From the very beginning, the individuality of an animal expresses itself best and is best-defined through its particular knowledge.* Almost certainly, this organism is not conscious in any remotely human way. Knowledge and adaptation are quite similar. Only living organisms (or machines programmed by man to pursue a certain goal) can have knowledge.

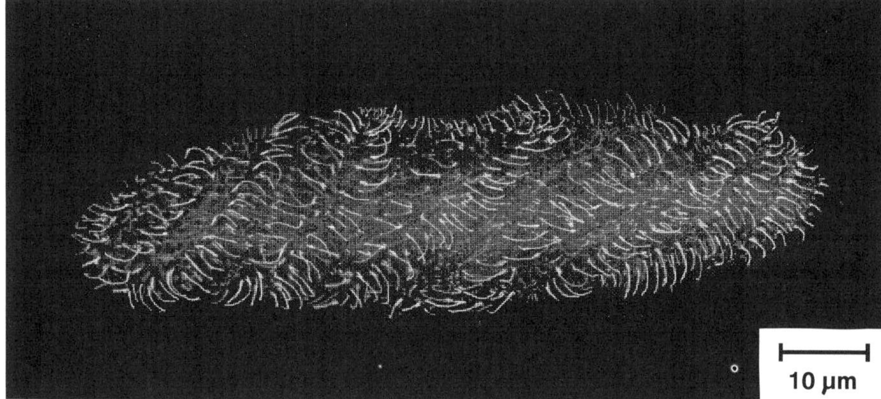

Fig. 2.1
Paramecium caudatum, after a SEM microgram

"Phobic response" is the proper name for the simple behavior of the paramecium. Even simpler is a reaction called "kinesis." This can be found in a wide range of other organisms, and even in mammals. It consists of a random zigzagging walk. Such a pattern of travel is effective in harvesting the food contained in a given area. The randomness insures that the organism slowly moves on to different regions. Now, if there is no food in a given area, or if being there is unpleasant, the organism simply increases its speed. In this way it will leave the undesired area faster. As soon as it reaches good pasture, it slows down again. Thus, more time is spent in rich areas.

Higher on the evolutionary scale is tropotaxis, or movement directed toward a desirable region—for instance, where there is more light, where it is warmer, or where there is more concentration of oxygen or food or less concentration of poison. This requires directional sensing. The organism has to establish the direction to the light source or the direction of a gradient of concentration before it can move.

Such low-level knowledge is limited to certain features only, and only within a certain environment. It is fully automatic. An example shows this: The above-mentioned paramecium has a tropism (attraction) toward H^+- ions of carbonic acid, since they are associated with rotting organic matter which contains food. If the experimenter introduces a drop of poisonous oxalic acid, the paramecium swims toward the identical H^+- ion of this acid and dies. Another example is a moth flying into the fire.

Of course, in our search for human nature, animals with a true nervous system will be more interesting than the paramecium. Nevertheless, the situational logic is similar even in protozoa, and there must be analogic solutions that insure survival. When the paramecium bumps into a wall and subsequently switches into reverse gear, there must be something that senses the impact and causes the cilia to reverse the sense of their wavelike motion. Sensory excitation must be received and transmitted from wherever it has been sensed to wherever a reaction will take place. Already, the protozoa possess an inner structure to accomplish such tasks. One finds tiny organisms that respond to touch, heat, chemicals, pressure, light, water currents, electric currents and gravitation (Westphal and Muehlpfordt 1976). However, a sensory area cannot be distinguished as separate from the motor area in all of the responses. For instance, in a weak electric current field, the paramecium migrates toward the cathode. In a stronger field the cilia at the posterior end start to beat in opposition, thus reversing the direction. This seems to be a direct influence of the external electrical activity on the mechanism of locomotion. Similarly, chemotaxis may be caused in a direct way, since some cilia in themselves are chemical sensors.

On the other hand, some very interesting sensory organelles do exist. Many protozoa have a light-sensitive "eyespot." In *Erythropsis richardi*, the red, light-sensitive spot is covered by a true light-collecting lens. The eye already begins a crude existence in the "first animals"! Another interesting organelle is the statocyst that has been identified in the ciliate *Blepharoprosthium*. It resembles our organ of balance in the inner ear. A particle of heavy mass is suspended in a capsule, exerting pressure on the side which points downward in the direction of gravitational pull. In other protozoa the same purpose is accomplished by sensing the direction of pressure from ingested food particles. This can be proven by feeding them small particles of iron. Organisms that swim against gravitation to reach the surface will now swim against the gradient of an applied magnetic field.

There is evidence for networks of fibers that seem to be conducting sensory excitation. In the ciliate *Euplotes* there is a central spot, the "motorium" from which such fibers extend to the cirri, which are fused, cilia-forming, leg-like protrusions. These are used for crawling-like locomotion. The organism is robust enough to continue moving after having suffered significant incisions. However, if the bundle of these "neuromotor" fibers is cut, movement stops (Dethier and Stellar 1961).

Some protozoa are highly differentiated in their inner structure; some are quite large. *Psammonyx vulcanicus* can reach six centimeters. A fossil form,

Nummulites gizehensis, was eleven to twelve centimeters. But their blueprint did not permit evolution into more advanced forms. They are "animals without tissue." Instead, they have spread out into virtually all moist environments, including guts of insects and higher animals and the bloodstream, where they continue to affect humanity with malaria, sleeping sickness, and other plagues.

Responses of protozoa are fully genetically determined and "hardwired." This is true even among the simplest multicellular organisms that have developed a nervous system. In some of them the wiring diagram has been almost completely described, and many, if not most, responses have been explained on the basis of such diagrams. Knowledge is contained in the cell's machinery and the wiring diagram. At the level of complexity of protozoa, this knowledge is mostly acquired through classical natural selection. It is surprising how quickly a small genetic advantage can spread through the entire population. For the interested reader I have included Vignette 2 (p. 293), explaining some quantitative details of natural selection and illustrating them with a numerical example of population dynamics. In the selected example, the genes of a small number of slightly more successful individuals take over virtually the entire population within fifty generations.

It has been said repeatedly that classical natural selection determines *local change* only and does not predict global evolution toward more complexity. *But there is so much to know and so much to adapt to* that in harmony with the principle of emergence, evolution toward more complexity and more knowledge is bound to occur.

Are there limits to this explosion of knowledge? Of course there are. As mentioned in the previous chapter, the medium of expression imposes a limit. Only certain architectures are possible under the laws of physics and chemistry. Knowledge can grow only as fast as the possibilities of expression; "niches" are found in assembling the raw material of life. But due to the complexities of protein chemistry, given enough time, we can be assured that many astonishing constructions will be found, having totally unexpected abilities.

Presently, the explosion of knowledge is actually seen continuing on a completely different substrate—silicon chips. Computer programs called "genetic algorithms" have already been mentioned. They include small random changes, mimicking mutations, and criteria for success in solving a problem. Successful versions survive and then evolve further. In this way, problems are solved in novel, unanticipated ways. Along the same line are the self-adapting artificial neural networks that will be discussed soon. Their natural counterparts live in human brains.

All this testifies to the universality of the principle of emergence, of which increasing knowledge is a significant part.

The Origin of Life, Lives of a Cell, and Cellular Automata

The complexity, the apparent purposive design, and the beauty of a cell's machinery are overwhelming. Can the cell really have evolved by mutation and selection alone? Common sense would deny this. Yet, thanks to Stuart A. Kauffman's ideas on self-organization, our imagination can be helped. The key, again, is an emergent process not unlike the one found in Conway's "Game of Life."

But, first, let me state to what extent conventional mutation and selection alone could carry us backwards in evolutionary time. For one thing, simple single-cell eucaryotic organisms, such as yeast, contain many less DNA base pairs. They have about $3\ 10^6$ or $1/300$ of what mammals have. Prokaryotes have only one-third of that. Moreover, there is evidence that primitive forms of RNA may have contained only the bases cytosine and guanine, which have a higher probability of spontaneous assembly. At that time, DNA was not yet in existence. At this point, the reader who is unfamiliar with the basics of cell biology is invited to consult Vignette 1.

Very likely, much simpler and shorter code sequences existed in the primordial soup. Gradual mutation quite believably may have covered most of the changes since that time, the construction of a cell enclosure being one of the more important ones. The evidence comes from the thousands of gene sequences in many different organisms that have been charted up to now (Eigen 1992), (Eigen and Winkler-Oswatitsch 1981). Along the evolutionary tree one finds that most ascending sequences show a tree-like pattern of branching of codes (speciation). However, some localized sequences show a remarkable degree of invariance in all species. Their small changes can be interpreted as a narrow, bundle-like divergence, whose point of common origin can be estimated as close to the first origin of life. The strongest evidence for this is found in tRNA (compare Vignette 1).

Thus, the mechanism of protein synthesis must have been established prior to any other variation, and essentially has remained in the blueprint ever since. Also, the further back one goes along the phylogenetic tree, the more predominant are G-C base pairs. The conclusion is obvious: very likely, the earliest codes only contained G-C and C-G pairs, a binary string of information. Such simplest schemes of replication and protein synthesis can easily be conceived to have occurred spontaneously. Events of this kind are presently studied with computer simulation (Langton et al. 1992), (Langton 1994). See also (Morowitz 1992). Thus, life must have begun in the primordial soup when a particular short and accidental chain of RNA, which could replicate itself, also replicated a chain of amino acids (a specific protein); this, in one of many possible ways, that may have helped in the speed (or accuracy) of further RNA replications.

Such positive effect back upon itself is called a "positive feedback loop" or, among philosophers, a "hermeneutic circle." Nature is full of such feedback loops for the following simple reason: the populations in such loops increase exponentially in time; that is, in every given, fixed time interval, the population doubles. If it doubled every hour, which is not fast for single-cell organisms, after sixty-four hours there would be 2^{64}, or about 18,447,000,000,000,000,000 individuals. However, without feedback loops there is no increase in numbers. Therefore, in time, most of the available material that can be utilized in systems with positive feedback will be found incorporated. Once established, the basic mechanism is open to improvement through mutation and selection. That is why our planet is teeming with life. Other planets in our solar system lack either the necessary material or the proper moderate temperature.

As already mentioned, the early self-replicating molecules existed near the threshold of chaos, and they had a large variability in their code. Consequently, they evolved rapidly and filled many niches. As organized complexity grew, methods of gene repair had to be invented (Eigen 1992). By then, a cell enclosure and a stable cytoplasm existed. Copying errors diminished, driving the systems toward slow, classical evolution by small, random mutations, selection, and retention. That slowness of mutation probably was the ultimate cause for the unchanged preservation of many species after a time of rapid initial evolution.

More difficult to understand is, how cells with identical genetic information can so easily specialize into the hundreds of different kinds necessary to make higher multicellular life possible. Here, it appears that the direction of thought taken by Stuart Kauffman (1993) again provides the best clue. It started from thinking about properties of electronic circuits called "random Boolean networks." The network considered consists of randomly interconnected logic gates (fig. 2.2). It has "AND gates" and "OR gates." Each gate has two inputs. Its state is transmitted to the inputs of two other gates. Such gates have the following characteristics: An "AND gate," at the time of the next tick, switches "on" (high voltage) if both inputs have been "on." Otherwise, it switches "off" (zero voltage). An "OR gate" on the next tick switches "on" if one or both inputs have been "on." Ticks are like the generations in the "Game of Life." In fact there are strong similarities to this game. A random Boolean network of this type differs from a cellular automaton only by the feature that a given pixel cell (or

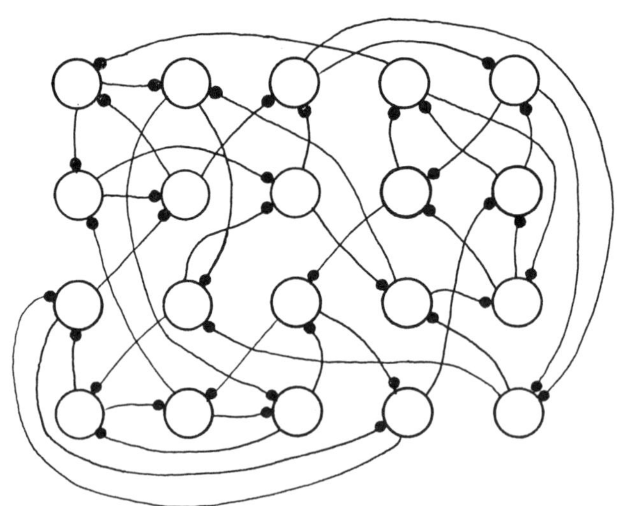

Fig. 2.2
Boolean Network that can have State Cycles

logic gate) is not responding to the states of immediate neighbors, but receives input from two randomly selected pixel cells that may be located at any distance. In other words, the difference is in the definition of neighborhood.

If such networks are left ticking away by themselves, they will settle into periodic cycles of patterns, not just patterns. This needs a more detailed explanation: Consider that "on" states of any gate are made visible by a light bulb connected to it. Now one would see a pattern of lights that changes at every tick. It turns out that a sequence of the same patterns will periodically reappear. Such a sequence is called a "state cycle." As computer simulations show, such random networks of 100,000 gates can settle into only about 370 different and stable state cycles, even though it has a supra-astronomical number of 2^{100000} possible different states (patterns of light). Only a very small fraction of them will ever be activated during a given cycle.

Now the connection to the mechanics of the specialized cell: Genes are expressed regulatory proteins. Each protein induces some reaction in the cell and in turn can trigger expression of another gene. Suppose two proteins are needed on the average to trigger the expression of a new third one, and you have the complete logical analog of the above Boolean network of 100,000 gates. Each "on" state corresponds to an expressed protein (gene). *A stable state cycle of gene expression could drive the life process of a specialized cell.*

But the most fascinating detail of that hypothesis is its numerical result: as stated, 100,000 genes would have 370 different stable cycles of expression. That means 370 different kinds of cells could be expressed by the same genome. In reality we have 254 different cell types in our body. Considering that the number of genes is likely less than 100,000, corresponding to less state cycles, the match seems to be excellent. *So each specialized cell is one stable state cycle of gene expression?* It's quite possible, even though the numbers have to be taken with a grain of salt. A mutation at a particular site makes the cell switch into a different cycle. It may become a fast replicating cancer cell.

The state cycle hypothesis provides even a deeper insight. When it comes to embryology, the most awesome process is how the cells are led to differentiate into the 254 types, in well-defined different regions of the body, in order to generate a functional machine. Here again it is worthwhile to reconsider the cellular automation called "Game of Life." Consider each undifferentiated cell of the emerging fetus to represent a "cell" of a cellular automation, albeit one that follows much more elaborate rules. Instead of just being "on" or "off," in response to chemical stimuli from neighbors, this cell could assume 254 different states, with each state itself capable of producing a number of different specific chemical signals, which in turn switch the states of neighboring cells. These would be complex rules for a cellular automaton, but it would still function exactly in the manner of the "Game of Life." After having seen this simple game evolve, I would not be surprised at all if the much more complex one would settle into one of the most complex, dynamic patterns known—our body. If different rules reside in the DNA sequence, the process may settle into an elephant or a hummingbird.

The wonderful machinery of embryonic development does not loose a bit of its beauty when viewed in this conceptual framework, but it becomes believable that, indeed, it is machinery.

In an accidental word game, the living cell has become the cell of a cellular automation, and this game called "life" has become the actual life of a multicellular organism.

PHILOSOPHICAL INSERT 1: Metaphysical Musings

What is this universe, that it can bring forth this immense complexity, functional to the smallest of details, mechanical on one side, and meaningful to us humans on the other? In this mechanism there is no hint to be found about anything that would point beyond universal and known physics. Yet there are many serious searchers who have not had the advantage of understanding the craft of physics, who say that all this reference to physics is a result of the way we do experiments and interpret them. That we ask the wrong questions, and therefore obtain invalid answers, and that, if there is any universal truth, it is different.

Every good physicist will understand these doubts and then categorically deny them. And that is not because of "indoctrination," but because the whole interconnected web of physical insights, down to every fine quantitative detail, makes such exquisite sense, is constantly proven correct through technical application, and has an aesthetic quality that convinces. It convinces as naturally as I am convinced when I meet an old friend, that this, indeed, is the same person I know.

Besides, any attempts to find "supernatural" events, "mind over matter," or the like, or ghosts, have failed miserably. An entire generation of "parapsychologists" at a number of academic institutions have tried and failed. Every single account of any finding has been discredited either as fraud, or as an incorrect use of statistics. Some of the "miracle workers" admitted their fraud, others were "debunked" by professional illusionists or hypnotists.

Not to be confused with the supernatural is the truly astonishing influence our brain can have on our body, especially in causing disease as well as healing. The brain also is the organ of central control of the body's physiology. Therefore, shouldn't one expect such influence? The control function remains unconscious, yet it has links to the conscious part. In this way our mind, indeed, can influence our body. But, as will become quite clear later on, the mind also is a function of the body, interacts with it, and depends on it. So the "healing mind" is not outside of the complex physicochemical control system that keeps us alive, but an integral part of it. The whole system is not different from the many self-healing physiological reactions known to medicine, except that we can influence the process through willed practices, including meditation and the like, based on empirical knowledge transmitted to us from earlier generations.

Where does this leave us? I think we simply have to accept that the universe is akin to a gigantic cellular automation, as described before. If we want to remain honest, that is where the current limit to human understanding has to be set.

In Conway's "Game of Life" there is an initial asymmetrical figure consisting of nine live pixels, called "Methuselah." When played, this explodes into an ever-spreading universe with chaotic fluctuations in population numbers; many "gliders" emerge and move out to infinity. Periodic patterns, short-lived beautiful crystals, are seen to flash, and it does not ever seem to stop. (Actually the population of live pixels becomes constant after about four thousand generations, but gliders continue to fly toward infinity forever, and periodic oscillations keep going).

Someone may be inclined to believe that in the gigantic cellular automation of the universe, there was a similar initial "God particle" that caused the explosion called "Big Bang," and furthermore, that a cosmic intelligence had put it there. I would not have any means to contradict this. Such a person would subscribe to the belief that someone knew beforehand what would come of this, in other words, had designed the Methuselah particle in such a way that life on Earth would emerge, even humanity. Here is the point where science ends and mythology begins. "I don't see any conflict between religion and science. Religion has to accept the science of the day and penetrate it . . . to the mystery," said Joseph Campbell (Campbell 1986).

I agree. But I personally do not find it interesting to engage in fruitless speculations. There is too much to do in this world, too many responsibilities to fulfill, too many quests to finish, too much to enjoy, and first of all—too much to understand that can be understood. I would suspect myself of trying to escape my obvious duty, if I indulged in such speculations.

But my friends from the humanities may feel differently. The door to future mythologies and helpful religions always will be open, because existential beliefs cannot be refuted. As said earlier, they do not expose themselves to the tests of reality. Campbell's statement above can be paraphrased: Every time there is progress in better understanding reality, religion has to retreat to where the mystery still remains. The "Game of Life" analogy exposes the current limits of our understanding of reality.

There already has been a retreat as well as a rejuvenation of Christian religion after Copernicus. I can see no reason why a still further retreated religion cannot become healthy and vigorous again, and provide psychotherapy and much needed moral prescription for those unwilling or unable to experience what this book tries to develop: that the true story of life can fill all the mythological needs ever experienced by humanity. As I hope to demonstrate later on, it also can produce a positive and more creative ethic that is natural to the species.

I personally, however, and other people I know, who have experienced the power of natural science and of the true story of life, cannot accept unfounded beliefs without feeling dishonest or, worse, irresponsible with regard to the rest of humanity.

In the meantime the challenge to find our real nature and our real place as part of this amazing enterprise, has still to be met. So we must return to study the web of life that has brought us forth.

Multicellular Organisms and the Need for Communication and Control

As mentioned earlier, it took six-sevenths of the 3.5 billion years of evolution of life, to mature single-cell organisms. But then suddenly a virtual explosion of multicellular life forms happened (Gould 1989), the so-called "Cambrian explosion." The quick evolution of all plant and animal phyla in the last one-seventh of life's history is certainly due to the fact that a very efficient cell chemistry was in place by that time.

If the view of embryonic self-organization is correct, the Cambrian explosion could find a surprising explanation. Might it be that at that time, after the invention of gene repair mechanisms, the number of base pairs in the DNA of single-cell organisms had reached a critical margin? Above this margin enough different state cycles, and thus different types of cells could synthesize, so as to permit the level of cell specialization necessary for a true multicellular organism? Now suddenly many new "experimental" multicellular organisms could begin to sprout. Subsequently, only the best functioning survived natural selection, as the paleontological record actually shows.

The breakthrough came when some single-cell organisms began to form colonies. These cells probably were less sophisticated than today's protozoa. When a cell divides, there is no need for the two daughter cells to quickly disengage. There may be advantages to stick together. Three more divisions in that mode, and there is a cluster of sixteen cells. Such, and still bigger, clusters have increasing volume to surface ratios. This has advantages of many kinds: Cell surfaces in open contact with the watery environment, teeming with parasites and unwanted chemicals, need to be maintained at the expense of energy. A small suspended particle is more prone to have to adhere to undesired surfaces. A bigger cluster will be able to move more efficiently on its own.

The most spectacular survivor of such early colonies actually belongs to the plant kingdom—the algae. There are several species of *Volvox* (fig. 2.3), hollow spheres formed by several thousand cells (Bold 1957). Most of the cells have cilia. The object is an actively moving plant! Already front and rear end can be distinguished. The smaller spheres, visible within, are daughter volvox. When mature, they will turn themselves inside out and leave the mother sphere, which disintegrates. The difficult inversion is necessary because the daughter's outer "skin" is adapted to the mother's internal chemical milieu, not to the water those algae live in. After inversion, when being released, they will carry some of the mother's internal liquid in their own insides. It is quite remarkable that a similar turning out still can be observed in all early embryonic development of higher animals. At this level a basic blueprint must have been laid down for all the future. The depicted *Volvox aureus* has fully developed sexual reproduction. Male colonies produce sperm packets only. Some of the cells in female colonies are specialized to become "eggs." Only they incorporate the sperm and divide into daughter colonies. Other species of *Volvox* follow simpler and less efficient strategies of reproduction.

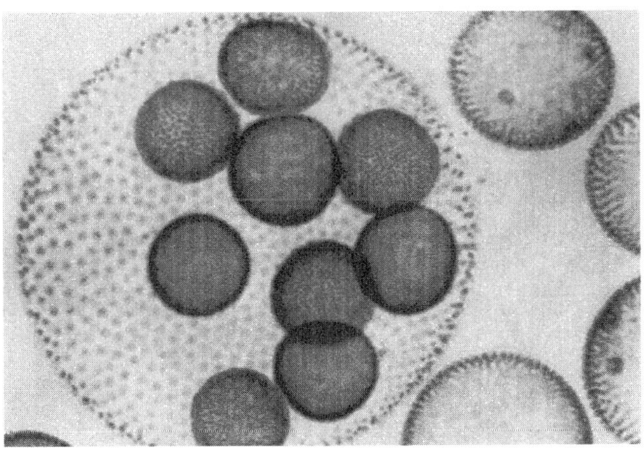

Fig. 2.3
Volvox aureus
[Reproduced with permission of Carolina Biological Supply Company,
Burlington, NC 27215]

Further advanced are the *Mesozoa* proper. To this class belong sponges and polyps. Coral polyps are among them. These are very simple sexually and non-sexually reproducing organisms, consisting of a countable number of cells. Even here the tendency to form "second order" colonies prevails. Corals are an example. The most exotic result of cooperation between primitive multicellular organisms is the Portuguese man-of-war (*Physalia physalis*). It looks like a large (about thirty centimeters) jellyfish with a sail above water and stinging hair of up to fifty-meter length below. But this is a colony of five different polyp populations. There is communication between the individuals because the colony can act as a unit. For instance, the lower part of the sail is periodically dipped into the water to keep it from drying out.

The advantage of multicellular organisms is division of labor. Cells can specialize at different tasks. Situational logic would now require that the tasks be coordinated. Communication becomes necessary. In those early multicellular organisms, communication is purely chemical. Some cells excrete substances into the internal space which are precursors to later neurotransmitters. Other cells react to these signal substances. Who produces what kind of juice, and when, and who moves what, is determined in this way. The release of a signaling substance into the internal fluid, and its detection at another location, is a very slow process. The problem was how to bring signal substances quickly to the target cells, and how to sense signals coming from other cells. To bridge the distance, specialized cells grew long extensions which became the axons and dendrites of neurons.

From the beginning, the nervous system had this double purpose: response to sensory input and control of internal vegetative functions. How ancient and fundamental the stabilization of internal milieu is can be seen from the fact that all cell-surrounding body fluids in us still have the same balance of electrolytes (sodium, calcium, potassium) as the oceans today and probably the early sea, in which the cells originally had evolved before five hundred million years ago. Multicellular organisms had to conform to this chemical need of their building blocks.

And with this an entirely new level of emergence is reached. Neurons build brains and ultimately human consciousness. In construction and function, the neurons of simple animals are already like human neurons.

Neurons

A generic neuron is represented in figure 2.4. The cell body contains all the necessary machinery for division and metabolism. What makes it spectacular are the multiple "branches" sprouting in all directions. The shorter ones are the dendrites. Most neurons also have one longer extension, the axon. In extreme cases, the so-called "pyramidal cells" can have an axon that is long enough to reach from the brain down to the lower end of the spinal column. As mentioned, neurons evolved from cells specializing in long-range chemical communication. Since diffusion of a dissolved substance over significant distance is slow, communication becomes faster, as well as more specific, if the cells grow extensions—the axons and dendrites. These are like interconnecting wires.

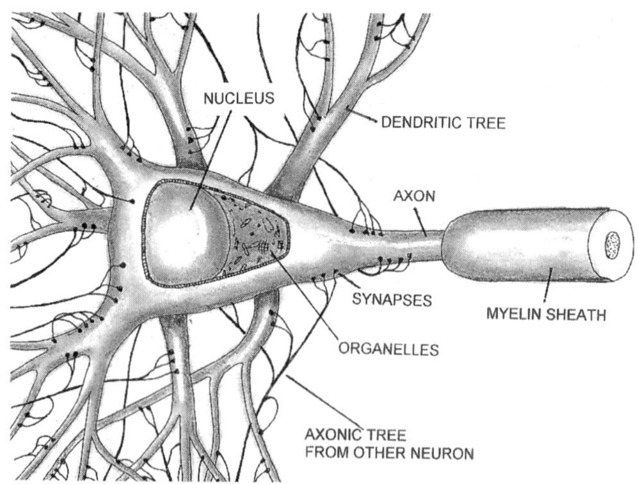

Fig. 2.4
A neuron near its cell body. Many synaptic bulbs are attached

Axons are output lines and dendrites are input lines. Long axons are enclosed in a myelin tube. This tube is a specialized cell itself. It acts like an insulator around an electric wire and vastly increases the speed of axonal transmission. At the end, axons branch out into a number of terminal fibers that connect to other neurons, predominantly to their dendrites, but also to any other part of the cell. The individual connection is a synapse (fig. 2.5). It consists of a bulbous extension at the end of a terminal fiber that attaches to the membrane of the target cell, but leaving a gap of about 0.2 micrometers, the "synaptic cleft." This is the last remnant of the original space separating chemically transmitting and receiving cells. The main transmission path has now become electrical in nature, and thus much faster.[9] The synapse has become the actual electrical connecting point. Its effectiveness to excite or inhibit the "postsynaptic" cell is variable. A neuron in higher animals can have one thousand to ten thousand synapses and can receive inputs from as many as one thousand different other neurons.

In the beginning, the neurons simply transmitted levels of excitation from sensory cells to motor cells. In a radical sense, this is still the most basic function, except that it has proven advantageous to make this transmission more indirect, to permit a global interaction of every sensory input with every other, thus making the response depend on the whole situation, and perhaps on previous sensory inputs. Internal control is similar. It also links sensors to effectors. In this way networks of neurons evolved.

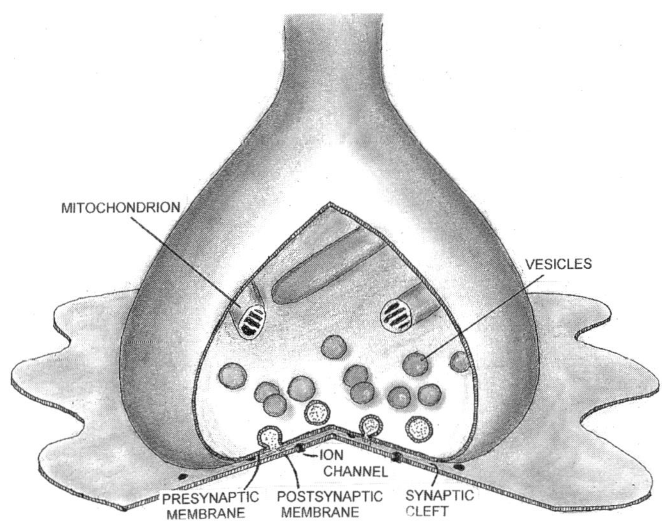

Fig. 2.5
Detail of synaptic bulb with cleft

[9] In electrical transmission, no actual transport of electrons or ions is needed from one end to the other. It is the repelling force between like charges that is immediately felt at the other end of this line. This translates into voltage (or potential).

A level or degree of excitation is mirrored in the frequency with which a neuron "fires." Some neurons can fire up to five hundred times per second. Each firing event is a short positive pulse in the electric potential on the inside of the cell. Normally the "resting potential" is about minus-seventy millivolts relative to the outside. The firing is an avalanche-like event caused by the gated sodium channels in the cell membrane. There may be a million such channels in one cell. Part of them is electrically gated. When some of them open due to some initiating event, positive sodium ions flow into the cell and rise the potential; this causes even more channels to open, and an avalanche of sodium ions follows. A sharp positive pulse is observed. After a few milliseconds, however, another event, the opening of potassium channels to the outside, makes the potential drop back to the resting level. Once triggered, the event propagates as a wave along the axon. Such a wave can travel with a velocity of up to one hundred meters per second and is not attenuated, since the axons also contain gated sodium and potassium channels, which reinforce (amplify) the pulse as it travels. Thus, a neuron has become a complex electrical as well as a chemical device.

The mechanics of synaptic transmission is as follows: The synaptic bulbs contain vesicles with a chemical substance, the neurotransmitter (compare fig. V 1.3). Upon arrival of the electrical pulse, vesicles burst (compare Vignette 1) and neurotransmitter is released into the synaptic cleft. This causes chemically gated ion channels in the opposite cell membrane to open or close, thus affecting a change in the potential of the target cell (postsynaptic cell). If the potential increases, the synapse is excitatory, otherwise it is inhibitory.

Altogether there are about thirty different neurotransmitters known today—some excite, some inhibit. The excitatory level of a given cell is a result of the concerted input from all synapses ending on it. If excitatory and inhibitory inputs balance out, nothing happens. If a motor neuron, i.e. one that has an axon leading into a muscle is excited, the muscle contracts.

It is wrong to carry the analogy between electronic circuits and neural networks too far. Unlike transistors, neurons are "temperamental." Their response depends on many more things than synaptic inputs. As living cells they are in constant chemical communication with the rest of the body. Their function depends on the state of the body, and in turn they can affect that state. The response also depends on the history of previous responses and excitations. Some of the neurons are naturally "silent" unless positively excited. Others constantly pulsate on their own. Others emit bursts of pulses at regular intervals. Still others specialize in the production and release of neuroendocrine hormones. But all activities are modulated by synaptic and chemical inputs. "The brain is a colony of pulsating polyps," in the words of Ramón y Cajal.

Long and short time changes in the strength of synaptic transmission happen as the result of repeated excitation or chemical influences. Besides weakening of synaptic effectiveness (habituation), strengthening (sensitization) can occur. Other long-term changes range from strengthening of a given synapse due to simultaneous excitation of two inputs, to the entire disappearance of synaptic connections. That is how our brain changes as the newborn learns to see, coordinate, walk, speak, and think.

Simple Neural Systems and Their Behavior

In 1948 Norbert Wiener staked out a radical claim. He said, "We have decided to call the entire field of control and communication theory, whether in the machine or in the animal, by the name 'Cybernetics.'" (Wiener 1961)

According to this, the function of neural systems would be similar to computer-like circuits: the brain—a computer. Among today's "cognitive scientists" there still are people who believe in this.

To biological thinkers, this line of reasoning does not ring true. A neuron has the individuality of a living cell. This individuality is defined by a particular knowledge. It individually responds and changes due to a multiplicity of external influences. Probably no two neurons are exactly alike. Rather than being a computer, a nervous system is a colony of simple individuals, collectively and individually adapted to specific tasks.

To obtain a first order impression of what neural systems are capable of, and how in the course of evolution they have learned to determine animal behavior, we have to turn to the simplest possible animals. Some mollusks have extraordinary large neurons; among them are snails, squids, and octopi. Such organisms have become convenient subjects for the study of function of networks of neurons—basic neuronal functions we share with all other organisms. It is only in number of neurons and complexity of interconnections where we excel. There seems to be one more difference: Simple organisms tend to have fixed numbers of neurons, with fixed, genetically determined synaptic connections. Yet the network of neurons in the human brain with its 10^{10} neurons and 10^{14} synapses[10] cannot possibly be determined by not more than 100,000 genes. As we trace the types of behavior through the ascending line of complexity, we find this difference mirrored: Behavior of lower animals is genetically fixed. With increased complexity, though, more and more individual adaptation is observed, until in the highest forms creative responses emerge, permitting survival in many different environments. This is accomplished by a high degree of initial randomness in synaptic connections. According to a hypothesis by G. M. Edelman (1987), adaptation now takes place by internal Darwinian selection (somatic selection) from variant groups of randomly interconnected neurons. The group that accidentally is best adapted for a given cognitive or motor task is selected for use, thus giving it occasion to exercise and strengthen the synaptic connections involved. Subsequently, the connections within the group fine-tune to optimize for the selected task. But there are other known mechanisms of self-adaptation. This will be discussed in chapters 5 and 6.

Presently we will concentrate on the example of a simple and hardwired nervous system—the sea snail *Aplysia*, also known as the "sea hare." On this level

---

[10] The exact numbers are not very well known. Depending on the author, the number of neurons is generally assumed to be between 10^9 and 10^{11}.

there is not a single brain-like organ, but several clusters of neurons, called "ganglia." *Aplysia* has twelve of them. Each contains one thousand to two thousand neurons. Best studied is the abdominal ganglion that controls circulation, respiration, excretion and reproduction. It also innervates areas on the back which protect the gills. One simple reflex, the gill withdrawal reflex, already can illustrate a variety of basic features of neuronal responses that are mirrored in broad classes of animal behavior (Kandel 1976, 1979). The gill withdrawal reflex serves to protect the tender gills from damage. There is a protruding area on the back, in the vicinity of the gills, the syphon, so-called because it helps circulate water past the gills. If the syphon area is stimulated by touch, the gills shrink away and become enclosed by the mantle. Figure 2.6 shows a simplified circuit diagram. The big circles represent neurons; the small circles are the synaptic bulbs attached to them.

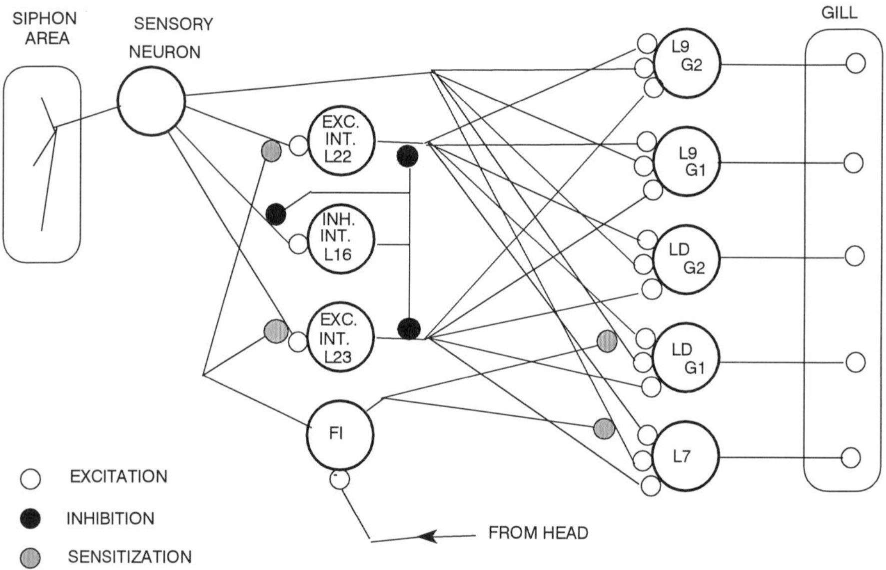

Fig. 2.6
Simplified diagram of connections relevant to the gill withdrawal reflex in *Aplysia* [After: Eric R. Kandel, "Cellular Basis of Behavior," W.H. Freeman and Co., 1976]

Three layers of neurons can be distinguished. The sensory neuron on the left is representative of an entire layer. This is the "input" layer. The second layer is represented by the four "interneurons" in the middle. The third layer is indicated by the motor neurons on the right. The main neuronal pathway is simple: The sensory neurons have input fibers in the syphon. The main reflex "arc" is as follows: Upon stimulation of the syphon, the sensory neurons stimulate the motor neurons of the third layer directly. These send axons into the muscles which cause the gill withdrawal. The other pathways probably are a secondary development. The direct pathway is supplemented by a "detour" through a middle layer of "interneurons" which also connect to other points of the system.

It is especially here that response modification occurs. For example, if the animal would crawl into a tight space, like under an overhanging rock or a large shell, and the syphon would be repeatedly or constantly touched, the creature might die from suffocation even though there was no real danger to the gills. The answer to such problems is habituation. The reflex disappears when constantly triggered. There is no need to go into the many details of how this happens on the molecular level. It all is based on long and short-term changes in synaptic strengths, particularly in the excitatory interneurons L22 and L23. In addition, L16 inhibits its neighbors and itself. This circuitry creates a precarious balance between excitation and inhibition. It determines how much excitation is passed on to the motor neurons from this layer. Small changes in synaptic strength can tilt it in either direction, rendering the system more responsive.

Instead of habituation, sometimes the opposite is required—sensitization. This is the increase in strength of the response. If the syphon of *Aplysia* is stimulated together with a noxious event to the head, indicating a more dangerous situation, the synaptic strength for gill withdrawal increases. This is accomplished by synapses coming from a "facilitating interneuron," abbreviated "FI," that connects to sensory input from the head. Its synapses either attach directly to another synaptic bulb or to the axon leading to it. As the diagram shows, these "presynaptic connections" appear before an excitatory interneuron or directly before the motor neuron. The neurotransmitter released by the presynaptic connection, in this case causes the synthesis of an enzyme in the cell membrane, which in turn causes a further reaction inside of the bulb, which enhances the uptake of calcium, probably by increasing the number of active channels. Such modulating *chemical synapses* will be found throughout all nervous systems. The human brain, for instance, is switched from being awake to sleep in this way.

All the described synaptic changes are said to follow a Hebbian rule (Hebb 1949). Hebb had postulated in a general way that synaptic changes must occur through use. In today's self-adapting artificial neural networks, similar synaptic learning rules are used. For instance, certain networks can learn to recognize the letters of the alphabet by simply being presented with random sequences of letters. In simple organisms one neuron and one synaptic modification may determine a response. In more complex brains it is rather a pattern of many sites within a huge network of neurons that cause more complex responses to more specific stimuli.

The described mechanisms of habituation and sensitization seem to be fairly general. Yet this is a simplified schematic, and it represents only a small sample of what can happen between and inside of neurons. For instance, the inhibitory or excitatory nature of a synapse can be determined by the state of the postsynaptic neuron. Some synapses change from excitatory to inhibitory after several firings. In rapid processing, such as in parts of the visual system, timing sequences of arrivals of pulses determine the overall output. Furthermore, the existence of more than twenty different neurotransmitters promises a large list of different functions of chemical control.

Among the many possibilities of synaptic changes through use is "local interaction" (Alkon 1988), (fig. 2.7). A synapse (*A*) conveys excitation to a postsynaptic neuron (N), via a dendrite. If downstream there is another synapse (*B*), that is excited shortly afterward, it will find its postsynaptic site already in an excited state. Such a simultaneous excitation on both sides of a cleft can cause an immediate and long-lasting strengthening of the site (*B*). This is nothing less than a memory trace, one of many that can reside on one and the same neuron. In simple organisms, one could call such synaptic change "conditioning": Stimuli (*A*) and (*B*) together cause a response, such as approaching the feeding site. (*A*) is the smell of food, (*B*) is a sound. After a while the sound alone will trigger the response.

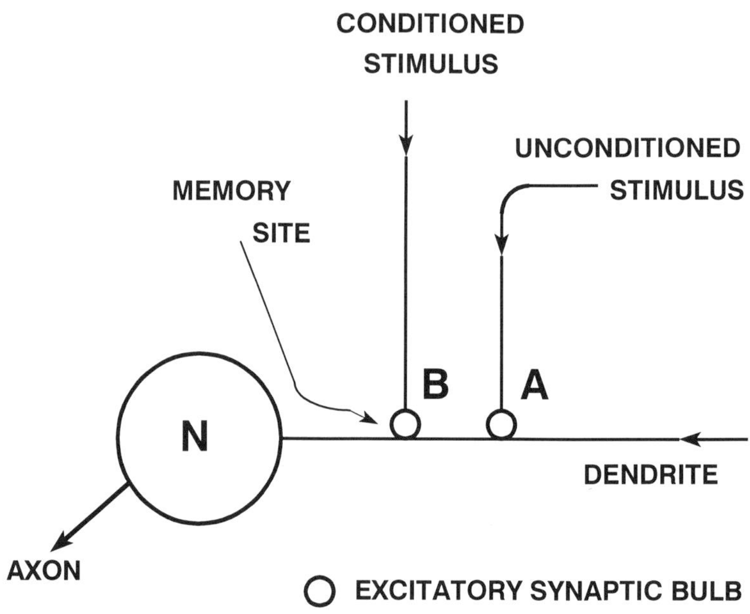

Fig. 2.7
Synaptic strengthening via "local interaction."

Thus, we note that already in the simplest possible reaction to the environment, the reflex, there is variability resembling learning and memory. The potential for variability is genetically preprogrammed. The term open programs has been introduced by Ernst Mayr for this. Soon after neural systems come into existence, they already begin to show flexibility of response. *The environment is multifaceted, and so must be the adapted behavior.*

No leading principle of organization of responses can be traced in evolution of the nervous system, except for a certain hierarchy of importance. The need to flee from a predator has to override the appetite to eat. Otherwise, the behavior of simple animals is more like a "bag of tricks." There is one trick for each external challenge. Even among higher animals, it seems to be rather our own preference for classification that imposes order on the description of behavior. Nevertheless, some standard attitudes of behavior do prevail, because some survival strategies are universally successful through their inherent situational logic. This is what makes ethological studies possible.

Before continuing any further, it seems useful to summarize some of the remarkable global features and capabilities of nervous systems:

1. There are command neurons which are active on their own, but generally modulated by the body's and brain's hormonal milieu.
2. There is complex synaptic connectivity with heavy mutual inhibition, which facilitates precarious states of equilibrium, ready to react to small sensory stimuli.
3. There are presynaptic changes, long-term and short-term, which facilitate habituation and sensitization.
4. Memory appears in individual postsynaptic sites which are strengthened by a long-lasting chemical change upon being triggered by two events in a prescribed sequence. But there exist other mechanisms of memory as well.
5. Each neuron has the ability to store a large number (in the hundreds or thousands) of such individual memory sites, each may be part of a different activity, response or "idea." But an idea, or such, probably manifests itself as a global pattern of many interconnected neurons.
6. The presence of a large diversity of neurotransmitters insures the existence of many variations to the themes struck in points 1 to 5.
7. Altogether the nervous system appears to be a fertile medium for emergence of new and unexpected characteristics of complex nonlinear systems, having an immense number of rules to be played out in the individual as well as in the phylum.

As far as adapted behavior is considered, we should always keep in mind these truly immense potentialities of neural systems. Whenever there is a survival niche to be conquered by behavioral adaptation, phylogenetic or individual, one can count on the ability of the nervous system to ultimately grab the opportunity.

The potential for complex "intelligent" adaptation can be studied with computer simulations of "artificial neural networks," or with more realistic models of interconnected real neurons. In this way, in chapter 5 it will become possible to develop a first order global model of human brain function.

As far as the evolution of behavior is considered, in higher and more complex behavior it becomes difficult to figure out the detailed workings of the networks of neurons. Therefore, until chapter 5, we will abandon neurological explanations and turn to the study of externally visible patterns of behavior—ethology.

A First Look at Ethology

If our own mind is the crowning achievement of Darwinian evolution, then the study of animal behavior throughout the ascending species must reflect our mind's ascent. More than that, the evolution of animal behavior, if compared with the simultaneous evolution of the brain, is one of the keys for understanding the human brain. The other, of course, is neuroscience and computer simulation of neural networks. Therefore we now enter a subject that will directly lead to what the first part of this book wants to accomplish—to discover what really makes us tick.

There are two approaches to the study of animal behavior. One is ethology, as championed by Konrad Lorenz and Niko Tinbergen. The other is Edward O. Wilson's sociobiology. Sociobiology strongly relies on genetic explanations. It even uses mathematical modeling of the population genetics of behavior. Ethology relies on detailed field studies of patterns of behavior, and tries to decompose them into components and compare between species. It often has been accused of anthropomorphizing, appealing to inner "flashes of recognition of the familiar" in explaining the behavior of animals.

Yet if there is continuity of evolution from a shrew-like mammal to *Homo*, as anatomical and paleontological evidence suggests, then such an approach certainly may have its value, particularly in a search for human nature. Today, a related field of inquiry calls itself "comparative psychology."

We will heavily rely on ethology, particularly on Konrad Lorenz's work, who has masterfully (and with proper caution) exploited the inner resonance that is caused when we look at other mammals. When looking at the behavior of other primates, this resonance of familiarity becomes strongest, whereas the genetic explanation becomes increasingly remote, even failing altogether, as in the case of altruism.

We now continue to follow the evolution of behavior. More elaborate than reflexes is a class called "fixed action patterns" or "fixed motor patterns." An example is cocoon spinning by the spider *Cupennius salei*. This is a sequence of about sixty-four hundred dabbing movements of the abdomen, as an oval container is weaved. Into this the eggs are laid. Then the cocoon is closed and detached from its base. This is an automatic sequence without any possibility of modification, as simple experiments show. If in the middle of action the spider is interrupted and the finished portion of the cocoon removed, after half an hour the spider may continue to work. But it builds the top, which now has no bottom. The eggs are laid. The opening on top is closed and the construction detached, letting the eggs fall out underneath. Also, if under the heat of the camera lighting the glands stop to produce thread, the movement still continues. The result does not influence the action in any way, no repairs are done, no modification is possible. One can almost imagine how somewhere a neuron fires sixty-four hundred times until it runs out of fuel.

What made the command neuron start firing and initiate the fixed action pattern? Some hormonal change perhaps, facilitated by the readiness of eggs, possibly in conjunction with the presence of a proper surface to attach the cocoon. Each fixed action pattern must be initiated or "released." The releasing mechanism, in this case, was partly endogenous (from inside of the body). Generally each fixed action pattern has its own *"key stimulus"* that releases the particular sequence.

A famous example of a fixed action pattern on a higher level is the egg rolling response of the graylag goose, studied by Konrad Lorenz and N. Tinbergen (1938). If an egg is placed outside of the nest of a nesting goose, or something that looks like an egg, she will use the underside of her beak to pull the egg into the nest. If the egg is removed after the action sequence has already started, the movement continues. Thus, the visual image of "egg outside nest" is the *key stimulus for the release of this fixed action pattern*. The pattern, by the way, already permits some variations. If the egg is near, the goose can remain on her other eggs. If it is far, she will walk toward it.

The egg rolling movement includes a component that apparently is on a more advanced level. This component uses sensory feedback. Since the beak is narrow, the egg could easily slip away sideways. So, while being pushed back into the nest, a careful balancing act takes place involving tactile feedback from the underside of the beak or from neck muscles. But there cannot be a neuronal connection between the two components because the balancing sideways movements stop when the egg is removed, the linear pushing, once released, continues. N. Tinbergen classifies the balancing component as "taxis," a directional orientation via sensory feedback. Simple taxis already can be found in protozoa. Frogs who want to catch a fly first use binocular vision to turn exactly into the direction of the fly. Then the tongue flicks out. Again, it's a two component action—first taxis, then fixed motor response.

Of course, frogs and geese are quite apart from *Aplysia* or a spider. They have real brains with millions of neurons, and they have a huge repertoire of behavior. Therefore, it is remarkable that still some of the early types of response persist. They persist in still higher animals. When dogs bury a bone in the corner of the living room by going through the movements of covering it with soil with their hind legs, we have a fixed action pattern. Likewise when they turn around three times before laying down, and no grass or snow is there to be trampled. Of course, humans can switch off silly actions by using conscious decision. Still, in the time when barbers used scissors, I often encountered "snipping away on air." Apparently it is easier to keep the automatic fingers going than make an effort to turn snipping on and off at the right moment. Of course, snipping away is not a fixed motor pattern that is innate. But once learned, synaptic connections are made on a level below where consciousness resides. Now we have a learned fixed motor pattern. No music could ever be played and no ballet performed without acquiring a repertoire of learned fixed motor patterns.

The above examples of taxis and fixed action (or motor) pattern shall presently suffice. Inventory of such components of behavior is being taken by researchers in ethology. They proceed through painstaking field observation, recording, and statistical evaluation, as well as laboratory experiments on animals. Journals and books are filled with an impressive accumulation of facts. One good introduction is the already cited text *Ethology—the Biology of Behavior*, by Irenaeus von Eibl-Eibesfeldt. The existence, throughout the animal kingdom, of repeatedly appearing types of behavior testify to successfully evolved patterns of neural connection and functioning. Like the liver, heart, and kidney, such networks are organs of behavior which, once invented and successful, are passed on to higher and higher species. They might become superseded by more complex forms, but often can be recognized as building blocks of the new.

After the 1960s, ethological thinking had to establish itself against some deeply ingrained objections from the behaviorist camp. Today we can safely say that the behaviorists have conceded. In order to avoid being accused of teleology by behaviorists, Konrad Lorenz and his school had stopped using the word "innate." It is much more to the point to replace innate by "phylogenetically acquired." The behaviorist postulate was that all behavior in higher animals, as well as in *Homo*, is learned by the individual. This erroneous view goes back to origins rooted in a psychology that was not based on biological thinking.

As a discipline of biology, ethology is closely related to morphology which is the study of animal forms and organs, their correlation and possible succession along the phylogenetic tree. Likewise, ethology is the study of animal behavior forms, their relation and succession along this tree.

How does one prove that there are behavior patterns which do not have to be learned? The king's highway to this is the deprivation experiment. For instance, a bird can be raised by artificial incubation, put in complete isolation, and restrained from flapping its wings. Yet when set free at the proper stage of development, it flies. It couldn't be otherwise. The species where fatalities from learning to fly are prevalent would have become extinct a long time ago. Other isolation experiments show that some bird species sing on their own; some need "tutoring." Whitethroats (*Sylvia communis*) have been raised in isolation in soundproof chambers. They spontaneously develop their species' complex repertory of songs (Sauer 1954). Chickens produce their typical calls under the same circumstances, even when surgically deafened. However, Oregon juncos (*Junco oregonus*) only acquire their typical song if they can hear themselves (Konishi 1964).

Chaffinches (*Fringilla coelebs*), when raised in isolation, develop an incomplete song, but one that roughly resembles their species-specific production. When recordings of different birdsongs are presented to them, including their own species' version, they "recognize" it and learn to imitate the whole sequence. If "strophes" are presented in a different order, the unnatural sequence is imitated. Other birds do lack their own innate song. Bullfinches raised by canaries learn the canary song and pass it on to their own young (Nicolai 1959).

Konrad Lorenz (1965) lists other clear cases of complex behavior that do not have to be learned—the starlings' navigation by the sun, which it has never seen before; the mantis' complex mechanism of exact aiming and grabbing its prey; the fixed courtship movements of the drake, which release a specific answer in the duck; the internal clock in many different animals, and all opto-motor responses. Many of the patterns can be recognized as phylogenetically acquired without performing any experiment at all:

> A young male salticid spider which, after molting for the last time, approaches a female must neither mistake another species for his own, nor must he perform the signaling of his specific courtship dance in any other way than the one to which his female responds; otherwise, he would be eaten by her immediately. He has no opportunity in his short life to gain any information about what a female of his own species looks like, nor what movements he must perform to inhibit her feeding reactions and to stimulate her specific mating responses. (Lorenz loc. cit.)

Even the human infant is born with a multiplicity of responses, as was demonstrated by J. Bowlby. (Bowlby 1958).

As mentioned, primitive building blocks of behavior may remain functional on higher levels. One example of this will become relevant in humans. It is the existence of *super-key-stimuli*. "Egg outside nest" was the key stimulus that released the egg rolling sequence in the graylag goose. Under natural conditions all the eggs encountered are of similar size. But if an experimenter positions an extra large artificial egg in front of the incubating bird together with a normal one, the gigantic egg proves to have "supernatural" releasing power. In figure 2.8 an oyster catcher tries to recover such an egg. The bird tries to roll and mount the super-egg, which could not be successfully incubated even if it contained a fertilized embryo. Such overriding *super-key-stimuli* do not occur in nature. That is why they can derail a well-established survival strategy.

Another example of a super-key-stimulus can be seen in territorial aggression of the male stickleback. This fish considers any object with a red belly a male intruder into its territory (Lorenz 1966). The attack is proportional to the size of the artificial red belly. Natural competitors are disregarded in the presence of an oversized stimulus. We will encounter the power of super-key-stimuli again in the second part of this book, in the artificial stimuli introduced through human culture. And humans will not be exempt from their power.

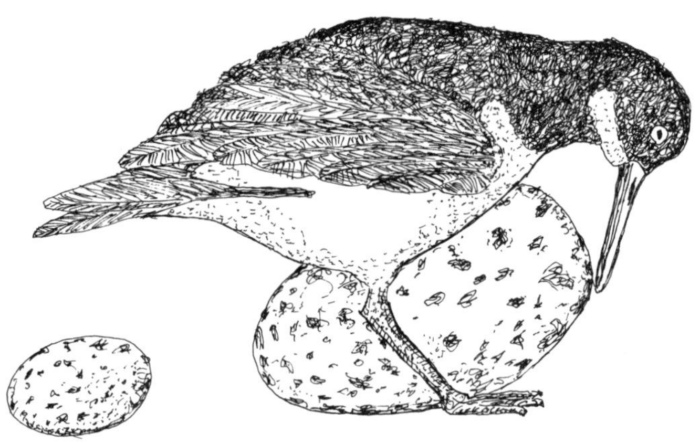

Fig. 2.8
Artificial giganic egg is an irresistible super key stimulus for egg rolling
[After N. Tinbergen]

From Motor Learning to Innate Curiosity

When a mouse is put into new territory it begins to explore, moving slowly, feeling its way with the vibrissae. But soon it learns to run quickly along well-chosen paths. It establishes *path habits* based on motor learning (mnemotaxis). This has extreme survival value: When pursued by a predator it can run with quick, well-trained movements, matched beforehand to all the obstacles. A good way to catch a mouse (according to Lorenz) is to first chase it off the familiar territory. How automatic the path habit becomes can be demonstrated by deliberately changing some of the path's features. Lorenz writes:

> When I removed from the path of the water shrews a little wooden box they had accustomed themselves to jump on and run across, they jumped into the air at the point where the obstacle used to be, and then sat there on the ground in a state of complete disorientation. Then they began to explore with their vibrissae, and turning back, they recognized a section of the path along which they had come. Gaining confidence, they turned to face forward again, rushed off, and jumped into the air again at the very same spot. They reminded me of children who get stuck when reciting a poem and start again a few lines further back in order to try and get over the difficulty—with mnemotaxis![11] (Lorenz 1973)

---

[11] Mnemotaxis, or "movement by memory."

Such motor learning can be understood as assembling a series of consecutive motor patterns. Unmodified "general" running is obviously phylogenetically acquired. But along a familiar path one runs with short steps here, then switches to a big jump there, and to a gentle curve at the end. One fixed pattern triggers the next. There seems to be a rhythm inherent in the movement. Indeed, isn't reciting a poem similar? Both activities are "Motor Melodies" (Uexküll 1936). John Eccles has shown that motor skills are coordinated in the cerebellum, a very ancient part of the brain. Even in humans, once learned, motor habits are hard to change. Athletic coaches know the difficulty of modifying wrong habits. Yet once a skier, always a skier.

The learning of path habits and other motor skills is a great step in the ability to acquire knowledge. This knowledge is about orientation in space. The highest skills in motor coordination can be found among the tree-dwelling monkeys and apes. In swinging from branch to branch and tree to tree, the quick succession of grabbing and releasing of hands and feet at exactly the correct moment must be somehow prefigured in the brain. There is not enough time to wait for detailed sensory input and processing. Such prefiguration must in some way be associated with a model of three-dimensional space and the temporal movements possible in it. The better the model, the more skilled the movement. It is natural that *Homo sapiens* descended from the best experts in motor coordination—the apes. In fact, in chapter 4, I will discuss the origin of thought in the pre-imagination of motor sequences.

The types of learning discussed so far—habituation, sensitization, conditioning, and sequencing of fixed motor patterns—were driven by the environment. And even though there is long-term memory involved, this memory is automatically accessed by events, such as a conditioned response to the end of one fixed motor pattern that triggers the next.

The conditioned sequence of fixed motor patterns gives birth to a new phenomenon: *the joy of exercise and play.* An animal that relies on path habits, or any other sequence of conditioned action or motor patterns, must be under the selection pressure to ever more perfect its skill through exercise. How will evolution solve the problem of motivation? Simple—it invents pleasure of exercise. The animal which likes to exercise will be at an advantage. Now this is a rather obvious solution. Pleasure really was already invented; from the beginning there was search for food, protective shelter, and sex. All such activities come under the name "appetitive behavior." When they find their goal or "consummatory act," the search ends, at least for a while, *as if* there is a feeling of satisfaction. The *as if* is important. We probably never will know if and how animals feel satisfaction. But obviously there must be something inside that reports: done, relax. F. A. Beach (1948) has performed an operation on male chimpanzees preventing chemical signaling when the seminal vesicles are empty at the end of copulation. The unfortunate victims continued to copulate until utter exhaustion.

This search, consummation, and rest, has to be given a name. If we call it "search of pleasure," everyone understands. This does not necessarily mean we ascribe human-like inner experiences to the animals. Nevertheless, after that mechanism has evolved to the human level, search of pleasure properly describes that mechanism at its "upper human limit."

Pleasure is caused by a simple neurochemical mechanism. There are "pleasure centers" in the upper brain stem. When excited, they release chemicals. Similar to drug-induced "highs," this causes a feeling of pleasure that is sought after. In fact, drug addiction is a chemical super-key-stimulus that causes super-pleasure that is very hard to refuse.

Coming back to the motivation for the conditioning of motor skills through exercise: Pleasure was there already. The novelty was merely to link it to exercise. Joyful exercise is play. Seeing kittens play proves the point: The high jump, coming down with extended claws, the imagined catching and letting go of the prey, all indicate playful pleasure. Cats, dogs, wildcats, and wolves play "rough and tumble." Cats "catch" the thread ball; dogs retrieve the stick and "shake it to death." Squirrels play "hide and seek." Even adult cats do actually catch, let go, and recatch their prey. How else can we interpret this, but by saying the cat enjoys exercising the skill of catching. Anyone who has seen dolphins perform cannot suppress the feeling that they thoroughly enjoy themselves, the reward being only part of the game. H. F. Harlow (1960) has demonstrated the same urge to exercise their acquired motor skills in macaques. They displayed their skills "for sheer pleasure," without any reward given. Joggers know the high spirit induced by running.

Expanding on playful exercise, Lorenz postulates the existence of a *"perfection reinforcing mechanism"* and speculates on the origins of art, especially dance, as a result of such endogenous motivation. The survival value of that mechanism should be obvious by now. Dancers are motivated by the joy of perfect movement, and spectators are charmed by it.

With ever-increasing memory, the stage is now set for another great emergent leap in the evolution of knowledge. If there is memory, and an action sequence has resulted in a pleasurable reward, this can be remembered: Do this and you will experience pleasure! *Temporal feedback from success has been invented.*

To fully appreciate the greatness of this new invention we have to compare it to the older scheme of purely genetic knowledge gain as it is represented in figure V2.1. Behavioral mutation in a primitive organism is a long-term experiment. If successful, there will be more offspring and the new genetic trait will crowd out the less successful traits within a number of generations, (fig. V2.3). But since in higher organisms mutations are random, most of them will not result in an increased number of offspring and will soon disappear again. Thus, at least in higher animals, progress of genetic evolution is slow and has to be measured by many lifetimes of a large number of individuals. Still it remains a form of feedback from success or failure, like the new kind of feedback via neuronal memory. There are two overwhelming advantages, though, to neuronal memory. The first is speed of adaptation. One individual can now sequentially adapt to many different environmental niches. The other lies in the fact that a wrong attempt does not result in death of an animal and its potential offspring, it only results in the death of an "idea." The individual animal can continue to experiment and soon can stumble upon behavior that leads to ever more consummation and thus to better fitness. The new mechanism amounts to nothing less than *the innate desire to learn*.

To summarize the chain of events that has led to the new level: At the beginning we think of the single-cell organism in the primordial soup. It simply feeds and multiplies. But the soup is not equally good everywhere. Those who can, move toward better pasture. Thus, *kinesis* and *tropotaxis* emerge. Meanwhile, symbioses and colonies prove to be advantageous, and the multicellular organism evolves. Among other organs, it needs neuronal networks to communicate between sensory and motor areas. As the animal searches and escapes danger in a changeable environment, better adapted, more complex neural systems are favored. Habituation and sensitization become the first mechanisms of individual adaptation of behavior.

Meanwhile, network complexity grows further. The "joy of exercise" becomes an innate reward system[12]. Its advantage is the learned acquisition of optimal motor skills—one of the most obvious advantages in survival. This further favors more neuronal complexity. Finally, out of the endogenous motivation to exercise and "play," aided by the increased number of memory sites, there emerges individual learning from success or failure. Whereas habituation, sensitization and conditioning were initiated and driven by external events, learning now happens through active exploration. Learning has become a species-specific, internally motivated activity—an instinct. We might as well call this instinct "curiosity."

In fact, curiosity can be widely observed, particularly in young animals. If a new object is placed into a rat cage, all will cautiously approach and sniff. They will try to chew on it, climb on it, and urinate on it. If the object does not contribute to any needs, it will be soon forgotten.

Not all agree that joy of exercise comes first, then results in play, and later the instinct to learn. Eibl-Eibesfeldt writes:

> One could even assume a separate drive to play, but I am inclined to believe that the drive to learn, which is the basis of all curiosity behavior, coupled with an excess of motoric motivation, will suffice to account for the phenomenon of play. Learning takes place during playful experimentation in the same way as during endlessly repeated movement games. (Eibl-Eibesfeldt 1970)

---

[12] A pleasure center is located in the brain stem (compare chapter 5). If an electrode is implanted there and a push-button is available for electric self-stimulation, rats will self-stimulate to exhaustion. Humans report pleasure with sexual overtones.

The truth may be that motoric motivation, curiosity, and play form one of the mutually reinforcing circular feedback loops so often found in the process of adaptation.

We now have come to a point where in rough outline the origin of certain mental accomplishments has been explained. Mostly the discussed achievements have been of the kind that on the human level would be called "cognitive skills." Essentially those are reactive. *They permit us to learn to react skillfully to the demands dictated by the realities of the external world.* Yet in the individual learning of such skills, certain innate or instinctual determinants were already found at work. These set and pursue an active goal. *On the human level such active endogenous motivation would be called "the will."* Our inner experience of this will amounts to much more than just seeking pleasure. Frustratied will gives rise to the strongest of all possible emotions.

With this we enter a world that is very different from cognitive skills. Here, we do not want to know, but we want to win.

Chapter Three

The Emotions of Being
or *The Roots of Will*

The will to live, internally, drives the perfection of motor and cognitive skills. But externally we may encounter other individuals having their own will. And that may spell trouble. Will against will is the fabric of social interaction. On the evolutionary time scale, the will to live must have preceded any cognitive abilities. In fact, these abilities only serve that original will to survive. The social world remains tough and violent, or at least charged with strong emotions. Animals and early humans developed elaborate means to escape this predicament. But even today we are plagued by the difficulties of living together in peace and harmony. Lately, it appears that we have not progressed, but have possibly even regressed in that art.

The Four Types of Aggression

Social skills also are a form of knowledge. They show how to treat one's companions in order to increase one's own success and the success of the species. But this knowledge is of an entirely different order. It remains largely instinctual even in higher primates. Social interaction was too important to be left to the liberties of individual modification. Of course, on advanced levels, individual learning also begins to enter into social behavior, but at a slow pace. Even in humans, where cognitive abilities are crowned by rationality and consciousness, social interaction often remains chaotic, bloody, and irrational.

A) Territorial Aggression

In many species one finds aggression between its members to be a widespread and most elementary form of social interaction. It all begins with territoriality, which according to Lorenz, might have originated with path habits. If an animal knows a territory well, it is at an advantage in it. It does well to exclude from it all rivals of the same species, because they all eat the same limited resource and use similar shelters. Many species do best if their members are evenly distributed over the available ecological space. A given square meter will only permit a certain amount of food gathering or hunting. If overgrazed, the resource will be destroyed. If underutilized, more individuals will be added until an equilibrium population density is reached. A simple way to insure uniform distribution is mutual aggression. It is even found that aggressiveness toward an intruder increases inversely proportional to the square of distance from "headquarters." There is obvious selective pressure for stronger aggressiveness: The more aggressive animal will have a larger territory, more resources, and more offspring.

Some now-classical observations of territoriality (Lorenz 1962) were made on populations of fish on coral reefs. The coral reef is a habitat of many niches. Different species can coexist, each specializing in a different resource. The striking coloration of those fish is like a flag. Aggression is only directed toward the same flag since it feeds on the same resource. In fact, the coloration is thought to have evolved so that peaceful coexistence of many species became possible. To mistake someone for a food rival is wasteful. Territorial aggression is widespread. It ranges from the little stickleback (a fish), through lizards and songbirds to grizzly bears and howling monkeys. And we certainly should not forget man. But many other animals do not own a territory and often prefer to live in herds, or schools. Sometimes a change between the two modes of existence is possible. Sticklebacks swim peacefully in a swarm until they find a suitable territory. Then they change their attitude. Even the color of the male's belly changes to red.

Of course, territoriality and reproduction cannot be independent. Generally it is the male who is more territorial and more aggressive. If a male red grouse is implanted with a pill supplying excess male sexual hormone, the bird becomes more aggressive and increases its territory (Watson 1966). A male owner of territory may admit one or more females, often after difficult instinct releasing rituals designed to initiate pair bond. Then breeding takes place. Now defense of the territory becomes indistinguishable from defense of the female or of the young.

In the natural habitat, territorial fights rarely result in damage. Inhibition to inflict damage has evolved together with territorial aggression. Damaging a member of the same species is counterproductive. In mammals, ownership of a territory is marked with all kinds of scents, urination, defecation, and with special scent glands. Such marked ownership often is not contested. Well-known is the territorial song of birds. Lorenz writes:

> It is remarkable that in many species the song indicates how strong and possibly how old the singer is, in other words, how much the listener has to fear him. Among several species of birds that mark their territory acoustically, there is great individual difference of sound expression, and some observers are of the opinion that, in such species, the personal visiting card is of special significance. (Lorenz 1966)

According to ethological theory, intraspecies aggression is an instinct—a phylogenetically acquired, species-specific, endogenously motivated activity. Aggression can be released by stimuli such as the presence of a territorial or sexual rival, and it is not learned. If not released, sometimes the threshold for aggression is lowered until it might erupt upon a minimal stimulus or no stimulus at all. But this is not true in general. Observations of instinctual aggression abound (Eibl-Eibesfeldt 1970). Rats and mice raised in isolation show all their species-specific aggressive behavior, occasionally with more than normal intensity (King and Gurney 1954), (Banks 1962), and (Eibl-Eibesfeldt 1963). According to Lagerspetz and Talo (1967), aggressive behavior in albino mice matures spontaneously around the age of twenty-eight days. Some strains are more aggressive than others, and crossbreeding shows the factor of inheritance (Lagerspetz 1964). Fighting cocks and the fish Betta splendens can be taught a task if the reward is a stimulus situation that releases fighting and threat behavior (Thompson 1963, 1964, 1969).

B) Mating Rivalries

Another form of aggression is associated with mating. In many species males accumulate and jealously defend a "harem." Or males fight among each other and the winner is accepted by the attentive female. This breeds stronger animals, a clear advantage. Yet the process may get out of hand. Sexual rival fights primarily breed rival fighting ability, which is not necessarily identical with survival fitness. That is how the gigantic antlers in some species evolved. Some can serve no other purpose but in rival fights and are shed and rebuilt every year, posing a tremendous load on general fitness. Sexual selection also may produce male birds with an enormous display of useless feathers. The peacock is an example and the bird of paradise, which is flight impaired due to its beautiful tail plumage.

C) Ranking Rivalries

Still another kind of aggression can be found in some animal species which live in social groups. Here, each individual has a clearly defined social rank that is defended. The term "pecking order" seems to stick to such behavior ever since it was introduced by Schjelderup-Ebbe in 1922, based on studies of domestic chicken. It is not clear whether social ranking has phylogenetically evolved from territoriality. But it is very likely that territorial aggression has been simply modified, now defending the territory of the entire group against other groups. The original aggression between individuals becomes tamed through the pecking order. If everyone's rank is clearly defined and known, wasteful infighting can be kept to a minimum. Social ranking can only establish itself in animals with true learning ability. After all, one's rank is learned by trial and error, an error being to start a fight with a stronger animal.

Another survival value of social ranking order can be seen when the group is threatened, for instance by a predator. The threat can now be met in an organized fashion. Generally, it is the experienced and strongest leader who takes action, either by leading his group to safety or, if possible, attacking or misleading the attacker. Do we see here the invention of moral responsibility? Washburn and DeVore (1961) once observed a baboon band that had entered open and dangerous territory. The group halted, and young and strong males protectively surrounded it. The oldest male went to explore. He located the lion, fortunately without being detected himself, then led the group along a safe path. All followed blindly without anyone questioning his authority. Why did the leader endanger himself? He was the dominant male with first access to food, females, and social attention. If a predator attacks, self-preservation would demand: Use your strength to run. Let a weaker individual perish. In fact, one might see here a selective advantage in the preservation of the strong. Why then is it that the leader takes action to try to preserve the flock? Does power bring responsibility? Later on I will try to resolve this paradox by positing the existence of a "similarity" bond between the animals.

Before considering this quasi-morality, we will have to look at one other broad area of social behavior—*Inhibition of aggression and ritualization*. But at first there is a fourth type of aggression to be listed.

D) Intertribal Aggression

Animals who live in groups and have tamed their rivalries by establishing ranking order generally defend their tribal territory against other tribes. It is not clear, however, whether this is a "redirection" of a basic aggressive energy from other individuals to other tribes. In all aggression there also is the genetic argument. Territorial aggression, mating and ranking rivalries, all favor the spread of more aggressive genes. In tribal aggression, "group selection theory" gives a fashionable explanation. The more aggressive tribe will have a larger territory, more food supplies, and, hence, the tribal gene pool will spread. In time it will crowd out other less aggressive gene pools.

Whatever the origin, the phenomenon itself is not contestable. The genetic explanation is strengthened by observations of rat and mouse societies. Here, ranking rivalries are absent. Probably there is not even individual recognition, but every rat with a communal smell is accepted as equal. If two different groups are forced together in a cage, mutual slaughter begins immediately. Males predominantly kill males, and females kill females. But this does not lead to the extinction of one of the groups. Rather, as soon a pair bond forms between two individuals from different tribes, this becomes the dominant couple. One by one, all other individuals are murdered as if by premeditation. A peacefully feeding rat may suddenly find its neck vein bitten through, and bleed to death (Lorenz 1966), (Steiniger 1950).

Here, we see the lower end of tribal aggression, where the tribe is a family. Where the tribe contains non-related individuals, no such extreme aggression is known. But the phenomenon prevails. In the next chapter we will encounter chimpanzee wars, and human political history up to now has been not much more than a succession of territorial wars.

The Taming of Aggression

Social ranking takes interesting turns at higher levels of evolution. Wolves are the most efficient killers. One would think that ranking or rival fights among wolves can become very bloody. But quite to the contrary, the animal with the strongest aggressive weapons, also has the strongest inhibition to inflict damage to its kin. A fight is instantly ended when one animal presents the vulnerable veins on the side of its neck to the other, in this way indicating submission. Other terms of signaling submission are derived from infant behavior (fig. 3.1) such as begging for regurgitated food and rolling on the back as if to present the anal region for cleaning as the young do (Eibl-Eibesfeldt 1970). The young are always protected. The wolf's social structure is extremely well-developed. A pack gives the impression of a family of intelligent and caring individuals (Lawrence 1986), (Fox 1971, 1980). Wolves are probably the most intelligent non-primate mammals, except for brain giants like elephants and dolphins. No wonder that the most intelligent primate has developed an early relationship with wolves—dog and man.

Fig. 3.1
Sumissive gestures of wolves. Food begging and presenting for cleaning
[After I. von Eibl-Eibesfeldt, "Ethology," Holt, Rinegart and Winston, 1970]

Aggression may be necessary, but in another context, fighting consumes energy (food) and time. Both generally are scarce commodities. Therefore selection has favored forms which minimize the effort but still serve the purpose. That is how *ritualization of aggression* came into being. With increasing mental abilities it became possible to estimate one's strength in relation to another animal without a real fight. That, of course, has happened in wolf society millions of years ago. Hoofed animals with horns or antlers have elaborate procedures to measure strength in a dignified way. A good example is the ritual neck wrestling between two Nilgai bulls (*Boselaphus tragocamelus*), figure 3.2, (Walther 1958). Each animal tries to press the other to the ground. The loser concedes without further ado. Such "matches" proceed in sportsmanlike fashion. Male elk behave according to strict and gentlemanly etiquette. They walk stiffly alongside each other, then step back, and when both are ready, they lock antlers and push. Often ritualization goes to the extreme, that a "meet" is avoided altogether. The weaker animal might be convinced to leave upon seeing or hearing the rival's obvious superiority—the bigger antlers, the vigorous song, the mighty puffed up posture.

Fig. 3.2
Neck wrestling of Nilgai bulls
[After: ibid.]

Emergence of ritualized fight reveals a new level of insight into certain characteristics of another individual animal and the proper reaction to it. Already, after unritualistic ranking fights, other individuals are personally recognized and remembered. *Now recognition and assessment can replace actual fighting.* By accepting this as an indication of a higher level of insight, we do not have to fall into anthropomorphic concepts. The new ability does not have to be conscious in the human sense, and the decision to fight or not to fight does not have to be called "rational." Yet the reality of the phenomenon is incontestable. Ritualization, once invented, will pave the way toward many new and more subtle forms of social behavior. Human competitive sport is only one of the late results of this development. The ritualized territorial wars among the Mayas of middle America was an earlier phenomenon of this type.

Aggression in Humans

When Konrad Lorenz's *On Aggression* was published in 1963, it stirred a vigorous dispute. On one side were the practicing ethologists and many biologists, who had become all too familiar with the ubiquity of intraspecies aggression as an instinct among animals in their natural habitat. On the other side were the proponents of "learning theory," for whom human aggression was individually learned through cultural tradition. This controversy touched on deeply-rooted philosophical convictions. Idealists did not want to give up the hope that humanity could be reeducated. The followers of Hegel and Marx found their masters contradicted. And to the liberal wing, for whom everyone is supposed to be born equal, it all smelled of racism since one could imagine aggression to be a racial characteristic. Most such opinions were not based on genuine understanding of the matter.

But is there an aggressive drive in *Homo sapiens* that is released on suitable occasions? Occasions would be seen in war, social suppression, nationalist propaganda, domestic disagreements, and general hopelessness. Even though the bloody history of our species and a brief survey of daily newspapers would confirm this, there are non-believers who would like to blame the cultural environment alone. Yet objective physiological studies on humans also support the view of our basic aggressiveness (Hokanson and Shetler 1961), (Eibl-Eibesfeldt 1970). *Recognition of such innate tendencies is morally very important, and to some it is a humiliating revelation.* One has to be careful so as not to fall into a simplistic trap. Intraspecies aggression may be a fundamental early development, but it is only one of the many instincts on higher levels that form "the great parliament of instincts" (Lorenz 1963). All of the mentioned inhibition and ritualization is present where aggression is present. In addition, I will suggest that a basic similarity bond exists underneath all aggression.

The instinctual underpinnings are an important ingredient of the human psyche, but we also have other emergent characteristics such as rationality, compassion, and an active center, the ego, with a certain power to reason. Rather than to conclude that man is an innate brute without hope for improvement, I would like to compare the emotionally healthy individual to an *orchestra of instincts, with a conductor firmly in charge of the beautiful music that is made by the many instruments (the instincts).* However, as will be illustrated in the second part of this volume, this kind of emotional health can easily disappear when the cultural support system breaks down. Then Serbs begin to exterminate Bosnians, or Turks slaughter Armenians, or German Nazis annihilate the Jews. Chimpanzees also have tribal wars of extermination.

The Rituals of Bonding

There are animal societies where being together is not a problem. Shoals of thousands, if not millions, of tiny silversides play the tropical seas. Diving through such a cloud of fish gives the impression of being surrounded by one diffuse organism.[13] They move, turn, and dance as a unit around you. There is no individuality in such societies. According to Lorenz, the survival value of shoals lays in the difficulty a predator has to catch a single individual in the presence of a hundred others zigzagging across his field of view. No predator attacks in a herd. Even lions first separate a prey from the rest of the herd of gazelles, zebras, or wildebeests. Shoals of silversides probably have roamed the seas for many millions of years exactly as they do today. It is territoriality and aggression that bring complications which have to be resolved by new emergent levels of social interaction.

We have seen how ranking fights and ritualized aggression lead to individual recognition. It is one of the most fascinating stories in the evolution of behavior how aggression is turned into a personal bond between individuals. Not unexpectedly, the beginnings of bonding are found in mating behavior and the care of the brood. Cichlids are a family of territorial fish known for their extreme aggressiveness. How does a female intrude into a male's territory for breeding? The answer is: very carefully. In the aquarium she might easily get killed.

Konrad Lorenz describes the ritual that takes place. She carefully enters the perimeter of the territory and is attacked on sight. Yet the male's aggressivity is at a minimum out there, and she can either quickly withdraw or present her side as an appeasement gesture. She tries this tactic again and again. Gradually the male habituates to her presence and she succeeds to enter the inner regions. But (in Lorenz's words):

> . . . the partner must always appear on the accustomed route, from the accustomed side; the lighting must always be the same, and so on, otherwise each fish considers the other as a fight-releasing stranger. Transference to another aquarium can at this stage completely upset pair formation. The closer the acquaintanceship, the more the picture of the partner becomes independent of its background, a process well-known to the Gestalt psychologist as also to the investigators of conditioned reflexes. Finally, the bond with the partner becomes so independent of accidental conditions that pairs can be transferred, even transported far away, without rupture of their bond.

In a further development the bond is firmly sealed:

> At first nervously submissive, the female gradually loses her fear of the male, and with it every inhibition against showing aggressive behavior, so that one day her initial shyness is gone and she stands, fearless and truculent, in the middle of the territory of her mate, her fins outspread in an attitude of self-display, and wearing a dress which, in some species, is scarcely distinguishable from that of the male. As may be expected, the male gets furious, for the stimulus situation

---

[13] The reason for this impression is that immediate neighbors move almost exactly parallel, with the same acceleration and maintaining exact formation. This is the only way to avoid a hopeless traffic jam.

presented by the female lacks nothing of the key stimuli which, from experimental stimulus analysis, we know to be strongly fight-releasing. So he also assumes an attitude of broadside display, discharges some tail beats, then rushes at his mate, and for fractions of a second it looks as if he will ram her—and then the thing happens which prompted me (that is K. L.) to write this book: the male does not waste time replying to the threatening of the female; he is far too excited for that, he actually launches a furious attack which, however, *is not directed at his mate but, passing her by narrowly, finds its goal in another member of his species.* Under natural conditions this is regularly the territorial neighbor. (Lorenz 1966)

One can assume that personal recognition now has occurred. But the key point here is the invention of a new "psychological tool"—*redirected activity.* The aggression is released by one object, but in the presence of a second instinct, the inhibition to hurt the mate, it is actually discharged somewhere else. The phenomenon of redirection is widespread. It must have emerged many times over in independent phyla. What an ingenious solution to a dilemma! The necessary territorial aggression remains in full order, or is even increased. It is now needed more than ever since the territory must be enlarged to feed two, and soon the hatchlings. But reproduction can proceed in peace.

A more evolved level of redirection is found in ducks, geese, and cranes. The famous dance of cranes is an example. The male opens its mighty wings, and with beak and gaze pointed straight at the female begins what looks like a dangerous attack. But abruptly he turns sideways, and now with wings still outstretched he presents to the female his vulnerable region in the back of the skull. Some species have a red mark there to accentuate the gesture. And now the bird runs an attack against another nearby crane. If none is available, it may choose a goose or even an inanimate object which it may pick up and throw into the air several times. The ritual is repeated often. It is as if the bird says (in Konrad Lorenz's words), "I am big and threatening, but not towards you—toward the other, the other, the other." Redirected aggression has turned into something new: a bond reassurance ceremony, or better, *a greeting ceremony.*

Much studied is the still further evolved ceremony in geese, particularly graylag geese. Figure 3.3 shows a diagram after Helga Fischer, one of Lorenz's coworkers. In this species the personal bond has become so strong that only little aggression is left. Yet the ceremony still prevails as a reconfirmation of friendship. Now the attack on the bad neighbor precedes the quasi-aggressive movement toward the female. The latter becomes a "coming home in triumph." Postures one, two, and three all have been well-studied by ethologists, and known to express aggression. Actual attack on a suitable enemy may ensue at this point. Four and five would indicate a serious attack on the female, except for the direction of beak and eyes, which point past the partner. As can be seen, both partners participate. The phases one, two, and three are accompanied by loud trumpeting fanfare. In the return phase, more subdued chatter or cackling can be heard, which Lorenz calls "passionate." The origin of this "triumph," or greeting ceremony in mutual aggression, can still be noticed when the partners have been separated for a while. Upon return, the greeting ceremony becomes particularly loud, passionate, and frequently repeated, as if the budding aggressive alienation has to be driven away. Or do geese enjoy seeing each other again?

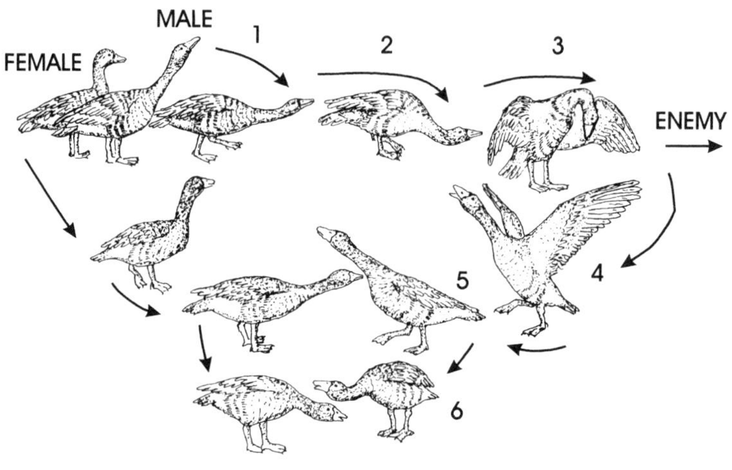

Fig. 3.3
Triumph-greeting ceremony of greylag geese.
[After: K. Lorenz, On Agression]

By studying homologies of this activity in various species of ducks and geese, Lorenz and coworkers have conclusively shown that the greeting ceremony evolved from redirected aggression. In hindsight, this is not a novel scheme of evolution: Swim bladders have changed into lungs and certain small bones of the jaw joint have become incorporated in the construction of the ear.

Goose behavior is remarkable in many other ways. They generally mate for life, and life can span fifty years. When one mate dies, the other goes through a period of severe physiological stress. Reduced tonus makes the face sag and the eyes recede. The bird loses strength and motivation and slides to the lowest rank in the group. In its restless search for the missing companion the goose becomes accident prone and may perish.

The lifelong association should be called the "triumph ceremonial union." Lorenz calls the triumph ceremony the "leitmotif" of a goose's life. Not only does the ceremony seal and constantly reinforce the bond; fragments and traces of it can also be found in most of the goose's behaviors and noises. In fact, it is more important than copulation. Males occasionally seek to copulate with strange females, but on those occasions the ceremony is never performed nor is the other female's nest guarded. The ceremony still remains exclusively monogamous. Occasionally a triumph bond is formed between two males. A female may become accepted and a "triangular" relationship formed. Such a group generally will become a high-ranking entity. Its breeding is highly successful since both males guard the nest.

Adolescent and "premarital" relations can strike a very "human" note. Unrequited love happens when one side already has begun to form a triumph bond somewhere else. Such a courtship does not last long, but Lorenz observed a hand-reared female whose ". . . faithful love for a happily mated gander persisted for more than four years. She never openly followed him and his family, she always 'just happened' to be where he was. Every year she proved her own fidelity and, incidentally that of the gander to his female, by laying an infertile clutch of eggs."

Konrad Lorenz does not shy away from using the word "love" without quotation marks. After all, he says:

> When cephalopods, like the octopus, squid, and cuttlefish, on the one hand and vertebrates on the other have invented, independently of one another, eyes built on the same principle as the lens camera, and when in both cases these organs have similar constructional units such as lens, iris, vitreous humor and retina, no reasonable person will object calling both the organ of the cephalopods and that of the vertebrate an eye—without any quotation marks. We are equally justified in omitting the quotation marks when speaking of the social behavior pattern of higher animals which are analogous with those of man. (Lorenz 1966)

Lorenz has been viciously attacked for such anthropomorphizing. Indeed he presents a philosophical testing stone of great strength. If we want to believe in our divine origin and God's moral law, we will cringe at the sight of his monogamous geese. But if we are content with our humble origin from the animal tree of evolution, we will delight in the beauty to be found in the reflections of our own existence in animals. Ultimately, I think, this more humble origin might be the more genuinely divine.

The conclusions to be derived from geese research are as follows: Human and goose emotional behavior in certain aspects is astonishingly similar. But we have not descended from birds. Our common ancestor is a very primitive lizard which certainly did not have any of the described behavior traits. Consequently the similarities are "analogies." In different phyla, evolution has found similar solutions to similar problems, such as the need to efficiently use resources and at the same time to reproduce, care well for the brood, interact socially on the background of preexisting aggression, and still keep the aggressive energy ready. It may be needed to defend the brood and to extend the territory.

What can be learned from Lorenz's geese, is a lesson on the nature of our emotions. Already Charles Darwin (Darwin 1873) noted the similarities in animal and human emotions. This does not mean they are inferior. Love's delight and pain can inspire us to the grandest achievements. Or should we say "redirect" us to the grandest achievements. Without Eros there is no art, no science, no joy, and no life. We should well recognize Eros' roots in animal nature. But the nature of man has integrated many additional developments. Human flirting, for instance, can be the most refined expression of our second nature—wit, the culture of emotions. As our rationality (logos) enables us to cultivate plants and breed livestock, so it enables us to cultivate our inner garden of emotions, to produce flowers whose beauty goes light years beyond goose love—or at least such can be our subjective belief.

The Similarity Bond and Beyond

We have found that intraspecies aggression generally is accompanied by inhibition to hurt severely. Moreover according to Lorenz, personal recognition and bonding is supposed to have evolved from the basic aggression. But there is one simple logic that would say that there must be more to the story than mere taming of aggression. *What would keep the group together if there were no positively attracting motivations?* This point has not been stressed much by ethologists, probably because overt aggression is much easier to observe than covert attraction.

When it was said that the pecking (or ranking) order makes living in groups possible, that did not inform us why the animals want to live in groups to begin with. Of course, there always is the genetic argument: Under certain conditions, living in groups improves survival. Yet ethology and comparative psychology believe in uncovering endogenous motivators. Following that philosophy, one is compelled by logical necessity to assume an underlying attractor. I will call this the "similarity bond." I think this basic communality already begins on the lowest levels of cognitive skills and first includes parents and siblings. The lower the level, the more uniform or similar are all reactions. Therefore, wherever learning begins, among the most constant features to be learned of the external world are the similar reactions of all the other animals of the family group, and later of the extended group. All this causes "familiarity" with the other animals. This activates one of the most ancient instincts: stick to the familiar, be it path habits, known territory, or familiar animals.

There is even a still lower-level neurological argument for the existence of the similarity bond. From the excited tweet and chirp of a flock of blackbirds to the chatter of a group of squirrel monkeys, one can literally sense the communality of emotion. Stronger even is this impression when diving through a shoal of the already mentioned silversides, even though their common world is not acoustic. Such phenomena certainly are associated with similar neuron firing patterns in the individual brains. Obviously there is communication. But instead of looking at it through anthropomorphic eyes, namely that here is sending and receiving of messages between individuals, on the most primitive level one should simply say: *There are no real individuals here; excitation of the acoustic medium of transmission is just an extension of internal axonal transmission.* Thus, a primitive group can be seen to have, in certain respects, a communal neural network, a communal brain. Every subsequent behavioral and cognitive achievement would have to build on this communal foundation. Hence the similarity bond may be so universal and so obvious that it tends to escape attention.

One can see the bond in action when a flock of birds suddenly alights after one bird alights. The hasty flapping of the wings of the first bird is the signal. But how do we interpret what follows? It cannot be a rational decision: "the hasty departure indicates that this bird has seen danger, I better follow suit." No, the danger reaction must have spread through a kind of contagion. It seems natural to assume that certain movements or noises evoke patterns of neural excitation in the other birds that are *similar* to the ones that have caused the movement (hence similarity bond). Is it a "feeling" of danger (for lack of a better word), that spreads through the flock? Whatever it is, it causes the flight reaction as if by imitation. What humans call "imitation" may well be a component here, but I think there is much more to the equation. *It is as if one animal gets under the skin of the other.*

Imitation produces a new phenomenon—local culture. French and American crows are the same species; they interbreed. Yet when sound records of American crows' calls are played to French crows, they are not understood. An American alarm call causes French crows to gather instead of to take flight (Frings and Frings 1959). French herring gulls do not react at all to American herring gull calls. Local birdsong dialects within the continent have been recorded on several occasions. Here is new emergence: cultural transmission—the first beginnings of culture can be found in animals! *Similarity bond plus learning equals culture.* The theme begun here will give a foundation to the second part of this book, and it will find its conclusion in the "Philosophical Insert 4."

On the level of birds, for us humans it is rather difficult to develop a meaningful understanding of this cohesion. But, at least among primates, the basic attraction can be well-observed in their regular efforts at consolation after an aggressive encounter (DeWaal 1989), and here it can be meaningfully understood by us. With a number of studies on chimpanzees, bonobos, baboons and rhesus monkeys, Frans DeWaal and his coworkers have uncovered a foundation of social cohesion underneath the aggressive encounters. After the fight, a tense period of stiff silence may follow, then the involved individuals "make up" by hugging, hand-holding, grooming and kissing, whereupon a visible wave of relaxation spreads through the entire group. One of the most pronounced features of chimpanzee groups is communal emotionality, as will be described in more detail in the next chapter.

On higher levels it seems to be proper to use the words "proto-compassionate bond" for the basic underlying cohesion. In the human species this will turn into compassion. Of course, even here the bond can be broken, and if broken by humans, it might be a much more serious and lasting affair (See cultural amplification in chapter 7).

The presented view has strong explanatory value in one hotly disputed area—altruism. In chapter 7, on cultural evolution, it will be explained that standard selection theory cannot account for altruistic behavior in mammals. If, however, the basic similarity bond and its further development into compassion are accepted, the problem disappears.

The early evolution of a similarity bond would make this a fundamental mode of communication between individuals. Even though apes lack language, there is no lack of communication of inner emotional states, leading to the mentioned communal emotionality. More than that, intentions and wishes of many kinds are immediately understood by all. We know the mode of such communication very well—it is *body language*. To some extent such communication can even jump boundaries between species. Every dog owner knows that, even though such exchange is a bit single-sided. The dog understands our moods, but we understand little about the dog.

Among ourselves we have lost the sophisticated use of body language. We can achieve some improvement by attending an acting school, whereas apes remain natural masters of that language. Any slightest cue lets one individual feel what is "under the skin" of the other.

Where this emotional mode of communication completely fails, however, is in cognitive insights related to the external physical world, such as manipulations with tools and gadgets by apes. This will have to wait for the invention of true language.

At the end of next chapter, I will have to further elaborate on the distinction between the two modes of communication. *In humans, art, poetry, literature, music, and dance are rooted in the similarity bond. Science and technology have to rely on cognition and language.*

To continue the quest of understanding our own nature, we now have to turn to the minds of our nearest relatives, the other primates.

Chapter Four

The Primate Mind

In the natural history of the human spirit, apes are the most instructive link between us and the animal world. On one side, they express all the emotionality of aggression and compassion they have inherited from their own predecessors, but in a form that begins to strike a chord in our own inner experience. The similarity bond can jump species here. On the other side, apes begin to show the first stirrings of rational insight that leads to language and reason, again demonstrating to us the roots of our own abilities to reason and to control emotions. As far as language goes, young bonobo chimpanzees pick up language almost like human children, without being actively taught. Having no voice box, they converse via computer buttons and electronic speech synthesizers.

The Uneasy Society of Baboons

Somewhere during the last twenty million years of evolution, mental skills reached one fateful new level of emergent integration—the mind of large apes. It is here that the natural history of humanity has its direct point of origin. At that time the brain of primates was no larger than that of other mammals. What made the difference was the ecological space. Members of the branch from which humans emerged likely had been dwelling high in the trees of primeval rain forests. To be successful in this environment, one needed skills that went far beyond what was required of ground dwellers. To climb and jump from tree to tree required a firm grip, excellent spatiotemporal coordination, short reaction time, and precise binocular vision. The binocular vision, aided by perception of color, measured accurate distance to the other side. The spatiotemporal coordination told how strong to flex each muscle when jumping, taking into account the sway of the starting branch. Finally the sure grip and quick reaction time had to save the animal's life every time it took hold of the other side. Four gripping hands evolved. Large areas of the brain became devoted to sensory input from the palms of the hands and the many muscles operating this complex mechanism, a feature we still inherit today. Thus, primates became master manipulators.

Gazelles and buffaloes live in the two-dimensional space of flat prairie. Their minds can remain two-dimensional. Birds follow a blueprint too ancient to be of much relevance here. The mind of tree-dwelling primates, however, evolved to be at home in three-dimensional space, another feature our mind still inherits, even though our ancestors left the rain forest a long time ago. Cognitive skills, once acquired, are too important to be lost easily. The point can be made that the *homo branch* is a result of tree-dwelling primates forced to leave their forest habitat for the prairie—thus becoming "overqualified" for the new environment, and therefore a commanding success.

The other lucky accident was that for the long-armed apes bipedal locomotion was the better way to increase running speed in the prairie. Now things could be carried along and a better panoramic view from a higher vantage point was obtained.

The baboons we encounter today in the open savannas of Africa are *not* our closest living cousins, but they have in common with our direct ancestor that important feature of having left the forest, without however, having become bipedal. Our closest living relative, the chimpanzee, unfortunately has remained in the forest. Therefore we will take a look at baboons first.

So many behavioral similarities can be found between apes and us, that female intuition and social sensitivity become great assets in ape research. The best and most devoted field work in ape ethology has been done by women. Jane Goodall, Dian Fossey, and Birute Goldakis became well-known. Less known is Shirley C. Strum, who researched baboons. In her book, *Almost Human*, Shirley Strum writes:

> When I looked at the baboon world through "baboon spectacles," I saw a complicated landscape, characterized by sophistication and social intelligence, populated by animals with long memories who, relying on social reciprocity, were of necessity "nice" to one another. In this world, males and females shared a complementary importance in the life of the group. (Strum 1987)

This impression describes well what happened to aggression and the ranking order when they reached the level of baboon society. The long-term memory lets them not only know each other, but also remember "individual social obligation." Aggression is not absent, but there are other social alternatives which render it less important. Social ranking is still there, but "political" moves between individuals have introduced a moderating effect. Yet, as we will see, matters may not be as smooth as they appear to be on the surface.

Strum cultivates a compassionate feeling toward the animals. Thus, a picture of baboons emerges which not only is objectively correct, but also has a humanistic depth dimension. Her baboons have anxieties, frustrations, and compassion. Yet what separates them from us becomes equally obvious. For instance, Shirley Strum was accepted as a non-threatening member of a group of thirty to forty animals. When observing certain rules of behavior, like avoiding eye contact and sudden movements, she could freely move about her friends. A juvenile tried to groom her, an adult male solicited her help. An adolescent male began to follow her, this being an activity that is typical and generally results in attempted or successful copulation. *Did they not notice that something was "different"?* When behaving in socially accepted forms, *for the baboons she was a baboon.* We remember the similarity bond. It can even build bridges between species.

Obviously, beyond the similarity bond that relies on body language, there is little insight into the realities outside of the communal social world. Injured individuals can be severely mistreated if they cannot show the required social responses. A male with a leg injury had to hobble along with great difficulty. This was perceived as threatening behavior, and the females "fled screaming." An injured individual may even be attacked. The similarity bond seems to break down when an inner experience is encountered that is not part of your own inner repertoire. It also remains nonfunctional when a screaming baby antelope is devoured alive. Dead infants are carried along until decomposition begins, then forgotten when occasionally laid down. A dead friend or family member is inspected and then left as a "social nonentity" (Strum's words).

But on the social side, even Shirley Strum was charmed. "The most difficult part of my fieldwork," she writes, "was adhering to my policy of not interacting with the animals. It took tremendous determination not to communicate with my subjects."

Everything a baboon does, on one level, is body language. Walking can be threatening or demonstratively relaxed. Position of tail is important. Facial gestures are known. Any direct gaze is a threat. A stronger threatening gesture is the "open-eyelid-flash," which consists of raising the eyebrows and hereby exposing the white of the eyelids, a gesture also known to human ethology. "Presenting" the rear end can be a greeting or appeasement. It is used independent of sex and also by the young. Early explorers of Africa have reported the same gesture in black women (Eibl-Eibesfeldt 1989). A "come-hither look" is known that consists of shaking the head, narrowing the eyes and smacking the lips. Probably this is an invitation to get groomed. A "fear grimace" and a "play face" have been described. All the signals convey an emotional state of the signaler and obviously are understood emotionally and elicit an emotional response.

The common emotionality becomes especially apparent when a new male wants to join the group. Exchange of males happens frequently, the more enduring center of the baboon troop are the females with their friendships and mutual child admiration. The new male patiently remains on the periphery for months. He tries to appear as inconspicuous as possible, but is viewed with tense apprehension by everyone. In Strum's account of such a situation, after a while he (Ray) begins to follow a peripheral female. *Confrontation with other males is avoided.* The female is scared for a long time. Everyone's attention is focused on the developing drama. Finally nervous grooming occurs. Jealousies of a previous consort arise, but the other male turns to some of his other "female friends," for comfort. Further contact is established through curious juveniles. The new male tries to build a network of grooming friendships. Finally a short and bloodless exchange with a high-ranking male becomes unavoidable. Yet still the predominant mode of social interaction remains through building friendships.

Among females, infants form a center of attention. It is a sought after privilege to touch or even handle someone's new baby. "Formal" visits are paid to the new mother and proceed under strict observation of relative rank. Lasting friendships between unrelated females are known. They may be interrupted by family duty, such as the arrival of a new child or grandchild, but are resumed as soon as time will permit.

How strong is their altruism? Children, even unrelated ones, are always helped. Friends help each other to prevail in ranking confrontations. Food, especially meat, is rarely shared among baboons, but is shared among chimpanzees. Most impressive is the self-endangerment of a "scout," as mentioned earlier (Washburn and DeVore 1961). Proto-compassion begins with an extreme tolerance for the young. Bellies of dominating adult males are used for slides. Children make big jumps in the general direction of adults, without any regard as to what body part they will land on. They stick fingers into the elder's noses, eyes, ears, and mouths, but tolerance prevails.

Figure 4.1 shows a typical marching order of a baboon band. This formation impresses by its appearance of functional group organization. The pregnant females and females with infants are in the middle, surrounded by the much bigger and stronger males. Juveniles and adolescents swarm unchecked around everyone, but can quickly take refuge within the circle. One strong male leads, and another mimics the rear guard. Yet this arrangement is not rigid. High-ranking females also have been observed to lead. It is not clear whether it is a "follow the leader" situation all the time.

Our powers of field observation generally end at the skin or fur. What goes on inside is hidden. Yet the work of Robert M. Sapolsky manages to provide a glimpse of the hormonal-emotional inner world of baboon experience. He

Fig. 4.1
Typical marching order of a baboon troop
[I. DeVore (1961)/Anthro-Photo, by permission]

combined ethological observation with studies of blood physiology in the observed animals. The animals are "darted" with an anesthetic through a blowgun. The primary object is to study stress for application in human medicine. Sapolsky (1987, 1990), and Sapolsky and Ray (1989) found a clear difference between dominant and subordinate males when they looked at the testosterone levels after a stressful event. Figure 4.2 shows the measured concentrations after the stress which is caused by the darting, and a short period of fearful disorientation before the drug takes effect. Testosterone is released at the end of a chain of reactions which are initiated in the brain's hypothalamus when it secretes cortisol. Testosterone primarily increases the amount of glucose that is supplied to the muscles. This is fueling muscular strength and endurance. But we know well the feeling that comes with this reaction. It is the stress anxiety we may experience when we face a tense situation, for instance, when we think a burglar is in the house.

High testosterone levels may be necessary for the animal to function in an emergency situation that requires all the strength possible, but those high levels are bad for long-term health. Therefore it is important to quickly reduce the level after the event. And here is where the difference between dominant and subordinate baboons is seen. The dominant animal shows a strong short peak in testosterone level, whereas the subordinate one actually responds by dropping the level. Nevertheless, *the average level of the subordinate is found to be higher*, rendering the animal to poorer health. These differences do not seem to be inherited or acquired during early development. If there is a change in "leadership," for instance when the alpha male ages or loses health, the new leader soon acquires the dominant physiology and the old one loses it! The effect is psychosomatic. It seems to originate in the social situation and is precipitated by the brain.

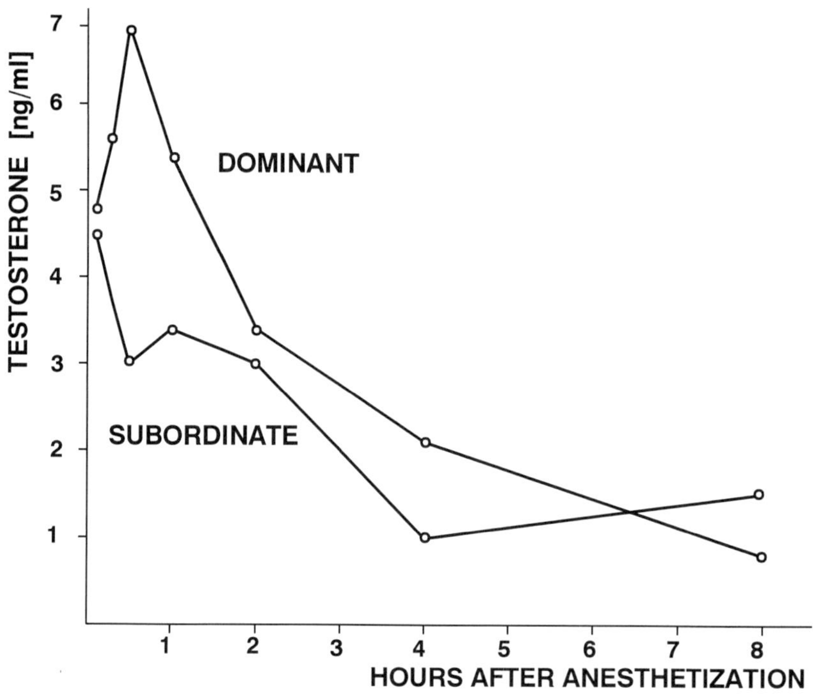

Fig. 4.2
Testosterone levels of dominant and subordinate babooon males after stress
[After: *Scientific American*, 262:1, (1990)]

The animal that remains in subordination has to live with constant stressful fears. His food can be taken away at any time by a more dominant individual, he cannot copulate in peace and must leave the safest sleeping place. That further increases the testosterone level. Sapolsky writes:

> . . . the olive baboons occupy a social landscape of Machiavellian dimensions. Alliances shift unpredictably; threats range from days of harassment to sudden outbursts of violence, and a baboon bent on avoiding the turmoil may still fall victim to another animal's problems. (ibid.)

Now we understand the endless sessions of mutual grooming. They are an antidote to stress. Shirley Strum, indeed, has described the visible relaxation associated with that activity.

Still, violent confrontations are rare, but now *we find the pecking order internalized*. Instead of overt fighting between bones and muscles, there is a struggle between psychological-hormonal inner "bodies." Aggression is still as good as new, but now it ravages on the dark invisible battlefield of emotion, frustration, and lonely stressful endurance. So, indeed, things are not as smooth as they appear to be. We feel some of our own traits mirrored.

Since Hans Selye's conceptualization of the stress syndrome in 1963, many studies have been conducted on humans. It was found that there are psychological defenses which can reduce stress. Among them we find: displacement by other challenging activities, self-hypnosis (a denial of the facts which precipitate the particular stress), and religious faith. If stress is caused by internal control of aggression, if not its frustration, one survival value of religion might be seen in the reduction of stress. More recently, the stressed manager goes to the workout or has a massage while his baboon cousin is being groomed by his most favorite female. But on second thought, the stressed manager may not be beyond a favorite female either.

Sapolsky and DeVore's descriptions of baboon social behavior include more problems of dominance than Strum's descriptions. Of course, there might have been environmentally conditioned local differences among the observed baboon troops. But maybe social intuition of a human male tends to understand the stress of dominance as akin to his own, and be more interested in it, whereas the woman might more readily recognize the tendencies of social cohesion. We have reached the point where animal observation turns into psychology. And psychology is a more uncertain subject.

Our Cousins, the Chimpanzees

"I thank them for all the times they worked so hard to teach me how to be a better chimpanzee." (Savage-Rumbaugh 1986a)

One can be taken in by baboon emotions, but one is awed by chimpanzee mentality. Chimps (*Pan troglodytes*) live in the forest. There always is a vertical escape route nearby and no need to stay under constant group protection. Most of the day they may forage independently, or in parties of only two or three. But they belong to a community and frequently stay in touch with each other through "pant-hooting" their position. This is a four-part vocalization consisting of wails, hoots, shrieks, and roars. Marler and Hobbet (1975) concluded that the vocalizations were individually recognized, so each knew where all the others foraged.[14] Moreover, with some practice even the human observers could recognize the voices. In good community spirit, pant-hooting, and buttress drumming can call the others to a discovered food site, such as a fig tree with ripe fruit—but only if there is enough fruit for many (Ghiglieri 1975). They do not only share food; also, in general, the intra-group relationships are more relaxed than among baboons. To compensate for this, however, there is a strong aversion of strangers (Goodall 1986). Apprehensive tension markedly grows as neighboring territory is approached. Territorial boundaries are fluent as in many other species and depend on the current relative aggressive strength and the numbers of the competing groups.

---

[14] An old Latvian woman once described to me how, in her country, the mushroom hunters in the forest indicated their mutual wherabouts. This was "yohooeing"–the human equivelant of pant-hooting

Most charming is the chimp's need of social contact, often expressed as physical touch, which includes embracing, hand-holding, kissing, and patting. This kind of contact is especially sought in situations of excitement. They have a strong emotional life and a need to share their feelings. Sharing calms them down. Only few examples will suffice to convince: If prey has just been killed, the onlookers become very excited; they jump, run around, stamp, embrace, and kiss. Similar commotion can be observed upon an unexpected find of large amounts of food, such as a large bunch of bananas (Goodall 1986). Figure 4.3 shows two males engaging in an open mouth kiss on such an occasion. When an adult male, Graybeard, first encountered his own image in a large mirror installed by Jane Goodall, he was extremely scared, "grinned widely in fear," hair bristling, and "reached out to four-year-old Fifi, who was nearby, and drew her into a close embrace. Slowly the grin left his face and the hair sleeked" (ibid.).

The dominant male, Goliath, once attacked another male who fled screaming. But then, still screaming loudly in distress, the attacked chimp slowly moved *backwards* toward the aggressor, looking cautiously over his shoulders. When he arrived at arm's length

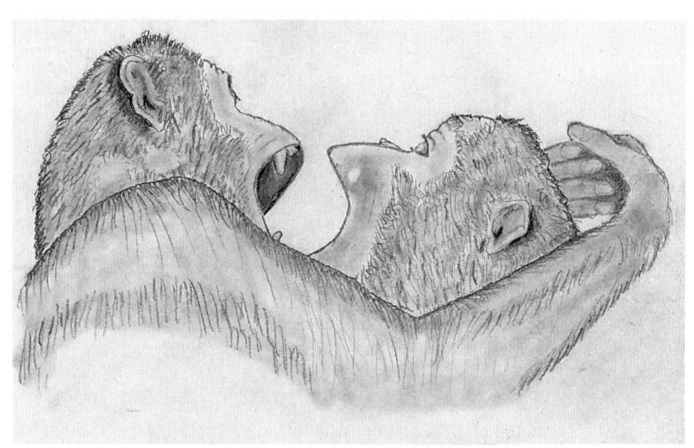

Fig. 4.3

Open mouth kiss of excitement between two males
[After: Goodall (1986)]

he crouched, still screaming. Finally Goliath began to pat him gently until both had calmed down (Ibid.). A fourteen-year-old male, after being attacked, went to his mother to hold hands (Ibid.). Figure 4.4 shows two females pant-hooting. This generally is an occasion of great excitement. In order to stay calm, they touch. Another exciting occasion is greeting, particularly after a long absence. This sometimes is almost too rough to be "cordial." A heavy male might "joyously" stamp on a frail little female, who screams at the top of her lungs, only to be embraced and groomed the next moment (Goodall 1986). Is there a trace of aggression that wants to be released at the reunion so that things will be smooth again? We remember the frantic greeting among reunited geese and the aggressive origin of the entire ceremony.

But an appearance of tenderness can also be encountered, as figure 4.5 shows. Embraces, kisses, and noises ranging from low grunts to excited screams can all be part of chimpanzee greeting. Erections may accompany any kind of excitement, and the males may touch their own genitals or that of others. Copulation, in contrast, can be rather businesslike. Temper tantrums can be thrown by all ages and sexes. One male had caught a small prey animal and had retired to a tree to eat. Another male asked to share but was ignored. So he threw an extended tantrum by screaming, stamping, running, and rolling on the ground. After a while, as if to demonstrate the compassionate bond to us, the first male could not take it any longer and delivered half of the prey to his opponent.

Fig. 4.4
Two females with infant pant-hoot while touching for reassurance
[From J. Goodall (1986), with permission]

Tolerance, when needed, can be exercised with great patience. Infants are notoriously bothersome. They often interfere with copulation. Particularly if the mother is involved they can become quite vicious. They climb on the female, grab the male in the face and generally try to get in the way as much as possible. But all is tolerated patiently and the act completed despite the obstacles. Apparently the juveniles are disturbed by emotions which do not yet find an echo in themselves. When the similarity bond fails, this seems to be scary.

Playing games and playacting is not restricted to children. Adults play chasing, tickling, and "play walking," a stiff, grotesque kind of walk. Females may be tickled under the chin by males. Mothers often play with children. If all this

Fig. 4.5
Male, left, is tenderly greeted by a female
[After: Goodall (1986)]

does not strike us as human-like, the bonobo chimpanzee (*Pan paniscus*) will. Figure 4.6 shows a pair of this rare species. Actually bonobos are not at all smaller than *Pan troglodytes*, but they are more elegantly built, and upright walk is seen more often. Also, they sport long hair on the head. They can swim and have been seen to walk bipedally in rivers to catch fish, whereas chimps are afraid of water and drown easily. Bonobos also are more intelligent. This will be discussed later. Frans DeWaal (1989) said, ". . . it cannot be denied that the bonobo is a key species for understanding human evolution." Their skeletal proportions are closer to our ancestor *Australopithecus* than any other ape.

Quite unexpectedly, here is the only other known species where the female is almost continuously sexually receptive. Thus, we humans are not the only "sexual athletes" among primates. In the bonobo, sex has been added to grooming as a means of peacemaking and relaxation. It does not only happen in the heterosexual way; pleasurable exchanges between females and between males are also common. Compared to *Pan troglodytes* there is a more stable association

Fig. 4.6
A Pair of bonobos (Pan paniscus)
[From: DeWaal (1989), courtesy of Frans DeWaal]

between couples, even though pair bonding per se does not exist. Copulation generally is frontal (fig. 4.7). Bonobo males do not cooperate among each other as strikingly as chimp males do. Consequently, the females can take a more central position in the tribe. Food is not shared among the males, but they share it with females.

In contrast to chimps, bonobos almost never resort to temper tantrums. Among males, aggression is typically subdued and may take on the form of one male rhythmically and lightly punching the chest of another, who accepts passively (fig. 4.8). A tense aggressive encounter between two males may end in sexual pleasure giving.

More about the bonobos later on, but presently some of the more disturbing aggressive behavior of *Pan troglodytes* will be discussed.

Fig. 4.7
A happy bonobo couple "making love."
[From: DeWaal (1989), courtesy of Frans DeWaal]

Like ourselves, our closest relatives have an overwhelming sense of tribal chauvinism. The proto-compassionate bond abruptly ends outside the tribe. Young females of other tribes are accepted as reproductive assets, but their previous infants are killed and often eaten. Old, nonreproductive females of other tribes are fiercely attacked and frequently die as a result of suffered wounds. Interestingly, the old female's offspring are not harmed. Sociobiological theory explains why the infants of *reproductive* females are killed: This increases the genetic offspring of a given aggressor male. Estrus of lactating females will return earlier by years so that they can bear out the aggressor's genes. Any behavior which increases the number of genetic offspring would be favored by natural selection.

But why are strange old females killed? Goodall (1986) thinks this may be to break the mother-daughter bond, which can be very strong. Mothers often are seen to accompany their daughters as they travel with males. This is a temporary "marriage," and who needs a chaperon? Social tensions are rarely absent, and two against one may be intolerable to the male. We know the stereotyped human mother-in-law. Still, inhibitions prevail if the mother is of the tribe. In this connection, Lorenz (1966) has observed in other species that aggression and sex are compatible in males, but not fear and sex, whereas in females, fear and sex *are* compatible. Thus, we may see here the beginning of innately caused misunderstandings between the sexes, so familiar to our society today.

Fig. 4.8
Ritual punching between two males
[From: DeWaal (1989), courtesy of Frans DeWaal]

The imperative to prevail within evermore complex social rules and contradicting emotional needs, may provide a strong selection pressure for increased brain complexity. Wherever there is selection, there are failures. Socially inefficient males will have no opportunity to mate. Females who fail to acquire infant-raising skills will cause death of their infant or produce an inept individual. Failure to adapt to the social environment is best described by the word "neurosis." Thus, neurosis, like in humans, may become a family trait that may be more culturally transmitted than hereditary. Jane Goodall (1986) had opportunity to observe a neurotic chimpanzee family. Passion was a female who gave birth in 1964 to a female infant, Pom. Passion showed an "extraordinarily inefficient and indifferent maternal behavior." Nevertheless, Pom survived and formed a "close, cooperative bond" to the mother. But this bond clearly was neurotic since mother and daughter as a team started to steal babies and eat them. Altogether they were seen to kill and eat three infants and are suspected to have devoured seven more during a four-year period. Thus, having neuroses is not a human prerogative. Human psychology understands that early life experiences can be conducive to neurotic malfunctions. And the same probably is true for chimps.

Chimpanzee Wars

Jane Goodall and her coworkers had unique opportunity to live through a piece of bloody chimpanzee political history. When she began her observations in 1960, it appeared that two peacefully coexisting groups came to be fed at her camp. Subsequently, however, only one group was to be distinguished. From 1962 to 1965, altogether nineteen mature and adolescent males participated in the feeding. The whole group consisted of up to fifty-seven individuals and in 1972 possessed a range of about fifteen square kilometers around the camp. Beginning in 1972, a smaller faction became gradually separated and began to occupy a separate range a few kilometers to the south. At first, friendly meetings and the exchange of females could be observed. But then males who only recently had been grooming friends began to attack each other. Between 1976 and 1977, five adult males of the southern group were brutally murdered by northern gangs with participation of females. An old southern female was also killed. Other females either emigrated somewhere else or joined the northern group. By 1977 the southern group had ceased to exist. Now a new strong tribe from still further south tried to take advantage of the weakened north and temporarily succeeded to extend their territory up to the camp area. Ultimately the old northern tribe had to remain content with a much-reduced territory, befitting their genocidally reduced size.

The northern invaders were extremely brutal. Jane Goodall writes:

> Humphrey grabbed Godi's leg, pulled him to the ground, then sat on his head and held his legs with both hands, pinning him to the ground. Humphrey remained in this position while the other males attacked, so that Godi had no chance to escape or defend himself. Figan, Jomeo, Sherry and Evered beat on Godi's shoulder blades and back with their hands and fists; Hugo bit him several times. Gigi raced around and around screaming loudly. Goblin kept out of the way. Finally Humphrey released his victim and the others stopped their attack, which had lasted ten minutes. Hugo, screaming loudly, stood upright and hurled a large rock at Godi; it fell short. (Goodall 1986)

On another occasion one of the attackers sucked blood from the victim's nose.

Such behavior is in stark contrast to the basic friendliness within the group and to what is observed with chimps who are habituated to humans in the laboratory. Obviously there is a very human-like characteristic at work here. Other members of the species can become "de-chimped" (Goodall's word) as other human nations in wartime can be stripped of their humanity. Cruelty is also possible against internal subgroups, as for instance by secret police gangs against "public enemies," not to speak of present-day street gangs.

As may be for many humans, going to war seems to be exciting for chimpanzee males. Jane Goodall's (1986) observations, and others listed by her, confirm this not only for *Pan troglodytes*. Males of many species enjoy patrolling the periphery of the territory and seem to initiate aggressive encounters with strangers. And so do the "militia" groups in the United States, who play war and invent enemies. All this, of course, is well in agreement with the basic and ancient territorial imperative. But knowledge of the formal genetic trait is one thing, to see the side of inner experience of that matter in our cousins, and to relate it to our own, is quite another.

The similarity to human street gangs is uncanny. Each gang has an anxiously defended urban territory that is marked, not by spraying urine, but by spraying paint. It has been reported that females have to undergo initiation rites by having sex with the leader. Alpha animals always have access to all females.

When we are stripped of cultural tradition, indeed, the distance to our animal forefathers shrinks considerably. We become apes with knifes and guns.

The Moral Imperative

Darwinian selection favors the strong, the efficient, and the clever. It seems to breed selfishness. Yet we have encountered acts of self-endangering altruism, and we will meet more of the kind. Altruism poses a problem to conventional "selectionist" theory. The individual who spends time and energy to help others will be less efficient to propagate his own genes. Thus, "altruistic genes" should gradually become extinct. To explain altruism, the theory has been made more sophisticated. It is argued that if an individual helps his kin, which have a similar gene pool, then even if he should perish by fostering his relative's genetic success, the genes he shares with them will be spread. Now, if one translates this into mathematics, there will be a competition between selfishness and altruism, and it is not entirely clear which will win.

But a simplified example can provide a rough idea of the effectiveness of kin selection: A child shares half of its genes with each parent. Siblings among themselves, on the average, also share half of their genes. So, if a brother is altruistic enough toward his sister as to completely forego his own reproduction, he must more than double the offspring of his sister through his altruism before the occurrence of his genes will be increased. In my judgment this is not likely to happen, but it is not impossible either. But first cousins have only one-eighth of their genes in common. If one of the two cousins is altruistic, he must therefore increase the offspring of the other more than eightfold, before he increases his genetic line. This would be almost impossible among mammals.

In parental care, the cost of more reproduction has to be compared with the cost of better care. It is found that in slow reproducing primates the advantage is on the side of better care. Therefore, parental child care can definitely be a result of selection of "caring genes." But it is already questionable whether altruistic cooperation between siblings can be a result of kin selection. Between first cousins this certainly cannot be the case. Therefore, except for the closest kin, among mammals kin selection theory does not explain the occurrence of altruism.

To understand general altruism, one may resort to something else. We noted that baboons remember help received from others. So when animals are able to recognize received altruism, remember it, and are moved to reciprocate in expectation of later rewards, one would find an altruism that may be more learned than genetic—*reciprocal altruism*. Of course, such altruism requires a rather well-advanced level of mental development. But animal groups which practice reciprocal altruism probably will be at a selective advantage against non-practicing groups. So one has to consider selection between groups. This seems to be reasonable. Yet what we see here is not so much selection for altruism per se, but selection for the increased mental abilities which make the practice of reciprocal altruism possible. The increased mental abilities, however, will certainly have many other unrelated selective advantages. So they might be selected anyhow, without the participation of altruistic behavior. On the other hand, one could imagine geniuses of selfishness to emerge if altruistic behavior would not provide a decisive advantage. So, above a certain level, altruism and cognitive abilities probably coevolve as a mutually reinforcing feedback loop[15] .

But all this appears to be far removed from genetic mechanics. I do not think that the concept of "altruistic genes" has much explanatory value on the primate level. Rather here we see again the emergence of new system characteristics when many cognitive abilities become integrated to create the primate mind. This of course happens on top of the already existing proto-compassionate bond. The new level is best understood with new concepts. It is like physics and chemistry. Chemical reactions are based on physics, but it is much more fruitful to understand them with concepts of chemistry. Likewise, all behavior is based on genetic selection, yet on higher levels it is much more fruitful to use a different approach. One would carefully compare behavior on different levels of evolution and try to find a logic or a system. On the level of primates we also may be aided by old-fashioned human introspective psychology. Here, generally we find many human emotions mirrored in animal emotions. Therefore this whole new enterprise, today quite appropriately is called "comparative psychology." The concept of reciprocal altruism is not a concept of genetics, but a concept of psychology, and how the new level of "psyche" may act back on genetic selection.

---

[15] Altruism can occasionally be observed on a much lower mental level, for instance in vampire bats (Wilkinson 1988). But this seems to be an exception facilitated by extreme selection pressure. These bats die of starvation if they do not drink blood at least every other night. Consequently, blood is regurgitated even to unrelated individuals.

In fact, here seems to be a good example of D. T. Campbell's *downward causation* (Campbell 1974b).

In that spirit, we now return to the chimpanzees. Jane Goodall describes a beautiful act of chimpanzee love:

> Madam Bee looked old and sick. Her arm, paralyzed by polio, dragged and several half-healed wounds were visible on her back, head, and one leg. It was very hot that summer, and food was relatively scarce so that the chimpanzees sometimes had to travel considerable distances from one feeding place to the next. Again and again Madam Bee stopped to rest. When soft food calls indicated that the two young females had arrived at the food site, Madam Bee moved a little faster; but when she got there, it seemed that she was too tired or weak to climb. She looked up at her daughters, then lay on the ground and watched as they moved about, searching for ripe fruits. After about ten minutes Little Bee climbed down. She carried one of the fruits by its stem in her mouth and had a second in one hand. As she reached the ground, Madam Bee gave a few soft grunts. Little Bee approached, also grunting, and placed the fruit from her hand on the ground beside her mother. She then sat nearby and the two females ate together.

And later:

> It was not the only time she (Little Bee) was seen to help Madam Bee in this way. On two other occasions she climbed down to the old female with her mouth and one hand full of palm nuts and laid the fruits from her hand beside her mother. Quite clearly she had some understanding of the needs of the old female. This ability to empathize, so highly developed in our own species, prompts much altruism in humans. If we know that another, especially a close relative or friend, is suffering, then we ourselves become emotionally disturbed, sometimes to the point of anguish. Only by helping (or trying to help) can we hope to alleviate our own distress. Was Little Bee, I wonder, motivated by a similar kind of emotion? Whatever the answer, it is evident that chimpanzees have made considerable progress along the road to human-like love and compassion. (Goodall 1986)

Here, the high level of integration of the chimpanzee mind becomes obvious. Little Bee seems to be not only *aware* of her mother's invalid state—something that the baboons were not capable of—but she also shows true compassion. I think this is the point in natural history where the "proto" can be dropped from the compassionate bond. It is not anymore "compassion without consciousness," something that is hard to visualize in human terms, but true compassion with awareness, that is to say, *at least with some consciousness*. This consciousness may be more fleeting than ours, or discontinuous, or vague, or whatever attribute we may use for want of better insight, but it is the real thing. And something else is born when an altruistic act is done consciously—*the moral imperative*. In human terms, "moral" is a word that belongs to the conscious domain of ego decision-making. *And when this is perceived as an imperative, imposed on us by an outside power like the unconscious emotional power of the compassionate bond, then, like Kant's moral imperative, it transcends our conscious ego.*

When raised among humans, a chimp will develop its bond toward humans. Lucy was an infant brought up in the family of Maurice and Jane Temerlin and their son. She used to hate the type of health food mandated on her. Surprisingly, it was found, that an appeal to guilt could be used to make her eat. That is how Maurice Temerlin accidentally discovered this:

> Without thinking about it at all I pleaded, "For God's sake, Lucy, think of the starving chimps in Africa!" For several moments Lucy did not move. She looked at me with a strange expression I had never seen before. Then she slowly began to eat. After a bite or two I added, "Take at least three more bites for your poor suffering father who loves you." Her tempo increased immediately. She then acted as though "eating everything on her plate" was a matter of great moral virtue. We did not believe this at first. But to this day, almost eight years later, we have verified the observation countless times. If Lucy does not eat what we want her to eat or as much as we think she should, I need only remind her of the "starving chimps in Africa" (it was the Chinese in my childhood) or whine, "How could you do this to me?" and Lucy begins to eat greedily. We have demonstrated this to our friends many times. (Temerlin 1972)

Of course, it is only the whining voice, not the meaning that counts. Guilt is often believed to be a result of repressive Judeo-Christian upbringing. But Lucy shows it to be of a much more ancient biological origin. Guilt is the discomfort you feel when you do not fulfill the imperative of the compassionate bond. If you are not nice to your fellows the distress partially will be your own. Poor Lucy did not notice that her foster parents played a cynical charade with her loving heart. When out in the forest, only one thing could summon Lucy back to the car—a (faked) shriek of distress.

The Origin of Thought and Language

By human measure, chimpanzees lack emotional control. Still, the first beginnings of a rational control function can be found in primates. Its nature is reflection and logic. And since many humans consider this to be the most important achievement of their species, naturally, many efforts are made to evaluate the capacity of primates to think, speak, and to solve problems. The 1960s and 1970s saw a well-publicized and popularized effort to explore language abilities. Less well-known are the much earlier investigations of the chimpanzees' insightful behavior by W. Köhler (1925). To me, the most impressive observation made by Köhler is the ability to solve practical problems "in the head," before any action is taken. A typical, and often filmed problem solution, is to obtain a banana that hangs from the ceiling but is too high to be reached. A box is available which can be used to step on and reach the banana. One can see the chimp sitting in the corner with a frustrated pouting expression. Its eyes wander around the cage. They chance upon the box. Soon the gaze focuses on the banana, then back to the box, then at the banana again, and back to the box. Then the chimp decidedly gets up, fetches the box, puts it under the banana and solves the problem. How could one explain such observations without concluding that in some internalized way the chimp pre-imagines in space-time the movements necessary to obtain the banana? Figure 4.9 shows some of the tasks Köhler's chimps were able to perform. To me this was an instant illumination of the origin of thought.

Many contemporary philosophers assume that thinking originates in language. Of course, this has been suggested by the fatefully unrealistic philosophy of Ludwig Wittengstein.[16] Also, this connection may have played a role in the interest for the attempts to teach language to primates. My own introspection clearly shows thinking to be related to spatial imagination. Of course, there may be two preferential types, logical and intuitive thinkers. Lawyers probably think logically, designers of machines think spatially. But it can be argued that spatial thinking is more ancient. Why would language otherwise be full of spatiotemporal images: *"Beyond recognition," "above suspicion," "pre-pare," "post-pone," "an-cestor,"* (from the Latin word *ante*, which means "before"), *"para-graph"* (the Greek word *para* means "alongside, aside from"). There is hardly a prefix in the etymological dictionary, whose origin can be traced, which does not invoke a temporal or spatial image. If a word is not spatiotemporal, it is invoking a human activity, which again is imagined

---

[16]Since I have professed to be a Popperian, I should mention that Popper and Wittgenstein were extreme antagonists. They are known to have engaged in a shouting match at a seminar given by Popper (Popper 1982); this included Wittgenstein threatening Popper with an iron fireplace poker. I found the best account of Ludwig Wittgenstein in C. J. Waddington's *The Ethical Animal* (p. 42), where he writes about his personal encounter with L. W.:

> Wittgenstein was, I think, so acutely aware of the otherness of other things that he never fully reconciled himself to the fact that words can have anything to do with them. I suspect that his intense concentration on the analysis and construction of languages arose largely from some profound feeling that for him at least all languages must remain almost totally inadequate for anything he felt is important to say.

And later on:

> No one who met him could easily doubt that Wittgentstein was a tragic figure of some magnitude. The nature of his spiritual situation will be discussed as long as that of Rimbaud or Jackson Pollock. My tentative diagnosis of it would be this: that Wittgenstein failed to realize that what he had to convey was essentially a poetic awareness of the otherness of reality; he was led, presumably by historical accidents of which I am ignorant, to enshrine his message in logico-philosophical jargon, whose very abstractness on the one hand served to exhibit the dislocation he felt between man and his surroundings, but on the other resulted in him being taken for a philosopher—an identification by which he was at first flattered, later filled by apprehension and a sense of guilt (Waddington 1960).

Fig. 4.9
Some of the problems solved by Koehler's chimpanzees
[Top after: Koehler (1926) Bottom after Yerkes (1943)]

spatiotemporal activity. Original nouns denote material things, which again are in space. Of course, it is possible to use language in a "sophist"-icated way without conscious spatiotemporal images. This is what the logical minds do. I think then the imagery remains in the unconscious background, like the complex mechanics of retinal image processing of which we are never aware.

Noam Chomsky has discovered a meta-grammar that is common to all languages. Unfortunately, he does not clearly make the connection proposed here. At one point he has even denied any relevance of ape research to his theory (Chomsky 1980), but his fellow linguist Steven Pinker seems to have moved toward a more evolutionary view (Pinker 1994). Could it be that spatiotemporal imagery is at the roots of grammar and even of logic? A brief look at traditional logic makes this likely. The three so-called "laws of thought" (Runes 1984) are identified as follows:

1. *Identity,* or "*A* is *A.*" Of course, the box in the cage is the box and not another box. It is there every time when looked at or imagined.

2. *Law of contradiction,* or "*A* is *B*, and *A* is not *B* cannot both be true." Again, this is clearly derived from objects in space and then applied to objects imagined in space.

3. *Law of the excluded middle,* or "*A* is *B*, or *A* is not *B*, are the only alternatives" *(tertium non datur).* Well, a banana and a box, real or imagined, are of different use. If a chimp would be shown a container in the form of a banana, it certainly would coin one of the often-reported new word creations: box-banana.

It is obvious that chimpanzees follow such logic when they solve their problems. Of course, this is purely operational and not conscious. Humans generally do likewise. In fact, these formal laws of traditional logic probably became conscious for the first time in natural history when they were formulated in Greek antiquity. On the everyday human level, spatiotemporal imaginations become instant, but may continue to dwell in the unconscious domain except in the trained minds of designers of gadgets and in specially gifted inventors.

Thus, very likely, the roots of thought are not to be found in language. But thought and language both are rooted in spatiotemporal imagination. To my knowledge, the promising attempt to base a universal grammar on spatiotemporal imagination still awaits a new Noam Chomsky. In such attempts one first should concentrate on sentences that describe material acts. On a higher level abstract terms follow the same structure. I will later argue that the study of chimpanzees, especially pygmy chimpanzees, verifies this hypothesis. They cannot speak because there is no acoustic machinery, but they are ready for symbolic speech since they have spatiotemporal images. As soon as given computer buttons to press, they speak through that medium. Thus, the organ of speech must have evolved later in order to communicate the imaginations. *The brain's images cause language and not the other way around.*

4.10
The compassionate bond and spatial imagination result in 'magical participation.'
[After: Koehler (1926)]

"The chimpanzee world is unequivocally dominated by the relations of objects in space . . ." writes Jane Goodall (1986). Spatial imagery, though, cannot remain an isolated facility. In emergent evolution, all acquired skills are immediately integrated to form even more advanced systems. How closely spatial imagination and the compassionate bond are intertwined is demonstrated by a photograph by W. Köhler (fig. 4.10). One of his chimps, Grande, is balancing on a ramshackle arrangement of shaky boxes, stretching to her limits to reach a banana. This critical moment is intensely co-experienced by Sultan, who is seen to tense both his hands in sympathetic action—*as if* his movements would support the attainment of the goal. Here, I am tempted to speculate on the origin of beliefs in magic—what else could precipitate the belief that one can influence another person's actions through the mind? The similarity bond, together with spatial imagination, lets us, so to speak, get under the skin of the other and experience his torment in trying to get the banana. Your effort becomes my effort too.

I am also reminded of huge mass events—spectator sports. The frenzy of "under the skin" or "magic" participation with the players of say, soccer, raises the passionate tribal spirit to such hysterical heights, that the aggressive competitive play may become translated into true tribal aggression against the other team's fans. The playing field is stormed, the umpire is beaten, players and fans of the other team are attacked. The aggressive spirit may even be released beforehand, in anticipation of the excitement, as occurred in May 1985 before a European Cup game in Brussels, where thirty-eight people died (*New York Times* 1985).

Thought appears first under subordination to the primate's social-emotional world. David and Ann J. Premack have conducted long-term studies of hand-reared chimpanzees under laboratory conditions. They devised a language that used various colored plastic shapes to designate words. A sentence could be formed by placing the shapes on a board in a proper sequence. To make the chimps understand and utter words in this unnatural way proved to be an arduous task. But it was possible. And it was also possible to prove with well-designed experiments that the apes, indeed, can use the language in a "symbolic" way; that is to say, they really imagine the objects or object categories designated by a given word, and they can properly understand and apply such general concepts as "similar" and "dissimilar." But even more revealing are the experiments to test whether they have "intentions" or recognize intentions in others. The Premacks' favorite subject, Sarah, was made to observe as food was placed in one of two boxes out of her reach. An experimenter was instructed to find the food and share it with Sarah. As soon as the experimenter entered the room, the ape began to signal with hands or feet as to which box contained the food. Now another "bad" experimenter was instructed not to share the food but eat it "greedily" in her presence. With this person the ape began to consistently point to the wrong box.

To point to the wrong box can be considered intentional deception. Deceptions of this kind have been observed by many other workers in the laboratory, as well as among wild chimps. Withholding information or giving misleading signals was observed by Jane Goodall (1986) many times. F. DeWaal (1982) reports a subordinate male limping, but only in the presence of the dominant alpha male, thus inviting mercy. All the above are instances of social manipulation. If the apes can use tools to manipulate the physical environment, such as twigs to fish termites out of small holes, or use screwdrivers to unscrew the lid of a box (Rensch 1967), or even open the lid of an electric receptacle to investigate the insides (and get shocked) (Temerlin 1972), they must be able to pre-imagine the result of their manipulation (not the electric shock, though). Similarly they must be able to pre-imagine the consequences of any of their social signals on the actions of others. That is how "thinking" comes into play here. Again, it is an "inner preview" of the familiar things to happen on certain occasions.

Thinking predates language, but once language is born, it may further facilitate thought. Language communicates thoughts to others. This is bound to become a powerful selective advantage to the better-speaking group, and via sexual selection, to the dominant individual who can keep his position because of better speech. If speech is exercised, thought is exercised, and stronger thought precipitates more sophisticated speech. Thus, a mutually reinforcing coevolution of thinking and language ability could have occurred. Indeed, the Premacks' apes have demonstrated that a language-trained ape is much better capable to solve problems (Premack and Premack 1983).

In the exercise of practical and social thought, conceptual generalizations force themselves on the thinker from the given facts. All material bodies posses "heaviness," "extent." Plants have in common leaves, growth, and rootedness which easily may result in the immediate recognition of any plant by its "plantness." Animals have an "animalness" which is easier to recognize than to abstractly describe. Conspecific animals must figure in the first and strongest concepts of all: "familiarity" and "other-sexedness."

When I say that the concepts force themselves on the thinker from the natural order of the external world, I merely state that they become automatic references, yardsticks in the practice of social manipulation. I do not mean that they are recognized in an abstract sense, *yet they are in existence and want to be named.* I think now language can begin to enhance abstract thought. When a concept is named, it assumes an enhanced reality. A motor-memory of certain sounds is formed. When the word is spoken, memory of the actual constituents of the generalization is activated, and now concept words can be manipulated in the mind in the same manner as concrete objects were moved around before. Now one can think, "All animals eat, but plants do not." Abstract thought is born, and its selective power drives the formulation of even more powerful language. But presently we will return to chimpanzee language.

A number of other apes have been successfully taught American sign language. Koko, a female gorilla, could draw on a vocabulary of six hundred words. Examples of her uttering are impressive (Patterson 1981). Upon hearing the fizz of a carbonated drink she signed: *Listen drink listen.* Or showing on her arm the place where the day before she had bitten another gorilla, Mike: *Bite there red.* When afraid of a big dump truck outside: *Afraid . . . close drapes.* Some of the spontaneous inventions were: *bottle match* for cigarette lighter, *eye hat* for a mask, *finger bracelet* for a ring, *elephant baby* for a Pinocchio doll, or *red mad gorilla* to describe her anger. Spontaneous humor was produced when she insisted that a white blanket was red—and then produced a small red lint stuck to it. Patterson and Linden write:

> Koko's humor can also be remarkably unsophisticated, and if a prank gets a rise out of someone, Koko is apt to repeat it. After noting with pleasure that blowing an insect on me produced a shriek and a jump, Koko did it again, this time laughing. Koko's laugh is a chuckling sound that is like a suppressed, heaving human laugh.
>
> Another of Koko's practical jokes is the attack with a plastic alligator. Koko will sneak up on some (supposedly) unsuspecting human with a little toy alligator hidden behind her back. When she is near her victim, Koko will abruptly spring up, brandishing the alligator wildly. The human is expected to assume a terrified look, scream and run. Koko knows that the whole thing is a charade, but she thinks it is vastly funny. It is wonderful that a gorilla might even for a moment think that it needs a prop like a toy alligator in order to scare a human. (Patterson and Linden 1981)

Here is a philosophical conversation with Koko. Drapes for her have a strong association with danger. Upon looking at a gorilla skeleton, the teacher Maureen asks in sign language whether this gorilla is dead or alive:

Koko: *Dead drapes.*
Maureen: Let's make sure, is this gorilla alive or dead?
Koko: *Dead good-bye.*
Maureen: How do gorillas feel when they die—happy, sad, afraid?
Koko: *Sleep.*

And later:

Maureen: Where do gorillas go when they die?
Koko: *Comfortable hole bye.*
Maureen: When do gorillas die?
Koko: *Trouble old.*

She has more concepts than words. When she learned "bracelet" she understood something "that goes around something else," this last word linked to a class of spatial imaginations, to a preexisting concept. So when she saw a finger-ring, knowing "finger" she could say "finger bracelet." Again, *concepts drive language.*

Most of the early studies have been criticized for lack of method. In order to pursue scientific ape psychology and language, more strict methods are needed. Sue Savage-Rumbaugh (1986a) does not teach sign language. Her chimpanzee's language consists of lighted buttons that carry symbols, figure 4.11. When any of her subjects presses a button, the statement is recorded in a computer. If an experimenter is present, the context of the event is also recorded. Like a natural language, the keyboard is always available. The chimps may chatter with themselves, as children do. Food may be automatically dispensed upon request, but one request on a lonely night was: "Please machine tickle Lana." The routine consists of daily language sessions and experiments to explore other mental abilities. Not all subjects do well. Like in humans, early childhood environment seems to foster or retard the mind.

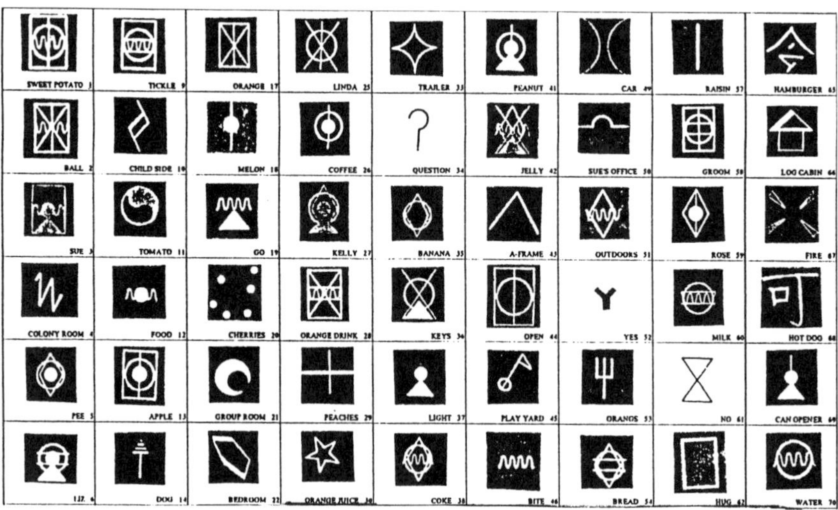

Fig. 4.11
Some of the lexigraphic keys used by Kanzi and other Bonobos.
[With permission: The Language Research Center, Georgia State University}

Here is not the place to review all the cleverly designed experiments. But the overall picture that emerges shows minds well on their way to take the last known step of emergent integration, the one into humanity. This is particularly true for the latest subjects: the rare bonobos (*Pan paniscus*). After having taught a number of common chimpanzees, Sue Savage-Rumbaugh started with a wild born female *pan paniscus*, Matata. This one did not prove to be an eager subject. In fact, her abilities remained far below the previous achievements. But Matata had an infant, Kanzi, who came to witness all attempts to teach his mother. He was not taught himself but was permitted to play with the keyboard if he wanted. When Kanzi was two and a half years old, his mother was brought to the *pan paniscus* colony at Yerkes Field Station in a further attempt to breed. Four months later she returned pregnant with Kanzi's sister-to-be. During her absence, recordings of all of Kanzi's utterances were begun. To the amazement of all of his caretakers it turned out that he had already learned to properly use the keyboard himself, without having ever been trained. He rapidly became more proficient than Austin, Sherman, and Lana have ever been after prolonged and laborious training. Unlike the other subjects, he did not confuse lexigrams when the keyboard was moved to a different place, or when they designated similar objects like apple, orange, or banana.

But the main sensation was this: Despite many attempts, none of the other apes had ever learned to understand spoken English. But *Kanzi was found to have learned the meaning of many words spontaneously*. For instance, when presented with three lexigrams, "green beans," "tomato," and "apple," and asked in English, "show me the tomato lexigram," he would point to, or take the tomato lexigram (no food rewards were given for reinforcement). Unintended teaching had occurred: The researchers used to accompany their teaching pursuits among the other apes with words. When the correct lexigram was pushed they may have said, "yes, right, tomato," or when a subject would indicate on the keyboard that she wanted to go to the tree house: "All right let's go to the tree house." Yet systematic attempts to teach English words to all the other apes had miserably failed. Kanzi, however, in time did match not only objects but also photographs of the objects to spoken English. In order to insure that no unintended clues were given by the experimenters, in other tests the speech was produced by an electronic voice synthesizer. This also was understood, despite the fact that the synthesizer was not a very clear one. Since Kanzi used earphones, the tester could not hear the words and give unconscious clues. A typical test may include most of Kanzi's vocabulary (sixty-six words at the age of four years). Examples are: bread, campfire, Coke, colony room, egg, grab, mushroom, open, refrigerator, TV, tickle, turtle, and umbrella. He was asked to match the object to the spoken English word; to match an object's photograph to the word; to match the photo to a given lexigram, and finally to match the lexigram to the word spoken by the voice synthesizer. Correct understanding was assumed when correct responses were given three times out of three trials. Under these constraints, Kanzi and later his sister Mulika could score better than 95 percent. Sue Savage-Rumbaugh writes:

> Kanzi and Mulika did well on formal tests from their first administration. They seemed to understand that the experimenters were

not communicating about something that was going to happen as they touched the symbol, but rather were posing a specific question. Sherman and Austin, by contrast, when first asked to label items in a test, appeared to anticipate that they would receive the item as a consequence of touching the symbol. When they did not, they became confused and consequently labeling had to be introduced slowly. (Savage-Rumbaugh 1986b)

Previous achievements with *Pan troglodytes* have often been belittled. H. S. Terrace, who had raised the chimpanzee, Nim, thought that their use of symbols does not really represent language. He noted that the symbols were only used to obtain something that was desired. The apes only could request an object, but not name it outside of the context of the request. Thus, he said, the chimp does nothing more than the chicken in a Skinner box. It presses a lever to obtain a drop of water. In that view, the chimp's achievement would be rather a conditioned response than the use of language.

Such critiques motivated Sue Savage-Rumbaugh to define criteria for a genuine ape language. (Savage-Rumbaugh 1986b). Those criteria were derived from studying language use of children. They are:

a) Appropriately request an item in the correct context.

b) Name a displayed item.

c) Correctly respond to a request for a particular item by selecting that from a group of things.

d) Demonstrate a behavioral correspondence between action and symbol.

e) Announce intended actions.

f) Respond to and make judgments about symbols in ways equivalent to responses and judgment about the actual thing (the referent).

Only point (a) had been readily mastered by common chimpanzees. The other points could be achieved marginally, but only after extensive training. The conclusion reached was that common chimpanzees do better than just produce conditioned responses, but they do not learn language in the same innately motivated way as human children do.

Therefore, Kanzi's and Mulika's achievements came as a complete surprise. Like human children, they spontaneously began to comprehend symbols *before* they began to produce them. Kanzi and Mulika had their keyboard connected to a speech synthesizer. In 1988 there were two hundred fifty-six symbols on Kanzi's keyboard, one hundred fifty of which he had demonstrated to master. So he can "'speak" audibly and converse with his human counterparts. The day was not spent with lessons of any kind. No formal training or food reward was ever given. In good weather there are walks taken in the fifty-five-acre fenced in forest that belongs to the laboratory. The forest contains stations where various food items are stored. Apes and humans hike together, gather food together and possibly share. Generally the apes take the lead and determine where to go, what game to play, and when to picnic. Among the games are hide, keep-away, chase, grab, tickle, and play bite. Kanzi and Mulika may spontaneously help to gather sticks for a campfire, clear new trails, pull weeds from the garden, or string up a hammock.

Often the humans are tired out by the ape's inexhaustible energy. Yet despite these conditions, language production and comprehension is noted and recorded. For instance, if Kanzi asks for a hamburger that is being cooked on the campfire, that is not a great achievement. But if he asks for one and points to a backpack that contains several foods, and then proceeds to pull out a hamburger as announced, that scores as proper use of "semantics." Obviously hamburger was clear in his mind *before* it was encountered, and the purposive action of selecting it followed. This would score on the points (d) and (e) of the above list. Or if he announces that he wants to go to the trailer and then proceeds to lead there, this scores on the same points. He would score on point (c) if asked in English as the table is being cleared: "Would you please carry the straw?" Whereupon, Kanzi looks over several items still on the table, then takes the straw and follows the caretaker to the next room. Or "Kanzi, would you give Panbanisha an onion?" And he looks around, finds a wild onion, pulls it out of the ground, and presents it to Panbanisha (his half-sister). Among other points, this clearly scores on (f). (Savage-Rumbaugh 1988). Figure 4.12 introduces Kanzi. Here, he is in a test situation, listening to recorded English utterances. The tester does not know what Kanzi hears and thus cannot provide subconscious cues. Kanzi satisfies all requirements of proper language use, and he acquires it spontaneously. His mental abilities and language acquisition skills are comparable to a human child.

How is it possible that without vocal chords the language ability has evolved, presumably laying dormant, ready to be used as soon as a medium of expression is provided? I do not see any other possible explanation than the one already suggested: Kanzi and his Pan troglodytes cousins all can already think in the sense of spatial imagination. Kanzi, though, is sharper by degree. In all the cited examples either a request from the caretaker triggers a chain of spatial pre-imaginations, resulting in action, or his own wishes trigger the same. Kanzi has reached a threshold of emergent integration. *His imaginations are effortless enough to link up with the ancient motivation of the joy of exercise. The result is motivation to learn language.*

We may ask, why did the voice box not evolve together with thinking? Because thinking primarily is linked to *individual* mastering of the physical environment. This does not have to be communicated to others. To the contrary, this may be against your own interest. On the other hand, all important social communication already happens effortlessly enough via the similarity bond and body language.

In that connection, a difference between human children's and the chimp's acquisition of manual problem solving is enlightening: chimps do not learn very well from a demonstration; they patiently experiment on their own until the solution is found. Children, on the other hand, immediately pick up from a demonstration. They can distinguish whether a hand movement is a demonstration of a technical manipulation, or is body language conveying an inner state. *The chimp's personal spatial pre-imagination can link up with language, but it is not strong enough to understand nonemotional manipulations of others.* Body language is exclusively reserved to covey the inner emotional states.

All this reconfirms the earlier view (chapters 2 and 3) that there are two ancient and distinct modes of mental function, and that these develop fundamentally different kinds of knowledge. In the next section I will call them "Knowledge I" and "Knowledge II."

Fig. 4.12
Kanzi with earphones, undergoing English comprehension tests
[With permission: The Language Research Center, Georgia State University]

PHILOSOPHICAL INSERT 2:
Conclusions for the Theory of Knowledge

The theory of knowledge, or epistemology, traditionally has been claimed as part of philosophy. Plato had already speculated on how it is possible that we can have knowledge about anything. In his spirit-matter dualism he had to conclude that knowing means remembering by the soul, which once had an existence in the eternal, supernatural world of ideas. Since that time the whole subject had not shed its mysterious flavor, until the time of Immanuel Kant (1724-1804). Kant was one of the last universal minds. He was equally at home with the fields of physics and mathematics of his day, and he shared the new objective spirit of enlightenment philosophy. He did not invent speculative first principles, but he uncovered functional roots of thinking that can be deduced by logical analysis of the way we arrive at conclusions. Up until very recently, his epistemology has remained the ultimate statement about the roots of our knowledge.

Thinking, feeling, and moral convictions, according to Kant's famous three critiques, rest on a number of "a priori categories," irreducible "first thoughts," or "ideas," upon which all thoughts and judgments are constructed "a posteriori"; that is, they are derived. The origin of the a priori categories had to remain unknown in Kant's time. He called them "transcendental" because they transcend into our mind from an unknown origin. *The most important a priori categories are three-dimensional space, time, and causality. Also, there is a "moral imperative" that is behind our ethical judgments.*

Yet Kant has been much misunderstood. Since the a priori categories are like "ideas," he himself used the misleading term "transcendental idealism" for his philosophy. This sounds much like Platonism, and often was understood as such, but it was not meant in this way. The categories simply are what is found in our mind by carefully inspecting how we think.

In our days, epistemology has made a great leap forward. Donald T. Campbell (1974a) has pointed out that the a priori categories can be derived from the evolution of mental abilities throughout the animal phyla. Konrad Lorenz (1940, 1973), the father of ethology, had made a first attempt to actually construct such an epistemology. The previous three chapters in this book were an attempt at such *"evolutionary epistemology."* This began when evolution was considered a process of knowledge gathering. *In chimpanzees, when thinking developed from "spatial pre-imagination," we witnessed the birth of the Kantian a priori categories of space, time, and causality. And when the compassionate bond developed from the ancient trait of the similarity bond, we witnessed the birth of Kant's moral imperative.*

Franz Wuketits (1990), a Viennese colleague of the late Konrad Lorenz, has pointed out the importance of evolutionary epistemology for the future of humankind. This new insight is compared to the Copernican revolution, with good reason.

Here, we are witnessing a process that has happened many times before: a philosophy has been succeeded by science, and a piece of the fog of mystery has been lifted again. Evolutionary epistemology became a natural science in its own right.

Unfortunately, evolutionary epistemology has not yet entered into generally accepted mainstream thinking. Rampant epistemological confusion still reigns about the place of natural science and physics in particular. Since these are important questions about the very foundation of this evolutionary synthesis, they have to be addressed. What does evolutionary epistemology say about the nature of physical insight?

Here, great contributions have gone by without much notice, especially in the English-speaking world. The German physicist and philosopher Gerhard Vollmer (Vollmer 1983, 1985, 1986) has produced a two-volume critique of physical knowledge on the grounds of evolutionary epistemology. I hope that after the insights of the previous chapters, my readers are well prepared for a summary:

1. The human mind has evolved by Darwinian selection in order to be successful in a world of medium-size dimensions, the "meso-cosmos" of the planetary surface. Here, through our acquired abilities of spatiotemporal pre-imagination (the Kantian a priori categories of pure reason), we can realistically understand, predict, and manipulate the world. The contents of what is known as "classical physics" became the final distillation of that process. This essentially had been finished by the year 1900.

2. Throughout natural history, having had no direct interaction with the physical reality of microscopic dimensions, "the microcosmos," and the gigantic stellar dimensions of the "mega-cosmos," we cannot be prepared to intuitively grasp the physics on the atomic scale, where quantum effects dominate, nor of cosmic dimensions, where general relativity dominates. Therefore, what has become known as "modern physics" must have an epistemological foundation different from the evolutionary one.

Every student of physics has experienced that direct intuition fails in the latter subjects, and that contradictions abound. Curved space, wave particle dualism, even reversal of cause and effect, none fit into any of the Kantian categories. It is remarkable that mathematics can still take hold of such things and even accurately predict. *Why is it that mathematics reaches farther than the human mind? That is a new and different epistemological question posted by the obvious success of modern physics.* Evolutionary epistemology cannot possibly have an answer to this. As far as I know, so far no answer exists by any approach. A first hint to such an answer may be taken from a previous insight in this book: the universe and mathematics, both, emerge from simple rules played out on a large scale. Both follow local situational logic. Might for that reason some "sympathetic correlation" exist between them?

But mathematical quantitative prediction, despite being a product of the human mind, is not that kind of knowledge that can truly push back the boundaries of what we experience as mystery. It can only give power, not unlike "magical" power, because we don't really understand what we are doing. We only see the results which justify the means. One can build a powerful technology on quantitative models of that kind. *We seem to encounter domains here that are of dangerous Promethean dimensions and remind us of Faust's bargain with the devil.* The second part of this book will address the corruption that this great "quasi-magical" power already has caused in modern technological civilizations.

The listed main points from evolutionary epistemology, together with the observation that to some extent we can reach beyond our natural limits of intuition, suggest that knowledge has to be more carefully defined. I suggest that there are two kinds of knowledge.

The first kind is identical with the knowledge we have traced in its ascent throughout the animal kingdom in chapter 2 (The Roots of Cognition). This leads to cognition directed primarily, but not exclusively, toward the inanimate external world. On the human level, such knowledge has built our technology. As with all knowledge, it improves fitness, and so far we have out-survived all other species, as well as other human tribes with less technological knowledge. *This knowledge is blind to anything but practical success. It shall be called "Knowledge I."* It ascends throughout natural history and then runs through both classical and modern physics.

The other kind of knowledge has to do with the innate will to live, which precedes any functional understanding of the external world. This was introduced in chapter 3 (The Roots of Will). Will in itself cannot be considered knowledge, but it poses problems to which special adaptation becomes necessary. Hence this spawns another type of knowledge gathering. Its first goal is how to handle the wills of other individuals. On more advanced levels it also concerns itself with the inner control and channeling of emotions. For instance, the ritualization of ranking fights is an early achievement of this kind. It is knowledge of personal social skills that also maintain inner calm (or peace), which frees energy for more important uses.

But on the human level there is an added complication. This arises from the acquisition of a conscious ego system. As the next three chapters will discuss in detail, this system develops on the cognitive abilities in the growing neocortex and their overwhelming success. *Seen from here, the worlds of will, our own and the ones of others, are foreign territory.* The ego finds itself in a situation where not only the external reality has to be adapted to (this was Knowledge I); also, the archaic, meddlesome, intruding, and illogical but powerful world of wills has to be included in the adaptation. *This area of adaptation I shall call "Knowledge II."*

Anatomically, the world of wills resides in brain stems and is experienced (by the ego) as external, mysterious, spiritual, and frightening.

Generally, Knowledge I is pursued by most natural scientists today, and especially by technologists. Knowledge II is the domain of theologians, artists, and humanists. Compared to the advanced state of Knowledge I, this domain today remains in the condition of Stone Age culture. As C. P. Snow has pointed out in his *The Two Cultures,* people in the pursuit of the two kinds of knowledge cannot communicate with each other; this results in a cultural schizophrenia. *It is one of the prime goals of this evolutionary synthesis to give an objective foundation to the pursuit of Knowledge II, so that ultimately the gap between the two domains can be closed.*

There are excellent physicists today who have not yet assimilated the message of evolutionary epistemology, nor are they aware of the difference between the two quests for knowledge that derive from it. While working on Knowledge I, they encounter the incomprehensible in quantum physics. Evolutionary epistemology could give their effort a solid footing by pointing out the limits of human intuition and stressing the further power of mathematics. But not knowing the difference, they switch to the pursuit of Knowledge II, and being especially unprepared for such task, they uncritically embrace primitive mysticism. Knowledge II has little to do with the external and objective world of physics. Fritjof Capra, in his *Tao of Physics,* is such a case. Also, today there is a large number of young students in the field of physical sciences who fall into the same trap of subjectivist mysticism.

A related case of epistemological confusion concerns Roger Penrose (1989, 1994), one of the best known theoretical physicists of the day. He sees a mystery in the existence of human consciousness and then tries to explain it with something equally mysterious: the unintuitive tools and concepts of quantum and field theory. Such interesting attempts, however, are bound to remain futile, because evolutionary epistemology already has clear and simple answers: the neural networks have evolved by Darwinian selection for their limited task. As the next chapter will illustrate in more detail, neural networks function according to well-comprehensible principles of classical (intuitive) physics. Nothing more is needed. The only thing that borders on mystery is their adaptive self-organization. But this already has been recognized as a rational process they share with the rest of the universe. Consciousness is not another mystery, but instead our subjective inner experience of the neurons firing away. It is experienced at the same level as feeling hungry. Having an existence in Popper's World Two only, it can never become an object of physics. Only the associated chemical and neurological processes can be explored by natural science.

With this epistemological preparation, we are now ready to enter the complex and philosophically explosive subject of the human brain. Its function is the crowning chapter in evolutionary epistemology. As this book continues, enough evidence will be uncovered about the function of the brain so that our mental processes and our willed goals can be firmly grounded in self-organization and emergent Darwinian evolution.

Chapter Five

The Brain and its Evolution

It was not possible to understand the brain, even in principle, until the power of artificial neural networks was discovered through computer simulations. They can self-adapt, for instance, to recognize letters of the alphabet, or to assign ill-defined objects to classes of similarity. Some can even read handwriting, pronounce English text, or reconstruct the picture of a face from a partial and garbled image. Here is another complex nonlinear system that can self-organize—it leads to what can be called "self-organized intelligence." Add to this the evolution of the brain as it can be studied today in surviving phyla, and the behavior that comes with it, and a first-order picture emerges of what the brain does. A basic antagonism (or duality) is found between the brain stem, with its ancient and inherited instincts, and the neocortex, which self-adapts to the external world of today. This dualism seems to be the most fundamental characteristic of the human psyche. It will be seen at work in cultural evolution and history, and it has the potential to explain the origins and value of art, literature, and religion. Consciousness is our inner experience of the working of some neural networks in the left temporal lobe. Their survival value is to assemble a picture of any current situation that represents a best guess from the available information. This may be partially wrong. But it is good enough to propel us to the number-one evolutionary success in this solar system.

Preface to Brain

When a measure of the current state of progressing emergence will be defined, brains will occupy a level of their own, and the human brain will top the scale, provided of course that one does not find a more developed neural system on a distant planet under a different sun. The progress from the "slimy to the sublime," as I have called it earlier, essentially is the progress of the brain. And if the brain is a machine—we are machines. Therefore, we are now entering a highly charged area of scientific pursuit, "the ultimate of ultimate problems," as William James has put it. Here, objective insight might become obscured even in the greatest of minds. When we look at the attempted syntheses by such monumental figures as John C. Eccles and Wilder Penfield, we find a superb command of the objective facts, but ultimately an escape into philosophical dualism: A ghost is put into the brain. This serves, so to speak, as a makeshift keystone to cap the otherwise physicalist vault of brain science.

For many, it is difficult to shed the belief in a divine connection, particularly if a physicalist keystone of human nature is not quite available yet. But I think such a keystone has arrived now. And this was one of the main incentives to write this book. Yet one is cautioned by Paul MacLean's words: ". . . it is impossible to say anything about such a complicated organ as the brain without being guilty of oversimplification." Therefore I have to plead guilty beforehand. Also, this chapter cannot be a specialist's technical contribution to the arcana of a rapidly evolving science. The goal is to extract from known facts a global model of brain function that is in natural harmony with the evolutionary perspective of human nature, as pursued in this book, and that will convey existential meaning beyond the mechanics of neuroscience. In pursuit of this, I will not shy away from citing some old experiments by Penfield because of their direct intuitive appeal. But the main insights will be derived from recent attempts of neuroscientists to form such syntheses. This science moves so fast that while this is being written, already new connections are being made. Nevertheless, as far as I can see, the proposed model tends to be confirmed rather than contradicted by the latest investigations and conjectures collected, for instance, in Gazzaniga (1995).

Unfortunately, there are many epistemological misconceptions that hinder progress in solving "the ultimate of ultimate problems" in a meaningful way. Many present-day "cognitive scientists" are a case in point. "Even the most naive observer can see that the nervous system is vastly different from a computer," writes Carver A. Mead in his book *Analog VLSI and Neural Systems*. Nevertheless, until recently there was a serious effort to model the human mind by sequential computers. It all had began with Norbert Wiener's *Cybernetics* (1961), which espoused the belief that living entities are simply control systems. This they are in principle, but on a much more sophisticated level than ever envisioned by Wiener.

Another early contribution had set the tone to somehow equate machine intelligence with the human mind. This was A. M. Turing's (1950) proposed test of machine intelligence: if someone would converse through a keyboard with a person and with a computer, and would not notice who is who (or what), this computer would possess "intelligence." Since the early days, Turing and Wiener had many followers. Most of them were non-biological thinkers coming from philosophy, psychology, mathematics, or engineering. Thus, after the demise of behaviorism, cognitive science was born as another attempt to reduce the human mind to an artificial construct that did not have to deal with the "too complex" reality of the real brain. Its two basic articles of belief have been well-summarized by Howard Gardner (1985):

1. ". . . it is necessary to speak about *mental representations* and to posit a level of analysis wholly *separate from the biological or neurological,* on the one hand, and the sociological or cultural, on the other."
2. ". . . there is the faith that central to any understanding of the human mind is the *electronic computer* . . . the computer also serves as the *most viable model of how the human mind functions.*" (Gardner 1986)

To anyone who has understood himself or herself as the product of biological evolution, and who has also understood the simple humanly programmed inner workings of a computer, manipulating "ones" and "zeros" forever and with forever increasing speed, such statements of belief must be appalling. Another critic of mainstream cognitive science is John R. Searle. Including himself among the cognitive scientists, not mainstream though, he concludes: ". . . we try to find out how humans might resemble our computational models rather than trying to figure out how the conscious human mind actually works." (Searle 1992)

Fortunately, since the invention and exploration of artificial neural nets, many of the creative minds have eagerly adapted to the new field, and have become "connectionists." That is, they have recognized that the solution lies in the complexity of synaptic connections, all neurons working simultaneously, or in parallel. This indeed, is what happens in the brain. But at this time the first article of belief is still alive and well.

In our pursuit of evolution of knowledgeable behavior, earlier we had to give up a detailed understanding of the neural correlate of higher behavior. This, indeed, would have been an impossible task. Yet, astonishing as it may seem, when we arrive at the human level, new insight into this difficult subject becomes possible again, not so much on small-scale neural detail, but on large-scale functional features. Today there are many different areas and methods of exploration to draw upon in order to understand the brain.

Positron emission tomography (PET) and magnetic resonance imaging (MRI) have been used to map metabolic activity in various regions of the brain while certain mental activities are performed. In apes, arrays of electrodes have been implanted to directly observe electrical activity. Studies of mental impairment of brain-damaged individuals have yielded a trove of information. Comparisons of brain architecture and behavior throughout the animal kingdom have provided great insight. But the most fascinating approach remains electro-stimulation of the surgically-exposed brain in conscious patients. This has been championed by Roger Penfield in the 1930s and occasionally is still being done today (Ojemann 1990).

By necessity this chapter will have to be rather technical, at least in the beginning. There is no escape from this. But think what can be gained: an understanding, in principle, of what makes us really and truly the intelligent beings we are, and why, despite all the cleverness, we often lose our way and make fools of ourselves.

Self-Organized Intelligence in Artificial Neural Networks

The level of excitation of most neurons is measured by the frequency of firings. As mentioned earlier, some neurons can fire up to five hundred times a second. Each firing sends a pulse down the axon. The axon branches out and ends in many synaptic bulbs attached to other neurons. Depending on a variable, the synaptic strength, or "weight," this excitation either helps to induce excitation in the target cell or contributes to calm it down (inhibition). Firings begin when the electric potential inside of the cell rises above a certain threshold, and each excited synaptic input contributes to a rising or a lowering of that potential. Thus, a neuron at a given time may have a characteristic as shown in figure 5.1: a "sigmoid" curve, where the sum of all synaptic inputs (positive or negative) has been plotted horizontally.

Such input-output curves are well-known in the art of electronic circuitry. Especially if the firing frequency is represented by a gliding output voltage, a fairly simple circuit can be built that has the input-output characteristic of a neuron. Hence it can model a working neuron.

Therefore, most insights on the dynamic of neural networks have been obtained from such models: artificial neural networks, or from their simulation on computers. What is reported in the

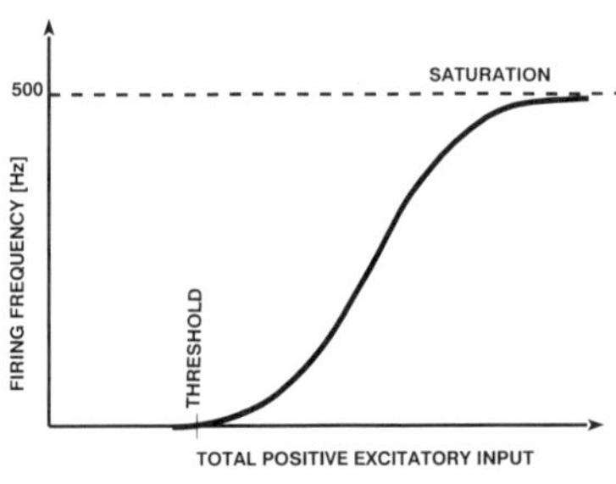

Fig. 5.1

Simplified excitation characteristic of neurons

following is equally applicable to natural, as well as artificial, networks.

In the neural networks studied in Aplysia (fig. 2.6), basically there were three layers of neurons: the sensory neurons (only one of them was included as representative of an entire layer); the "interneurons," and the motor neurons. If, for the moment, the sideways inhibitions in the middle layer are disregarded, as well as the separate inputs from the head, one finds that each neuron of a previous layer makes synapses to every neuron of the next layer.

Such networks are called *fully connected*. In figure 2.6 there also is one axonal tree that jumps over the middle layer and goes directly to the motor neurons. Networks with this feature are said to be *trans-connected*. They are prevalent in natural systems, but rarely considered in artificial ones.

The fundamentals of self-organization can already be discussed on the simple two layer, fully connected network of figure 5.2. Although this particular representation is only four neurons wide, everything that follows is applicable to much wider networks. Consider a network that is 120 neurons wide. This would have 120 x 120=14,400 synapses in total. Now think that the input layer is connected to the 120 pixels of the image represented in figure 5.3a. This layer may consist of, for instance, the light sensitive cells in the retina, thus receiving excitations from the lighted pixels of the letter, A.

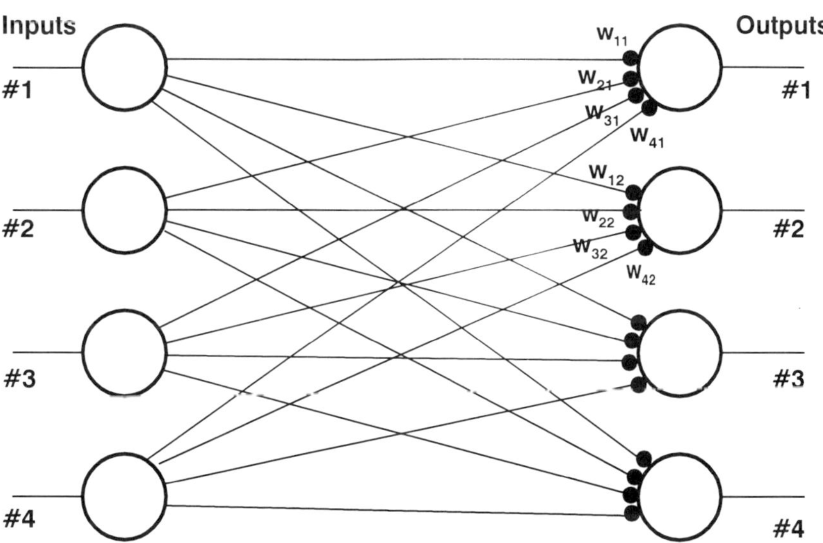

Fig. 5.2
Simple, fully connected, neural network.
Large circles are neurons. Small black circles are synapses

Now suppose that all the weights, w_{i1}, connecting to output neuron #1, which come from the twenty-two lighted pixels of the letter are made large and excitatory, and all other weights leading to this neuron are made small. Obviously the #1 neuron will receive strong excitation and will produce a strong output. Now project a letter, *B*, upon the retina. A quick comparison between the two images in figure 5.3 shows that *B* will light up only six of the previously lighted twenty-two pixels of the letter *A*. Therefore, excitation of #1 will be much weaker, possibly not even reaching the threshold, so that no output whatsoever will be seen. Similarly, all other symbols that are sufficiently different from our letter, *A*, will produce no output from #1, or a much smaller one. Obviously with that distribution of weights the *#1 output neuron is a recognizer (or detector) for the letter A.* But also patterns that have a deviation of only a few pixels from the standard will show a strong output. Therefore, one could say *#1 is a recognizer of a class of symbols that closely resemble the letter A. More generally, it is a classifier of patterns of excitation from any origin, including hidden patterns deep inside the brain,* whose meaning presently may elude us.

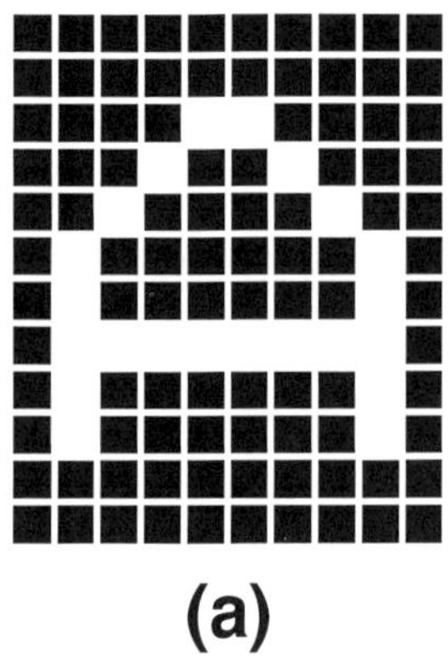

(a)

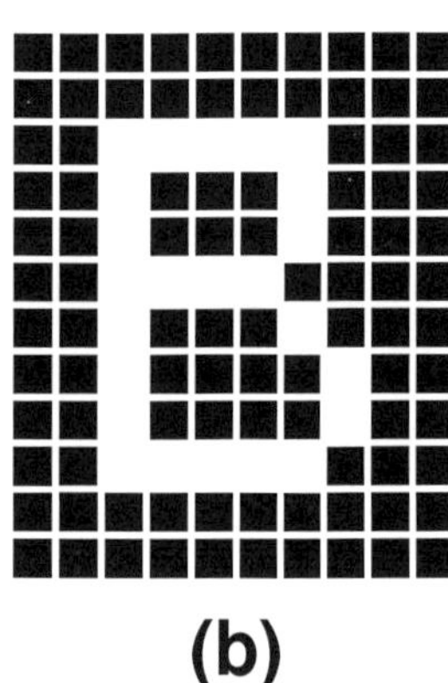

(b)

Fig. 5.3
Two images to be recognized by the network.

Similarly, #2 can be made a recognizer of a class of symbols resembling the letter, B. Ultimately all 120 outputs could be made to recognize different symbols, even though there are restrictions which presently need not to be addressed.

But the most surprising ability of such networks is yet to come. As mentioned before, synaptic weights, w_{ik}, can undergo changes when excitation is conveyed through them (Hebb 1949), in a way comparable to muscles becoming stronger through exercise. There is not one particular Hebbian rule, but a class of them. Suppose the rule for all synapses in the network is as follows: *any synapse is strengthened when it conveys positive excitation, while the target neuron also is in a state of increased excitation.* Stated differently, a "post-synaptic" excitation has to be present simultaneously with a "pre-synaptic" one, in order to cause an increase in the synaptic weight.

Suppose initially there is just a random distribution of synaptic weights. If now the letter, A, is presented to the input layer, due to the randomness, by necessity, one of the output neurons will receive an excitation that is stronger than all the others. This will be a post-synaptic excitation for the synapses leading to that neuron, consequently, its active synapses will increase their weights more than any other synapses. *Therefore this neuron, say it is #3, will further increase its sensitivity for, A, ultimately becoming a detector.*

This simple self-adaptation to recognize a symbol suffers from a certain lack of contrast. Since neurons other than #3 also will receive significant excitation, even though a lesser one, they too, will move toward becoming a detector for, A, but at a slower pace. This effect can be eliminated by sidewise inhibition in the output layer, as in figure 5.4. Each neuron makes inhibitory synapses to all others in the same layer, and an excitatory one upon itself (in the figure only the tree of #3 has been included). The initially strongest output (for instance #3) will now suppress all the other neurons in that layer, while reinforcing itself with the excitatory synapse. This arrangement may be called a "winner takes all" circuit, and it, indeed, is found in natural neural networks. Now no other output neuron will have a post-synaptic excitation, and no other synapses will be strengthened.

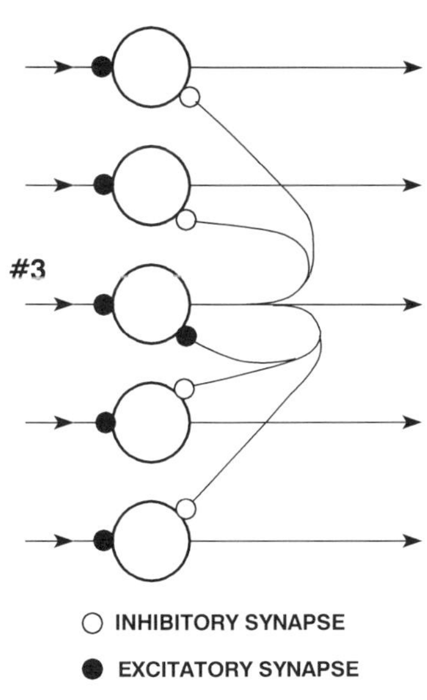

Fig. 5.4
"Winner takes all" through sideways inhibition.
Axonal tree of only one neuron has been included

After the first adaptation, projecting a different symbol will cause another neuron to adapt itself, and so on. In practice, however, such *self-organized intelligence* exhibits a peculiar limit. When too many different symbols are presented, the networks have the "human" tendency to forget previously learned symbols in favor of new ones. But there are schemes to correct for that (Carpenter and Grossberg 1987). They include nothing more than global feedback control loops.

In terms of phylogeny as well as ontogeny, such networks are rather simple organs. Their genetic blueprint merely says: connect to each neuron in the next layer and make sure all the synapses become stronger with use when the postsynaptic neuron is on. *That is all that is needed to cause self-organized intelligence.* That kind of intelligence may seem a bit limited, yet it can be a building block for much more advanced function.

When a third layer is added, the network becomes a "classifier of classes," particularly if not too much sideways inhibition appears in the middle layer. Then not just one neuron of that layer will represent a given symbol, but a pattern of several neurons, excited to varying degrees. Deviations from the standard will be better represented in the variations of the middle layer patterns, and consequently the third layer can much sharper perceive boundaries between classes. The fabric of networks found in the neocortex, indeed, consists of many successive layers in neighboring areas.

But we humans do not simply recognize and classify things, we also are active on our own. We may, for instance, speak or play the piano. How can such tasks be accomplished by self-adapting neural networks? In fact, two networks together can do that. Suppose we learn to play the piano. We look at the music sheet and see a symbol, say a chord C-E-G that is to be played. We can assume that a network of the above kind, connected to the eyes, previously already has adapted to recognize the symbol for that chord. So somewhere there is a neuron, or a small group of neurons, that will light up whenever a chord C-E-G is seen on the music sheet. Now suppose this excitation goes to input #2 in figure 5.2 (it is irrelevant which input that is). Also, suppose the output layer consists of motor neurons driving muscles of the hand and arm.

In learning the necessary hand position for this chord, direct self-adaptation cannot do the trick. We need to be taught the arbitrary convention. Suppose with help from a teacher, or using a book, in ways I am presently not ready to discuss, we succeed to move our fingers to the correct position. The only way to do this is to activate certain motor neurons in the output layer, say #1, #3 and #4. Because this is a teaching effort outside of what this network does, these motor neurons must have additional synaptic inputs from somewhere else. But what counts is this: they are excited so as to move the fingers to the proper position. Now comes the adaptation: the synapses from input #2, namely w_{21}, w_{23} and w_{24}, will find a postsynaptic excitation already in place. Therefore they will grow stronger as in the previous adaptation. Next time when a chord C-E-G is read and the #2 input excited, the same muscles will be strained and the chord played "automatically," without conscious effort, as any accomplished piano player does. A learned motor pattern has been established.

The architecture for this learning network consisted of two of the two-layer networks cascaded. Actually this can be reduced to one three-layer network, because the output layer of the first network also can serve as the input layer of the second one.

Of course, the above demonstration only explains the principle. There are not just three, but at least thirty muscles to be activated to a proper degree, before the fingers are even placed in position. But that does not matter. There are at least 10^{11} neurons available and 500 to 1000 times more synapses.

These principles are extremely enlightening because now, in principle again, one can begin to understand how functional knowledge about the world can enter into our brain from this world itself.

These two examples should suffice to generate enough respect for the possibilities of self-adapting neural networks. Now we can approach the mysteries of the brain with more confidence. Vignette 3 provides a more detailed introduction and more examples of the power of these fascinating complex systems. Today the above mentioned networks, and many more, have all been implemented in computer programs. Some of them are commercially available. They move robots, read handwriting, pronounce written English text (by driving sound synthesizers), evaluate the creditworthiness of bank customers, try to predict the stock market, and solve many more problems previously considered unsolvable.

Yet the real networks of the brain certainly contain many more variations of the theme struck in this section, and many more tricks of which at this point we know very little.

The Brain's Architecture and Evolution

In the following[17] I will try to avoid as much as possible the labyrinthine complexities of anatomical nomenclature. So as not to confuse the non-specialist reader, only the absolute essentials will be introduced. Figure 5.5 shows a vertical section of a human brain. This slice is taken parallel to the face from the middle region. At the bottom it includes an upper portion of the spinal cord. The darker areas show the *gray matter*. This contains densely packed neurons and their local interconnections. The lighter areas show the *white matter*, which primarily is a mass of thick bundles of axons—the cable basement or the wire harness. On the outside is the convoluted cerebral cortex (rind), which is gray matter. When stretched out, the human cortex covers about 1.5 square feet (0.14 square meters) of surface, and on the average is about 0.12 inches (3 millimeters) thick. It contains 70 percent of all neurons. A microscopic examination of a properly dyed specimen superficially reveals a structure resembling a layered classifier, as in figure 5.2. Except that not two, but three or four layers with an increased population density of neurons are present.

Figure 5.6 is a drawing by S. Ramon y Cajal, the father of neuroanatomy, who discovered the layers in 1888. He used a staining method invented by Camillo Golgi in 1875 which shows a small fraction of all neurons, but these in their totality. Today much is known about the fine structure of connectivity within the cortex. Besides the horizontally layered structure, a vertical columnar organization can be found. Each column has a cross section of 0.1 square millimeters and contains about ten thousand neurons of different types. Among these, the number of large pyramidal cells with an outgoing long axon is in the hundreds. Each column seems to be a functional unit. In a given neighborhood of the cortex all columns are of a similar construction. However, on the basis of

---

[17] The main sources used in this section are (Nauta and Feirtag 1986), (Nieuwenhuys, J. van Voogd, and C. Huijzen 1978), and (Popper and Eccles 1977).

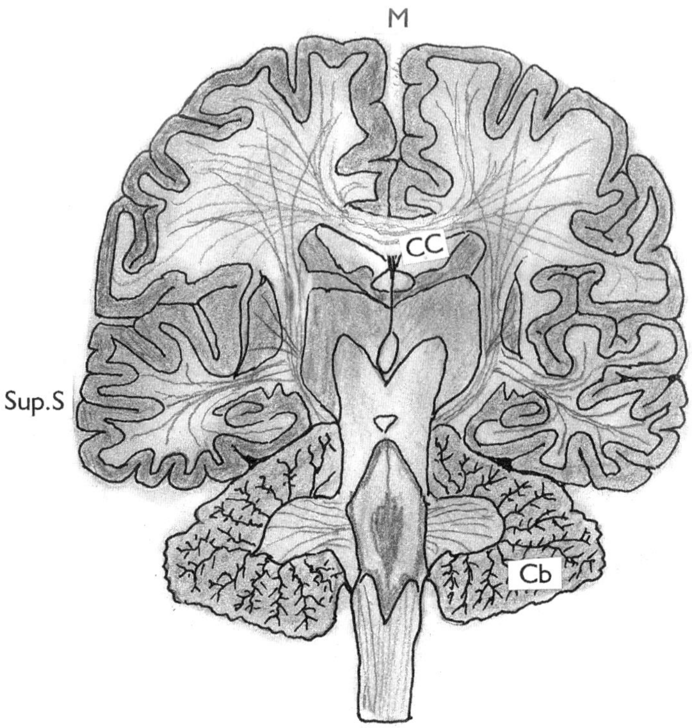

Fig. 5.5
Vertical section through human brain
[After: S.J. DeArmand et. al. (1989)]

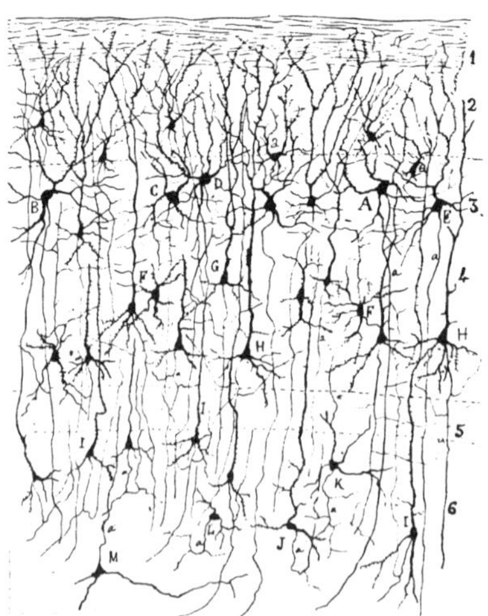

Fig. 5.6
Golgi stained crossection through cortex of ratsketched by Cajal in 1888, indicating
the layers

[With Permission: The Cajal Institute. CSIC. Madrid]

the predominant cell types and connective organization, more than forty different regions of the cortex can be distinguished. These are the areas of Brodmann[18] (Krieg 1953). Hundreds of millions of axons leave the cortex, and a similar number enters as input.

It is, of course, hopeless to try to map out a complete wiring diagram of the subcortical fibers. The white matter looks like a homogenous fatty mass and not like a wire harness at all. But there are methods of preparation of the brain by alternately pickling and freezing that make the axonal fibers separate from each other. If such a brain is carefully torn apart with tweezers, direction of tear will follow the fibers, and the structure becomes visible (Ludwig and Klingler 1956). Today the main axonal bundles and general tendencies of the wiring diagram have been traced by this and other techniques.

Figure 5.7 shows such a preparation by tearing into the inner (medial) surface of the right hemisphere. The medial plane is indicated by M, in figure 5.5. One notices the fibers emanating from an upper swelling of the brain stem, the thalamus (Th), and radially fanning out toward the cortex. The black areas indicate edges cut away in order to expose the fibers inside. The oval ridge, visible halfway between stem and cortex, is the crosscut corpus callosum, which consists of about two hundred million fibers crossing the medial plane to link both hemispheres. In figure 5.5 this is indicated by *CC*.

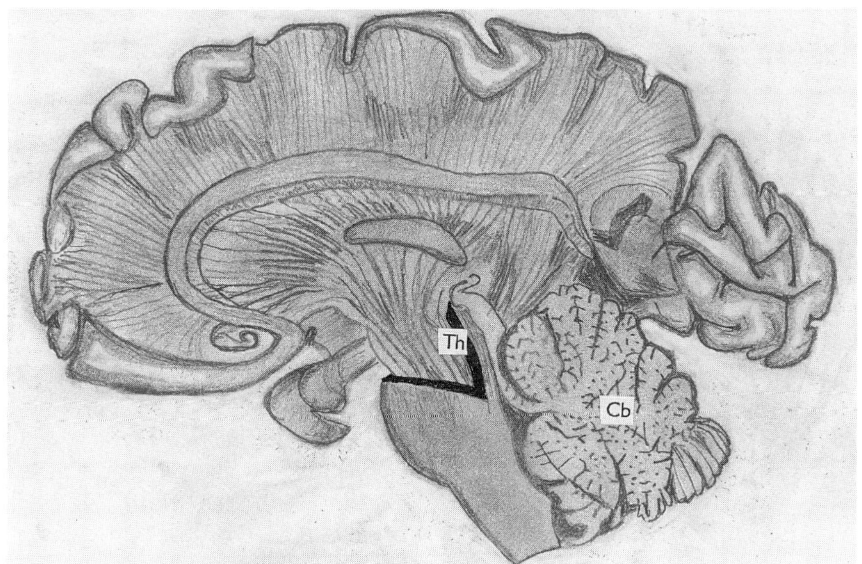

Fig. 5.7

Human brain prepared with the Klingler method. Right hemisphere seen from medial side, surfaces ablated to show radial (cortico-thalamatic) fibers

[After: Ludwig and Klingler, Atlas Cerebri Humani]]

[18] This classification is somewhat subjective. Many more distinct subareas are being recognized today.

Besides the predominant radial (or cortico-thalamic) fibers, there are tangential (or cortico-cortical) fibers that link various areas of the cortex. The bundles of the corpus callosum are most pronounced. But, as figure 5.8 reveals, there are many bundles running from one neighboring fold (or sulcus) to the next as well as over long distances from front to back. In this preparation, again the ablation is from the medial plane down into the (left) hemisphere. All of the cortex has been removed to show the bare wire harness.

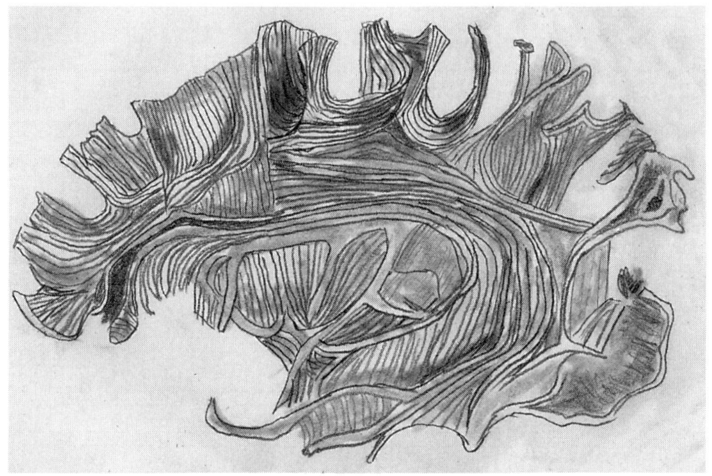

Fig. 5.8

Human brain prepared with the Klingler method. Left hemisphere seen from medial side, most surfaces ablated to show tangential (cortico-cortical) fibers.

[After: ibid.]

We now return to the microscopic detail of the cortex itself. It is quite impossible to trace all the connections even within one of the mentioned cortical columns. But representative partial circuit diagrams have been obtained (Eccles 1984). In figure 5.9 I have redrawn the diagram presented by Eccles in the schematic form introduced in Vignette 3. The letters near the cell bodies indicate particular types of neurons. *PY* is a large "pyramidal cell" that sends out a long myelinated axon, *MA*–a "power driver" so to speak. The outgoing axons, *ND*, go to dendrites in other columns in the neighborhood.

CC is an input from such neighborhood, and *TC* comes from the thalamus. Surprisingly, we find this sample of the total columnar network to have the simple architecture of a *partially trans-connected feed forward network* of four layers of the type sketched in figure 5.2. No feedback is seen within this column. The network appears to be only partially connected, but here is only a small sample from many more neurons and connections of this column. There is heavy connectivity within the column. But connectivity between columns is limited to a neighborhood of two hundred to six hundred columns, containing a total of a few millions of neurons (Szentagothai 1978). Since each column sends its axons into every direction, the whole cerebral cortex looks like one huge locally connected layered neural network. In addition to the local connections, which reach out sideways 1.5 to 4 millimeters, there are the already mentioned tangential (cortico-cortical) "association" fibers which span the entire area of the cortex. These are the express lines of communication between cortical provinces. Full connectivity is not necessary and is not possible in large networks. If all neurons of the brain were fully connected, our brain would probably have to be at least one hundred times larger in volume.

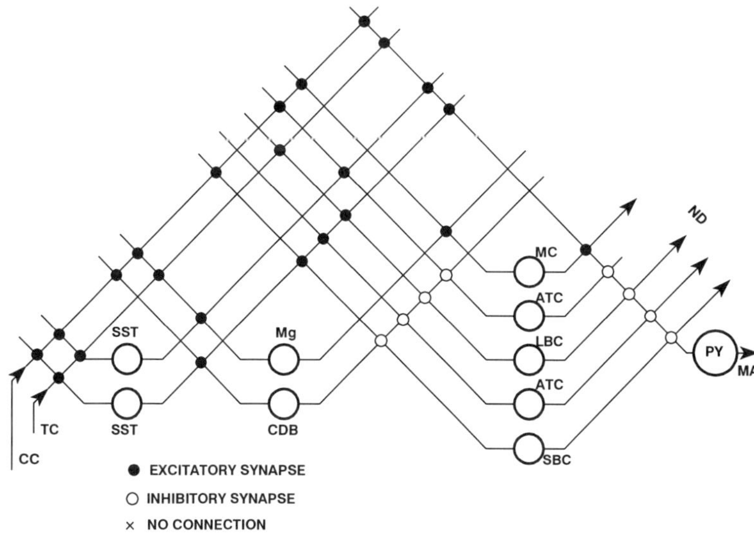

Fig. 5.9
Circuit diagram representaion of a sample from a cortical column

The cortical network certainly has features that go far beyond what artificial neural nets of comparable size have. One would expect that the various types of neurons are adapted to particular tasks specific to their position within the network—special tasks, of which at present we have no good understanding. *The network certainly is self-adapting as to whatever its function locally may be.* Known examples of self-adaptation within the visual cortex will be presented later on.

The huge network's inputs and outputs are the hundreds of millions of radial fibers as well as some connections at the edges of the cortical sheet. In evolutionary perspective, it is important to note that the area of convergence of all the radial fibers is part of the brain stem. This is the oldest part of the brain, a mere knob at the head of the spinal cord. It can be found in all vertebrates. It also is the first part of the brain to be formed in fetal development. Thus, in phylogenesis as well as in ontogenesis, the cerebrum (cortex and underlying wiring) is an outgrowth of a much earlier central nervous system.

Figure 5.10 shows two views of the human brain stem. Structurally this is the most complex part, "a wearisome labyrinth," in Ramon y Cajal's words. It is the center through which all sensory inputs and all motor outputs are routed. It consists of a crowded heap of nuclei–accumulations of neurons which can be recognized as individual units, functionally as well as anatomically. On the surface, some of the nuclei become visible as "colliculi" (little hills), best visible in the dorsal (rear) view in figure 5.10. Many of the major nuclei connect through radial axon bundles to their own "private" areas of the neocortex.

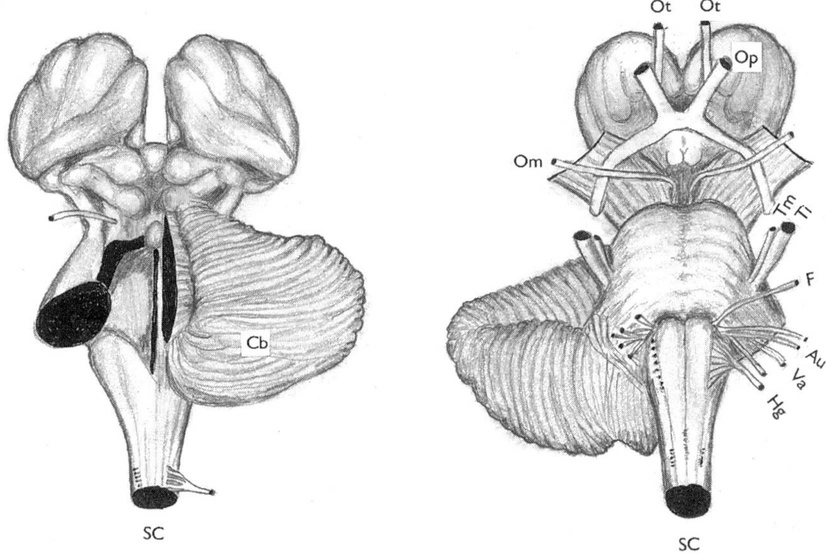

Fig. 5.10
Human brain stem
[After: Nauta, W.J.H., Felrtag, M,(1986)]

In the figure, we see on the ventral (front) side, on top, the two cut optic nerves (*Op*), entering from the eyes. Below, the oculomotor nerves (*Om*) exit to effect eye movement. Further below, on the side, the facial (trigeminal) sensory input lines (*Ti*) are visible and the facial motor output (*Tm*) is next to them. They serve the upper face. Below is a starlike outburst of many nerves: another facial nerve (*F*) for the lower face, the vagus nerve (*Va*) which is the only nerve that descends down into the body outside of the spinal cord to control inner organs, and two auditory inputs (*Au*), one from each ear. Further down is the hypoglossal nerve (*Hg*), conveying taste sensations from the mouth. And ultimately all the somatic-sensory input and motor output lines of the spinal cord (*SC*) join in from below.

Thus, all sensory input, except olfaction, enter through the middle brain stem and find the first synaptic connections inside of it. But the nose, too, is connected. In all vertebrates, two olfactory bulbs are positioned directly on the tip of the brain stem. Sometimes they are fused together. In mammals, as the cortex grew, the stem had to grow an extension, the olfactory tracts, in order to keep the bulbs near the nasal cavity. The olfactory tracts (*OT*) are seen at the top of this drawing. On the dorsal side, the two hemispheres of the cerebellum (*Cb*) are attached. One has been removed for better view.

The cerebellum is like a small, highly-convoluted cerebrum, with its own cortex. In figure 5.5 a cross section of the cerebellum appears in the lower part. Compare also figure 5.7. This also is an ancient organ. It facilitates feedback motor control. Motion systems, whether natural or artificial, all need feedback loops: a muscle contracts while its antagonist stretches, the current angle of the elbow is sensed, and the muscle is told to stop flexing or to exert more counter force when the desired angle is reached. That is one of the functions of the cerebellum. Another is to keep a memory of preprogrammed automatic movements. Lorenz's water shrews with their motor memory had a cerebellum as big as their cerebrum.

It is instructive to look at the evolution of the vertebrate brain as it is reflected in today's surviving phyla (Igarashi and Kamiya 1972). One of the simplest vertebrae is the lamprey, an eel-like creature, not yet considered a fish. Its brain is about 1 centimeter (less than 0.5 inches) long and depicted in figure 5.11a. In front of this swelling of the spinal cord are the two olfactory bulbs (*B olf*), with an extension called the olfactory lobe (*L olf*). Behind this is another bulb, the epiphysis (*Ep*), which is the control center for inner organs, their chemistry, and for whatever homeostasis[19] this simple creature has evolved. Further back are two more bulbs which are connected to the simple eyes, the optical lobes (*Lo*). And finally a ragged extension on both sides follows, which is a budding cerebellum (*Cb*). In appearance as well as function, the analogy to the human brain stem is striking. Of the basic ingredients, only one is missing—there is no cochlear (ear) bulb. The ear will be invented later.

With this simple accumulation of nerve cells, this creature can effectively swim, escape, search food, feed, smell (and taste?), see, and reproduce. We will never know what it can see, but obviously it sees enough to have survived a few hundred million years of struggle. Compared to Aplysia, the lamprey is more advanced. There are more neurons, and they are efficiently concentrated in one control center. The dark cavity, *V*, is the beginning of what will become the fourth ventricle in the human brain. It carries the nourishing liquid that surrounds all cells in all brains.

For the next level, representative of true fishes, the carp brain is shown in figure 5.11b. This is about three centimeters long. It has sprouted a number of

---

[19] Control of the inner milieu, such as temperature and composition of fluids

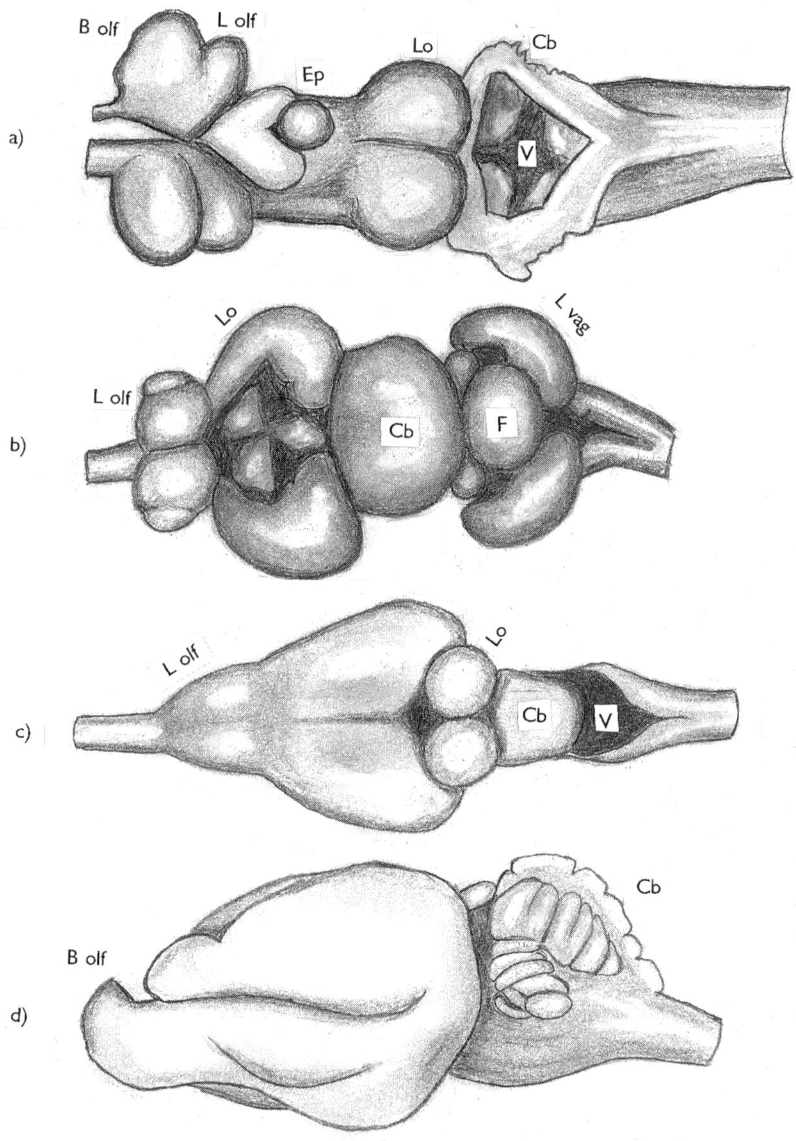

Fig. 5.11
Ancient forms of vertebrate brains. (a)Lamprey (top view) (b) Carp
(top view), (c) fapanese turtle (top view) (d) Oppossum (side view)
[After: Igarashi, S., Kamiya T., loc.cit.]

new bulbs or nuclei. The largest one is the cerebellum. Obviously the carp has a more sophisticated mode of locomotion, using several fins. The optic lobe is much enlarged because the carp has true eyes. The new facial lobe must have to do with food intake (taste and chewing). The lamprey only had a sucking intake hole. The vegetative somatic machinery also grew more complicated, requiring two new vagus lobes (*L vag*). Not seen, on the underside, one finds the two mammillary colliculi which will continue to be visible throughout the entire line up to *Homo sapiens*. In fact, the morphologist can trace most of the centers, nodes or "bulbs" throughout all ascending phyla of vertebrae. New ones may be added, but the functions of the old ones are so fundamental that they continue, even though possibly growing larger.

Next we take a turtle, the Japanese terrapin (*Clemmys japonica*), figure 5.11c. This brain is slightly smaller than the carp's, yet as percentage of body weight, it is bigger. Behind the olfactory lobe there is a beginning of the two cerebral hemispheres. Like in all brains to come, the two hemispheres are mirror images of each other, separated by a medial fissure. Bordering to this groove are areas which will become the hippocampus (compare fig. 5.16). The area further out will give rise to some other regions which in higher mammals are the immediate neighbors of the brain stem (the striatum and the cingulate cortex), and like the hippocampus, have a distinctly simpler microanatomy of neural connections. Being the most ancient parts of the cerebral cortex, they are sometimes called the "paleocortex" or "archi-cortex." The cerebral hemispheres have further grown on the level of the American opossum (*Didelphis sp.*), figure 5.10d. This is a side view in order to show how the hemispheres have wrapped themselves around the brain stem. Even a first inward folding has occurred. The cerebellum has gained much volume and complexity; mammalian locomotion is much more sophisticated. But underneath all, the brain stem nuclei are not only intact, but new ones have also been added and old ones enlarged.

Going on to figure 5.12, we can see how throughout cat, chimpanzee, and man, the cerebral cortex (or neocortex) has further proliferated. In this process of cortical expansion, the hippocampus, which is on the upper side in the Japanese turtle, has been pushed around the back, and is now located underneath (compare fig. 5.16). Altogether, each hemisphere has become like a shrunken balloon made of neocortex, whose open mouth is on the medial (inner) side. This mouth is hemmed in by a ring consisting of the oldest parts, generally referred to as the "cingulate gyrus" (from "cingulum," meaning "the band or girdle"). The top swelling of the brain stem (the thalamus) is halfway protruding into both of these openings. From here the radial fibers enter and connect to the inside of the balloons. The corpus callosum also uses these openings to link both hemispheres on the inside.

When comparing the neural systems with mental achievement, one can distinguish three distinct levels of emergence.

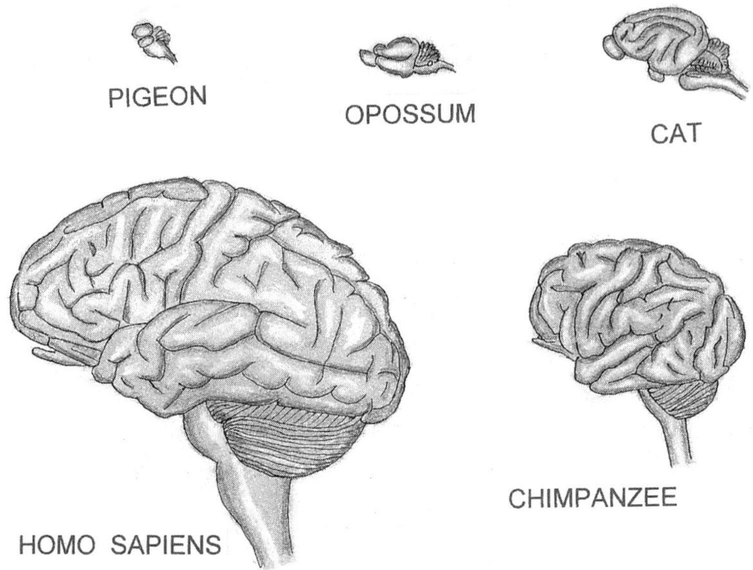

Fig. 5.12
Brains at various levels of emergence, drawn to scale
[After: *Scientific American*, September 1979, p.47

Emergent Level I: Representative of this level is Aplysia. The nervous system is distributed over large areas of the body. More than anything else, this is a straightforward control system.

Simple key stimuli trigger hard wired responses. Yet, in a limited and preprogrammed sense, adaptive modification can take place. Generally this is restricted to habituation and sensitization.

Emergent Level II: The control system now has acquired some memory; it is concentrated in the head and has sophisticated sensory organs for vision, olfaction, hearing, and body posture. A cerebellum is present for advanced motor control and motor memory. This is the basic brain stem animal. Fishes, lizards, turtles, and birds would be representative of this level. Their success is mostly based on improved locomotion and sense organs. Motor learning takes place. Yet otherwise adaptive modification beyond habituation and sensitization is very limited. And when it happens, it is probably associated with the first beginnings of the cortex—the paleocortex. Birds are the most developed among the species following this blueprint. Such animals are as ancient as they are successful.

Emergent Level III: This breakthrough came with the proliferation of the neocortex—a basically unspecialized network of neurons, forming layered classifiers, destined to self-adapt. On this level learning became an instinct. The basic brain stem animal, however, is still intact underneath. In fact, it must have developed quite a few new tricks in order to motivate the process of cortical self-adaptation through the joy of exercise, and generally to utilize this new organ. This last step is the most spectacular one. With the neocortex, intelligent life becomes possible.

Whether there is a fourth emergent level in a biological context, has to remain undecided. The transition from the general mammalian level to human consciousness looks more like an increase in quantity only, not in quality. But certainly there is a fourth, supra-biological level of human culture and ego system that is based on human consciousness. This will be discussed in the second part of the book, "Beyond Human Nature."

The control of the body's internal functions forever remained in the brain stem. The mammal with surgically removed cerebral cortex continues to live and to display some basic reflexes. The brain stem is the organ of control for somatic survival and well-being. Pursuing this goal, it will be found to also control and modulate cortical function.

If one keeps in mind the need of all neural networks to connect each point to almost any other, as in the neocortex, and the need to connect to one central organ of global control, then the overall design of the advanced brain makes exquisite sense. In order to connect each point of a surface to every other point, and to one central organ, and also to minimize the total length of axons involved, one would have to bend the surface into a sphere, surrounding a center, and then further shrink it by adding many inward folds. I am reminded of the layout of Cray supercomputers, even though there the geometry is not spherical but cylindrical. Yet it follows the same idea.

Cortical Localization and the Basic Dualism

When the interaction between brain stem and neocortex is considered, one cannot talk about one without the other. They are two focal points of a multiplicity of interacting feedback loops. But they contribute in different ways. I will try to sort this out with help from some old, but still fascinating, experiments by Wilder Penfield. He was one of the most celebrated early neurosurgeons of this century. In the twenties it was already known that epilepsy sometimes could be linked to scars in the brain tissue. Sometimes the scars were surgically removed. But the method was considered questionable and of limited success. In 1934, Wilder Penfield and some of his colleagues at McGill University established the Montreal Neurological Institute. There he was to spend a lifetime of intense research and to have a large surgical practice. Electroencephalography was not yet an established tool, and the knowledge of localization of brain function was in its infancy, yet the success of the operations depended on removing as much as possible of the diseased regions without cutting into any areas that would leave the patient mentally impaired.

To improve chances of success, Penfield used a technique exposing the brain under local anesthesia of the skull. Fortunately the brain itself does not have sensory cells. Then, in order to localize the patient's important areas, he used electrical stimulation with a hair-like electrode. When the motor area of the cortex was stimulated, the patient involuntarily moved a limb or finger. When the sensory area was stimulated, he or she told that for instance tingling was felt in the corresponding limb. Also, when the area of the scar was stimulated, sometimes the patient felt an epileptic seizure coming. In this way Penfield mapped out the areas around the scar before the critical incision was made.

This became a gold mine to brain research. But what does the record show? At first there are some fairly straightforward observations. Figure 5.13 shows a principal localization chart of the left hemisphere. On each hemisphere there are the two most pronounced deep inward folds, fissures or sulci, which define the main geography: In the middle, running at an angle of about $20°$ backwards with respect to the vertical, is the central fissure of Rolando. It separates the frontal lobe from the parietal lobe. Under the temple is the third main lobe, the temporal lobe. In evolutionary perspective, this is the youngest part of the brain, which has become especially pronounced in Homo sapiens. It is separated from the rest of the hemisphere by the deep "fissure of Sylvius." The rear of the cortex also is classified as a separate lobe, the occipital lobe, which contains the *fissura calcarina*, which is the center of the visual or calcarine cortex.

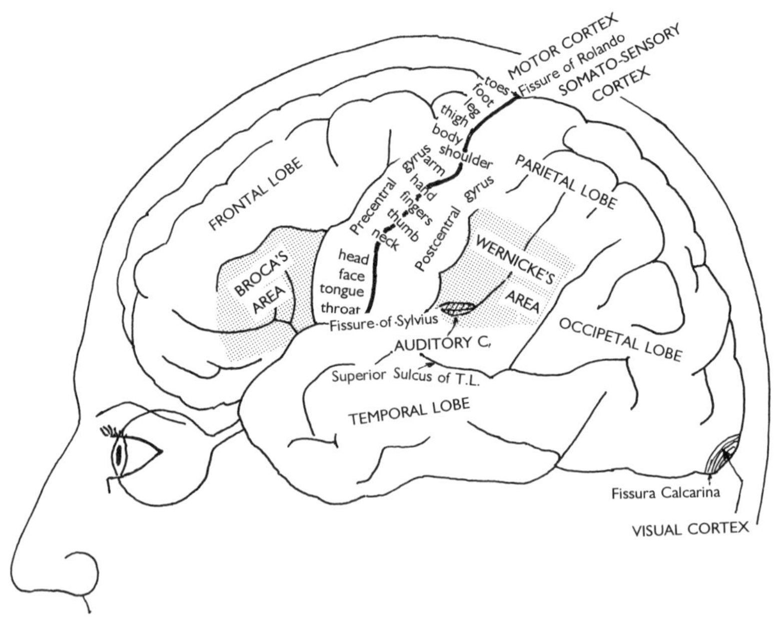

Fig. 5.13
Orientation map of left cerebral hemisphere

Along the frontal shoulder of the fissure of Rolando (the "pre-central gyrus") and extending into the fissure itself, is the motor area. Here, stimulation at any given point is directly followed by observable movement or twitching in a well-defined muscular field of body or limbs. From here long axons lead into all body muscles. The mapped motor areas are indicated in the figure. On the other side of this fissure on the "postcentral gyrus" is the corresponding somatic sensory area. The map here roughly repeats the one on the opposite shoulder of the fissure, but stimulation in this area generally results only in a perception of twitching, or of a movement, not in an actual movement itself. Examples of stimulation of the motor area are (Penfield and Roberts 1959) and (Penfield and Rasmussen 1952). Pulling of jaw to right, quivering of the jaw in a "sidewise manner," "a feeling in my throat that stopped my speech." (All of this stimulation was at the lower end of the gyrus). Vocalization, as well as arrest of speech, can be obtained. Penfield writes:

> Such vocalization is a vowel sound, a cry without words. One might consider that it resembles the cry of the infant if allowance is made for the difference in structure of the larynx of the adult which produces a heavier tone. It also resembles the cry of some epileptics at the onset of a seizure. (Penfield and Rasmussen 1952)

Stimulation in the appropriate upper areas on the right side leads to, for instance, closure of the left hand and bending of the forearm, arching of the left foot, et cetera. The right hemisphere is connected to the left side of the body and vice versa (contralateral). The patients are well aware of the movements, but when asked "Why did you move your foot?" the answer may be, "I don't know," or "I didn't move it, you did." One vocalizing patient who was asked to suppress the sounds ". . . observed humorously that if he could just come around to the other side of his head so that he could get his hands on the operator, he would be able to stop it." The movements are never skilled, and in Penfield's words ". . . not more complicated than those the newborn infant is able to perform."

Not all motor responses are elicited from the frontal side of the fissure of Rolando. In Penfield's observations about 20 percent came from the predominantly somatosensory area on the other side. The same movement can be elicited from both sides of the fissure. Salivation and mastication, in fact, are more often elicited from the sensory area than from the motor area. This shows the close connection between the two areas. Understandably each movement needs somatosensory feedback for its proper control. Is the topological mapping really exact and unique? Penfield could produce neck movements and movements of the eyeballs and eyelids from almost any point along the middle region of motor cortex. And single fingers could be moved from points well outside of the area devoted to hand and fingers. Today it is generally accepted that accurate topological maps do exist, but electrical excitations yield blurred maps due to the mentioned lateral connections between cortical columns, including connections across the fissure of Rolando.

Returning to figure 5.13, we notice in front and adjacent to the motor area for head, face and tongue, a dotted field, the anterior speech area or Broca's area. This is involved in the fine-tuned motor sequences needed for speech production. Another higher speech center is found on the other side of the central fissure, the posterior speech area, or Wernicke's area. Excitation of any of the speech areas will not produce speech. Not even any understandable single words are produced. But different kinds of interference with speech can be observed—hesitation, slurring, distortion, and repetition of words or syllables. Total arrest of ability to speak can be induced from many points, even outside of the speech areas. A subject was asked to count, and under electrical excitation he counted: "five, six, twenty, nine, . . ." After the electrode was removed he counted normally. Another was asked to name objects from pictures presented to him. A picture of a foot was shown while stimulating: "Oh, I know what it is. That is what you put in your shoes." And upon removal of the electrode, "Foot." A case that particularly impressed Penfield is the following. The patient had to name objects pictured on cards.

> Then, before the picture of a butterfly was shown to him, I applied the electrode where I supposed the speech cortex to be. He remained silent for a long time. Then he snapped his fingers as though in exasperation. I withdrew the electrode and he spoke at once. "Now I can talk," he said. "Butterfly. I couldn't get that word 'butterfly,' so I tried to get the word 'moth.'" It is clear that while the speech mechanism was temporarily blocked he could perceive the meaning of the picture of a butterfly. He made a conscious effort to 'get' the corresponding word. Then, not understanding why he could not do so, he turned back for a second time to the interpretive mechanism, which was well away from the interfering effect of the electric current, and found a second concept that he considered the closest thing to a butterfly. He must have presented that to the speech mechanism, only to draw another blank. (Penfield 1975)

Penfield and his associates reported many similar cases of induced speech block. But this patient with the initials C. H., who could not say "butterfly," has made history. In an intuitive flash, Wilder Penfield came to believe that there is a mind active and separate from all the cortical areas that can be blocked or stimulated. This mind observes the body's involuntary movements. It tries and cannot stop induced vocalizations. It can joke about its temporary incapacitation. Penfield continues:

The patient's simple statement startled me. He was calling on two brain-mechanisms alternately and at will. He had focused his attention on the cards and set himself the purpose of recognizing and naming each picture as it came along. At first each picture was inspected in the stream of consciousness. It was identified, named, and recorded. He was using areas of cerebral cortex that, at birth, had been uncommitted as to function. Evidently, the highest brain mechanism, impelled by mind-decision, can carry out these transactions, calling upon previously established, conditioned reflexes one by one. When I paralyzed his speech mechanism, he was puzzled. Then he decided what to do. He reconsidered the concept "butterfly" and summoned the nearest thing to butterfly that was stored away in his concept mechanism. When the concept "moth" was selected and presented in the stream of consciousness, the mind approved and the highest mechanism flashed this nonverbal concept of moth to the speech mechanism. But the word for "moth" did not present itself in the stream of consciousness as he expected. He remained silent, then expressed his exasperation by snapping the fingers and thumb of his right hand. That he could do without making use of the special speech mechanism. Finally, when I removed my interfering electrode from the cortex, he explained the whole experience with a feeling of relief, using words that were appropriate to his thought. *He* got the words from the speech mechanism when *he* presented concepts to it. For the word "he," in this introspection, one may substitute the word "mind." Its action is not automatic.

Large areas of the cortex can be surgically removed, and the "mind" is still there. When it feels impaired, it might switch to satisfy its wishes in some other way, employing other parts of the brain. Penfield concluded for himself that this uncannily persistent mind must reside outside of the physical world of neural mechanics. Obviously he was a dualist at heart as many other of his great contemporaries, and René Descartes of long time ago.

But for me this case represented a different kind of instant illumination. There is no need to invoke the incomprehensible, I concluded, considering what ethology has to say, and understanding the power of neural networks, Penfield's observations lead to quite a different interpretation. No matter where this "mind" resides, it is simple enough to consist of low-level innate biological determinants. It does not speak itself, but merely turns on the speech mechanisms in the cortex. "It" does not have to have consciousness in itself, but can make use of the mechanics of consciousness and speech, among other systems. *The important thing is that it has certain wishes.*

The picture of Konrad Lorenz's water shrew flashed before my inner eye, as it stopped dead in its track when the expected box in the water was missing. Its little active self found itself deprived of the usual surface to jump on. It couldn't quite stop the jump that would have propelled into unknown waters, but it managed to convert the impetus into a grotesque upward leap that so much amazed Konrad Lorenz. This animal's choices were not many. A cerebellum for running was available, yes. But almost no memories could be found in the tiny cortex, and no solution in sight. So what does it do? *Like C. H., the water shrew tried the next best in its simple circumstances. It backed up for a distance and tried again, with the same negative result.* Lorenz does not tell how it ultimately resolved its quandary.

I did not see much difference in *quality* between the self of the shrew and what Penfield called the "mind" of C. H. It was a matter of *quantity*, though.

When the persistent self of C. H. wants a neocortical system to recognize and name the contents of the picture, and this system is blocked by an electrical shock, the result is emotional frustration, and nothing more sophisticated. We have encountered emotions in animals without much cortex, such as geese. It is quite natural to assume that the most basic "will" to take action relates to the instinctual sphere, and resides mainly in the brain stem. This "will" may also be elaborated in more adapted detail by the right cortical hemisphere which remains untouched by Penfield's electrodes. As will be seen shortly, the workings of the right hemisphere mostly remain unconscious. The left neocortex, on the other hand, must contain the cognitive apparatus that (if not blocked) can recognize the picture of the butterfly from memory and associate the previously learned name with it.

When we ask the question, why does the cortex try to name the contents of the picture and not one of many other things it would be capable of doing, like noticing the predominant color or that the picture of the butterfly is a cheap print, the answer seems to suggest itself: the somatic self has made up its will previously by reading some simplified cortical output that was tailored to its needs and produced during the initial conversation with the doctor who had set up the experiment. It "set himself the purpose of recognizing and naming each picture as it came along" (see above). But "it" did not have to understand any details of the task *as long as there was assurance that this was desirable.* The details of execution have remained on the cortical level of neural networks.

In this view the relation of brain stem to cortex reminds us of the relation of an animal to its environment. Already Jakob von Uexküll (1934) had understood how each species, in a way, lives in its own private and restricted environment. Due to limited knowledge and sensory apparatus, the animals receive and react to only a fraction of the total reality of nature around them. To the rest they remain blind. *Similarly the basic shrew-like self may be blind to most of the cortical activity, except to some selected output.* Nevertheless, it can well satisfy the basic survival needs by issuing simple orders for cortical processing for subsequent, well-informed action. *In this view, neocortical activity simply enlarges the action space of the basic brain stem animal.*

Instead of having to physically explore the world every time, and risk being eaten, some of the action can be staged beforehand in the virtual world of the cortex.

An insert about the usage of "high" and "low" is now appropriate. In computerese, a "high level command" designates a compact and simple statement that is translated by the resident program into a large sequence of "low level" steps compatible with the hardware. In connection with brain function, "high cognitive level" is more appropriately associated with the cortical systems that do all the detailed processing, whereas "low level" would apply to simple instructions originating from a brain stem that does not have direct cognition of the complex external reality. Such usage seems logical and natural. This shows once more the incommensurability of computers and brains. Thus, in organisms, the ultimate decisions may be made on a low cognitive level. Life's important decisions generally turn out to be more emotional than rational.

My hypothesis was reinforced when I learned that Penfield had localized his "brain-mind interface" in the brain stem, particularly in the reticular formation, the most ancient organ that deserves the name "brain." Primitive animals succeed well with very little cortex on top of the brain stem and the cerebellum. In the sense used throughout this book, they "know" by instinct how to make good use of their ecological niche, how to interact socially, how to reproduce, and how to avoid being eaten. Consequently, since the cerebellum is already occupied with motor control and the cortex remains small, the instincts that determine behavior on the ancient level must be primarily located in the brain stem. And there is absolutely no reason to believe that human instincts are centered anywhere else. After all, the main internal organs do exactly the same in the water shrew, the bonobo, and in *Homo sapiens*. Also, the location for control of homeostasis and of the endocrine glands is in the hypothalamus. This, too, is a part of the upper brain stem.

The need for close proximity of the centers for physiological needs and the hedonic wishes of the self is obvious. These wishes must be in tune with the current condition and needs of the body. When satiated, it might want to sleep and digest. When hungry, it will look for food. At the proper time of the year, following a hormonal circadian rhythm, it will want to build a nest and reproduce. The circadian clock, by the way, is also nearby in the *supra-chiasmic nucleus.*

The somatic self can be understood in terms of neural networks. Add to a small system like NETtalk (see Vignette 3) the refinements that natural neural nets have above artificial ones, especially hormonal modulation. Add also some neurons emitting excitation on their own, again under hormonal control, then multiply everything by ten thousand and you may have something that remotely resembles the physical substrate of the somatic self of a mouse. Multiply this again by one thousand and you may have your own somatic self's central machinery. The resulting 3.10^{10} neurons are still not more than a small fraction[20] of the total number of possibly 10^{12}. This network can send signals to most parts

---

[20] Just reduce all numbers, in case the stated total number of neurons should prove too high.

of the cerebral cortex, and it will receive back messages tailored to its simple modes of somatic knowledge. The received messages may contain highly processed sensory input already classified by comparison with memorized contents. In fact, as artificial neural networks demonstrate, processing sensory input is inseparable from forming memory. Since childhood, a model of the external world is built up in the cortex. The key to processing of sensory input is classification with the aid of this model of the world which resides in the acquired distribution of synaptic weights. *The source of this model is the world itself.*

The somatic self requests sensory processing, and in response to processed sensory input it will give low-level commands for action which are translated by the motor cortex and the cerebellum into complex, well-executed and fine-tuned "motor melodies." The resulting behavior thus originates in somatic needs and can realistically exploit given external possibilities. *But the somatic self is not more than a bag of tricks of phylogenetically acquired knowledge evolved to survive in the real world.*

Where, most likely, is the central organ of the somatic self located (or Penfield's "mind-brain interface")? Since Penfield, many researchers have placed it in the reticular formation (Moruzzi and Magoun, 1949) and (Magoun 1963). The reticular formation is named after the characteristic grid like microstructure of neuronal connections. Figure 5.14 shows the location and extent of the reticular formation in our brain stem. The reticular "nuclei" in fish can be individually traced throughout phylogeny and are still present in *Homo*. Around them an extremely complex web of specialized new nuclei has evolved, which ultimately, we may say, have been capped by the cerebellum and the cerebrum. But the reticular formation remains at the most strategic location that can be imagined, at the bottleneck of all axonal traffic between body, sense organs, and the rest of the brain. It is also located at the front door of the body's chemical control center.

Toward its upper end are many areas which produce regulatory hormones that enter the bloodstream and influence

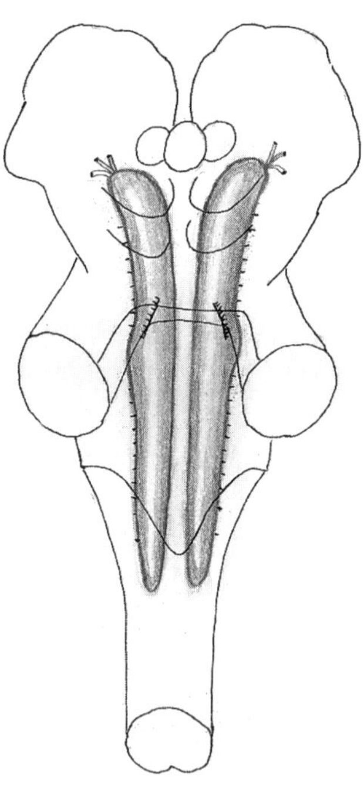

Fig. 5.14
Location of the human reticular formation within the brain stem

everything from blood pressure to ovulation. In turn, such areas themselves are receiving chemical feedback from anywhere between genitals (estrogen) and the olfactory bulb.

Today there is much hard evidence for the posted model. Using a technique developed by David Hubel (1959), micro-electrodes have been inserted into animal brains that probe single neurons. Multiple probes in the brain stem or any other area of the brain have been used to follow electrical signals from their origin to wherever they are sent. Activity of single neurons has been continuously recorded for days. Although the exact location of a probed neuron is not known at first, this can be established after the experiments are completed: A burst of current through the electrodes is applied. This causes a small local burning. The animal is sacrificed, the brain is dissected, and the burned spot(s) microscopically identified.

For instance, giant neurons in the pontine (or middle) area of the reticular formation have been found to initiate rapid eye movement (REM) sleep. Intense traffic of signals, the so-called "pontine-geniculate-occipital cortex (PGO) spikes," have been found to originate in the brain stem and follow the radical fibers to the cortex. It is known that during REM sleep the visual cortex is stimulated in this way, resulting in dreams. In the waking state, exploratory behavior is accompanied by a theta rhythm of around six Hertz (cycles per second). This is a global cortical activity which can be picked up with electrodes on the outside of the skull. This synchronizing rhythm is so powerful that even the rat's whiskers move with it. The origin again is in the brain stem. Among the effects that the rhythm has on the cortex is formation of new memory. A neuron can acquire long-term synaptic changes when excitation occurs during the peak of a theta period (Winson 1990).

In the final analysis, the will to live emanates from the brain stem. It also controls the mode of functioning of the brain, or stated differently, *it modulates the cortical activities to the animal's own hedonic end.* In the absence of a neocortex, to the same hedonic end, the primitive animal directly takes advantage of the possibilities of the real world.

Intelligent Cortical Systems

Like any executive decision maker, the somatic self is helpless in the absence of well-abstracted data, and of a well-organized staff to skillfully carry out the decisions. This branch of the brain's business is made up by the overwhelming majority of all neurons. They form the huge self-adaptive neural network of the neocortex.

In this section, I will try to explain in bold lines how the various areas of the fully-developed and adapted cerebral cortex receive sensory input, interact among each other and with the somatic self in order to form that marvelous machinery that can make a realistic model of the external world, move about it knowingly, and become aware of it.

To us humans, the three most important sensory systems are the visual, the auditory, and the somatic. The sensory tracts from eyes, ears, and from the body, all at first pass into the upper brain stem. From here the excitation is conducted via massive radial bundles of fibers to three primary sensory areas on the cortex. In figure 5.13, the primary visual and auditory areas are stippled (visual cortex and auditory cortex). The somatic sensory area already has been mentioned. It runs along the central fissure of Rolando. The primary visual area is at the dorsal (backwards) end of the occipital lobe. Only a small part of it is visible. In humans, the major part has been pushed around to the invisible back side of the lobe (the medial surface), and much of it is inside of the calcarine fissure located there. The primary auditory area also is barely visible on the upper rim of the temporal lobe. Most of it is hidden in the fissure of Sylvius. Those three areas are major sources of cortical activity.

It is illuminating to trace the succession of the main thrusts of tangential (or associative) fibers originating from the primary sensory fields. Although many investigations with different methods have been conducted in this matter, I will center on one using rhesus monkeys (*Macaca mulatta*) (Jones and Powell 1970). The technique is not very appealing and cannot be repeated on humans. Selective areas of the cortex were killed, whereupon the decay spreads along the outgoing axons. After a few days the monkey is sacrificed and the affected areas are found by microscopic examination. Fortunately the basic construction and even the fine details of columnar structure are similar in both monkey and man. Similar Brodmann areas have been identified in both. The human brain, however, has some new areas added on.

Figure 5.15 shows the left hemisphere of the monkey brain. The first diagram, (A), includes the succession of pathways originating in the visual areas 17 and 18 of Brodmann. There are two main thrusts. One goes to the temporal lobe, first to area 20, then 21, and ends on a narrow field (dotted) around and inside of a fold (the superior temporal sulcus). These areas have special significance because the two other sensory systems also converge here. This thrust also continues to the frontal end of the temporal lobe and to the frontal lobe, area 46. Another direct thrust goes to area 8 of the frontal lobe.

Looking at the sensory somatic pathways, (B), originating in areas 1, 2, and 3, we see a main thrust going into the motor cortex, 4. This, of course, is part of the feedback loop needed for motor control. The other main thrust jumps through areas 5 and 7 and, again, ends along the superior sulcus of the temporal lobe. But there are projections into the frontal lobe as well.

Finally, in diagram (C), part of the auditory cortex is visible on the upper rim of the temporal lobe. It also sends fibers to the areas of convergence in the temporal lobe and to the frontal lobe.

What can be made of this wiring diagram? Let us first extract some common features:

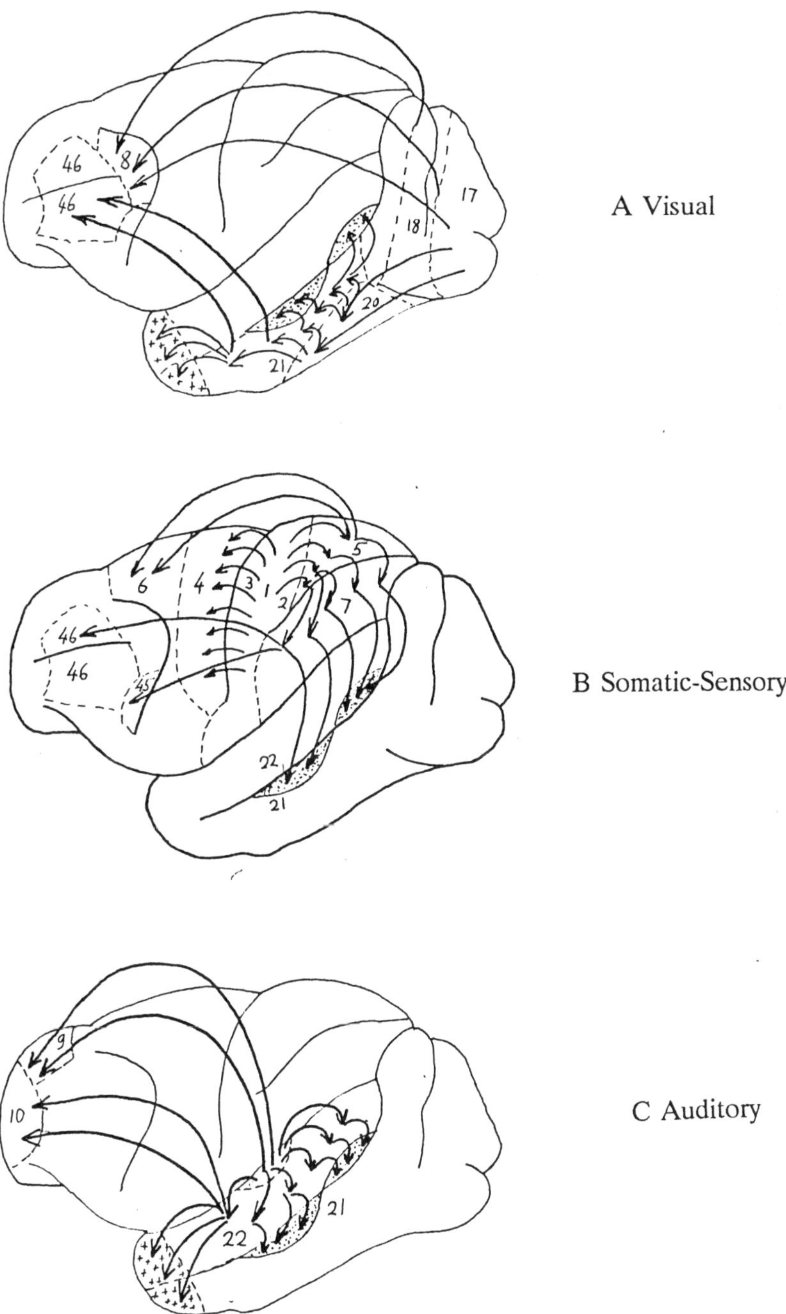

Fig. 5.15
Neural pathways in a monkey brain
[After: Jones and Powell (1970)]

a) To a first order of approximation, all areas of the cortex seem to be sequentially involved in the processing of sensory information of one kind or another.

b) The sensory pathways all have the same area of convergence on the temporal lobe.

c) A separate area of convergence of visual and auditory pathways only is in the frontal end of the temporal lobe, as indicated by the little crosses.

d) All sensory areas have projections into the frontal lobe. From Jones and Powell's work alone, no single area of this lobe can be strictly declared an area of convergence. But in a wider sense all sensory input arrives here. In humans the convergence is evident.

All the primary sensory areas are maps of the kind introduced in figure 5.3. Therefore the functional scheme associated with points (a) and (b) seems to be obvious: sensory input is sequentially processed by various pattern recognizers and classifiers in the cortical neural network. In the human brain, the processing chain must be significantly more powerful than in the rhesus monkey. There are a number of new Brodmann areas along the pathway, in particular huge new areas (18 and 19) have developed in front of the visual cortex, crowding it into the medial plane. At the end of this processing chain, which covers a large part of the cortex, a suitable abstract is delivered to an area in the superior sulcus of the temporal lobe.

What is the importance of this area? Inspection of figure 5.5 reveals that the depth of this sulcus (Sup S) is about as closely connected to the brain stem as possible for any piece of cortex. The short lower edge of the heavy sheet of radial fibers, the *capsula interna*, runs directly into this area. And furthermore, another link between this critical area of the cortex and an ancient system is made by the nearby *hippocampus*,[21] the main part of the paleocortex. The preparation in figure 5.16 shows the brain from below with the brain stem cut and some of the cortex removed to expose the *hippocampus* (HC) (carved surfaces are black). The two hippocampal lobes wrap around the brain stem. Each is merging with the temporal lobe all around its periphery, just opposite of the deep end of the superior sulcus. The two ends send thick bundles (fornix, cingulum, and others) into the brain stem, some ending in the ancient mammillary nuclei.[22] The hippocampus gives the impression of an organ collecting (and sending) signal flow broadside from the edge of the cortex, as well as communicating via a highly converged output bundle with the brain stem. It is hard to avoid the conclusion that the hypothesis of the circuits of the somatic self is strongly supported by points (a) and (b). Most of the cortex is involved in the processing of sensory input. When finished, the cortex presents the results to the somatic self.

But it is not the somatic self alone that profits from all the intelligent processing. The cortical areas closest to the described region of convergence are Brodmann number 20 and 21. As we will see shortly, here is the suspected seat of the conscious central interpreter.

---

[21] Greek for sea horse, which is quite recognizable in figure 5.16

[22] This name comes from the shape of this nuclei, as they appear in figure 5.10b, resembling female breasts.

I will leave points (c) and (d) in suspense for the time being and first explain in more detail the best known intelligent system—vision. This will serve as an example and prototype for the understanding of the other, less explored systems.

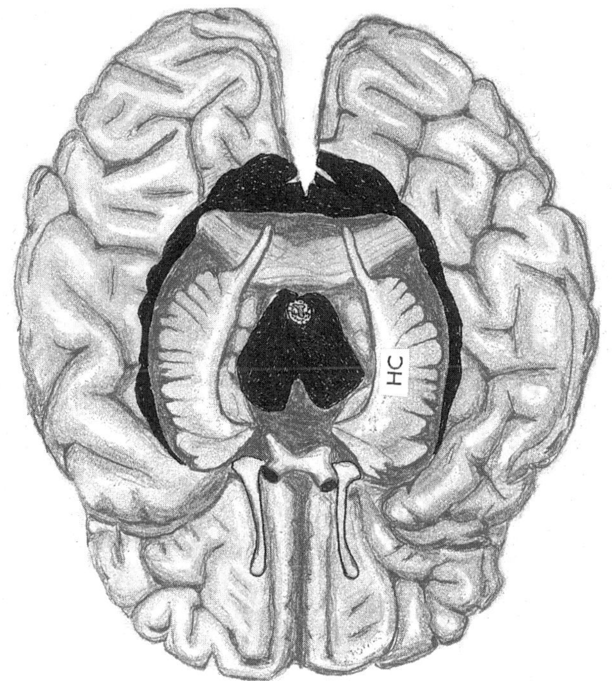

Fig. 5.16
Kingler preparation of human brain seen from below. The brain stem is cut and the hippocampus exposed.
[After: Ludwig and Klingler, Atlas Cerebri Humani]

The first stage of the visual processing chain is the eye. The evolution of the mammalian eye can be inferred with relative certainty. First there are light sensitive cells on the skin of primitive organisms. The ones that are located at the bottom of recesses help to crudely resolve the direction from which the light comes. Rattlesnakes still have such recesses as secondary eyes for infrared. Higher optical resolution would require a deepening of the recess, ultimately resulting in a pinhole camera eye. But that is not what happened (Poliak 1957). Pinhole cameras cannot provide a sharp image and good sensitivity at the same time. Only a small amount of light can pass through a small pinhole.

But the pre-fish, which might have developed such a low grade cyclopean eye, was partially transparent, as many sea organisms are. And here comes the emergent accident: The naturally curved epidermis (outer skin) right and left of the cavity acted as a lens on each side. *Two much brighter and clearer images appeared on the back side of the primordial retina!* Thus, the two-lens camera eyes came into existence by accident. Moreover, now the view was (accidentally!) panoramic, not just forward looking. Remnants of that sequence of events are the retinal neurons, which are on the wrong side—in front—thus obstructing the optical path. A comparable sequence must have happened at least twice on the phylogenetic tree. Among mollusks it resulted in the almost human-like eye of the octopus. Focusing, however, was solved differently. In mammals the transparent body at the entrance became stretchable with muscles, thus creating a lens of variable focal length. In the octopus the distance between retina and a rigid lens became variable like in photographic cameras.

With a high resolution retinal image in place, the attention now can shift to interpreting the image's contents. The most natural, compact, and economical way to grow bundles of axons toward the brain is to make them all parallel. This, in itself, results in the approximate, one on one topological maps of retinal excitation to be found in the lateral geniculate nucleus and further on in the visual cortex.

Now a chain of classifications and decomposition of that image begins, ultimately ending in my self-realizing, for instance, that an automobile is rapidly approaching and that I will be run over if I don't get out of the way fast. At first the image falls on an array of about one hundred million (108) light sensitive cells in each retina. Most of them, the "rods," are only sensitive to intensity. Other cells, the "cones," can distinguish color. A comparison with a television image is useful. Let me consider a picture that has three hundred thousand (3×10^5) "pixels"(this is more than the American standard has, but less than the French). Comparing the numbers, the eye should be far superior to a TV camera. Yet one finds only about one million (10^6) axons in each of the optic nerves leaving the eye. Is then the actual resolution of the entire visual field only 10^6 pixels, which is the content of detail in not more than three TV images? Strictly speaking, yes. But evolution has found ways to see better through better utilization of the given tools.

At first, the density of pixels is much higher in the center of the visual field, the fovea centralis. Since our eyes can turn quickly to any desired point, for all practical purposes we can see everything our self wants to pay attention to, with that maximum resolution. Secondly, processing of the outputs from the light sensitive cells already begins in the retina. There are neurons in the retina which convert the 10^8 light inputs into 10^6 axonal outputs. In that process, important features such as edges or movement are reproduced with an effective sharpness that surpasses what a simple 10^6 pixel array would be capable of.

Figure 5.17 shows a Klingler preparation of the brain seen from below. Much cortex and white matter have been removed in order to exhibit the visual pathways (again, carved surfaces are black). After the optic nerve has left the eyes (E), something strange happens which is a result of an ancient and unalterable blueprint of all vertebrates: The nervous system has to have mirror symmetry between right and left. Yet only both eyes *together* provide full stereoscopic vision. This problem of joining signals is solved as follows. Axons from *both left halves* of the retinas go to the *left side of the brain,* and axons from both right halves go to the right side.[23] Thus, the two images on both left half-retinas can be compared at the same cortical location, and vice versa. This requires crossovers from left to right and vice versa in the so-called "optic chiasm" (OC).

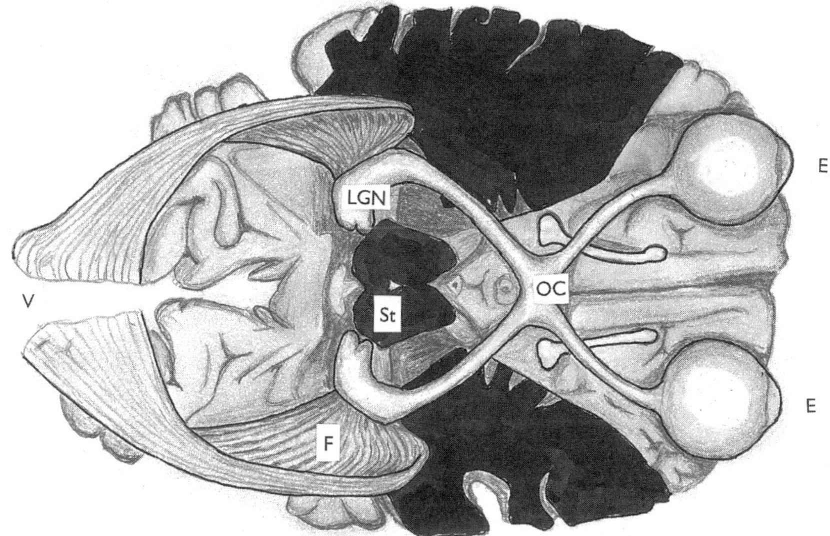

Fig. 5.17
Kingler preparation of the visual tracts
[After: Ludwig and Klingler, Atlas Cerebri Human]

After the crossover, the two mixed optic nerves enter the upper brain stem (St). Here, in areas called *lateral geniculate nuclei* (LGN), they make synapses, and signals are exchanged with other areas of brain stem and cortex. Connections to the *reticular formation* (the center of the somatic self) certainly are made. In this nucleus, also a low resolution, one on one (topological) map of the retinal area can be found (Marocco 1986). From the *lateral geniculate nuclei,* axonal fibers fan out toward the visual cortex (V), forming the so-called "optic radiation" (F), in figure 5.17. On the visual cortex, a much more detailed topological map of retinal excitation appears.

As can be seen in primitive fish, eyes are almost as ancient as the reticular formation. They predate the cortex. Therefore it is quite appropriate to assume that in association with the crude map of the visual field, in the LGN (upper brain stem) an older visual system has survived and still is functional. That, indeed, it

---

[23] Nevertheless, the vision system is "contralateral." The imaging lens projects the right visual field upon the left half of the retina, which projects to the left brain

is functional has been shown through studies in lesioned primates (Ungerleider and Chistensen 1977). According to these authors the *lateral geniculate nucleus* is involved in initially directing attention to various areas of the visual field by causing the foveal area to move to where it is most interesting. Here, again, we see the old little somatic shrew-self at work.

Another piece of evidence for the old visual system comes from so-called "blindsight." If parts of the visual cortex are damaged, the patient will experience corresponding blind areas on the retina. Yet the retina remains intact. Extensive evaluations of such cases (Weiskrantz 1980), (Weiskrantz et al. 1974) reveal that patients insist to see nothing when light or an image are projected on the so-called "blind retinal areas." Yet when asked to guess whether there was a flash just now or not, or whether the image of a grating is horizontal or vertical, they regularly achieve 75 to 95 percent correct guesses. But all the time, they insist that nothing has been seen and are baffled when told of their scores. The explanation, in view of the hypothesis of the circuits of the self, is of course that the self can see via the crude visual map in the *lateral geniculate nucleus*. That this information remains unconscious, points to the origin of consciousness somewhere along the pathway *well after* the visual cortex.

If the ancient visual system is crude, how did progressive corticalization improve its abilities? David H. Hubel and Torsten N. Wiesel were given a Nobel prize for their functional exploration of the visual system. First of all, the primary cortex was found to have a striking pattern of connections to right and left eye. Figure 5.18 shows a pattern of eye dominance on the surface of the visual cortex. The black areas are connected to the right eye, the white ones to the left eye. The width of each stripe is about 0.5 millimeters. For this pattern there is an explanation at hand: *Stereoscopic processing begins right here*. The images of right and left eye are slightly different. In this difference lays the three-dimensional nature of space.

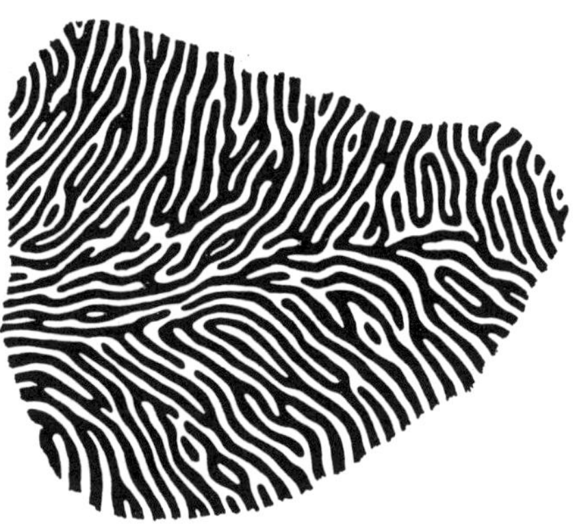

Fig. 5.18

Eye dominance pattern in a macaque visual cortex. This was obtained by autoradiography after injecting a radioactive chemical into the eye. The chemical (deoxyglucose) travels along the axons to the visual cortex.
[From: Hubel, D. H., Wiesel, T.N. (1979), with permission]

The local interconnections between cortical columns must be instrumental in figuring out distances and orientations of objects, edges, points, et cetera. But there is even more detail to the understanding of that process. Within a cortical column, the cells in the entrance layer (IV) best respond to a circular light spot on the retina of a certain small diameter. But Hubel and Wiesel (1978, 1979) have demonstrated that in the surface layer the best response of a given neuron is elicited by a thin line of a certain angular orientation. The angles vary according to the pattern given in figure 5.19. At first sight this cortical representation of image properties seems to be strange, until one realizes that it is exactly these differences between the eyes in angular orientation of lines which mirror the three-dimensionality of space. To prove this point, a complicated treatise in analytical geometry can be replaced by a simple experiment:

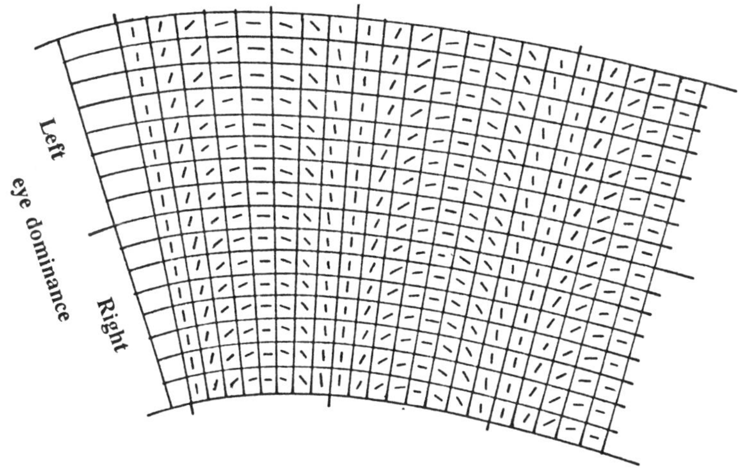

Fig. 5.19
Graphic representation of the surface of the visual cortex. The individual columns are visable. Arrows indicate the directional sensitivity of each column.
[After: Hubel and Wiesel, Ibid.]

Sitting upright, put a pencil in a vertical position about ten inches in front of your nose. Incline its top by about forty-five degrees towards yourself. Now, if the pencil is exactly in the middle and you close the right eye, you will see the pencil inclined to the right. If you close the left eye, it will appear to be inclined to the left. With both eyes this information is interpreted to mean inclination toward me.

This experiment begs for the following conjecture. These networks are adapted to solve the problem of stereoscopic vision in the most efficient way: extract lines and then compare the angles seen in each eye. If lines are not naturally present, as some experiments with optical illusions show, our system will tend to assume connecting lines between significant features of the presented image in order to be able to understand spatial relations. In a given region all neurons which are sensitive to a given angle are connected via the short-range axons. If a fraction of them is excited, all will receive some excitation, resulting in the mentioned illusions.

I cannot insist, however, that this conjecture has been proven correct. Paul Churchland has a different one (Churchland 1995): both retinas are projected on the same input layer of a classifier, in such a way that each neuron receives excitatory synapses from one eye and inhibitory ones from the other. If related retinal areas in both eyes see exactly the same image, all excitations cancel out to zero. In the second layer are "fixation cells" that detect the absence of excitation, indicating that both eyes converge on the same point in three-dimensional space. Now, because the angle of convergence ("vergence angle" in jargon) of the eyes is known to the networks from the muscular sensory cells in the eyeball muscles, one might conclude that stereoscopic vision works by triangulation, like the split-image devices in some older photographic cameras. But that is not what Churchland has in mind. He says that there are detectors in other layers that detect how far the images are split everywhere, thus eliminating the need to constantly refixate when looking at a three-dimensional scene. This seems to be quite within the possibilities of "classifiers of classes," as introduced here.

At the present state of affairs, however, nothing is verified for sure, and the line-direction sensitive cells may very well play a role. Whatever the actual truth, the conclusion is that there is more than one way for stereoscopic vision to work.

But how can any such intricate pattern of connectivity come into existence? Here is the point where principles of organization begin to appear that certainly go beyond genetic determination. Real neural networks even have a stronger ability to self-organize and self-adapt than artificial ones. Several researchers—(Miller, Keller and Stryker 1989)—have explored such processes by computer simulation. Stryker et al. assumed that the mapping of the retinal area upon the visual cortex is approximately topologically correct, and that there is arborization, meaning that any incoming axon branches out over a neighborhood of a certain size. It is also assumed that synaptic strengths are initially random but are increased by a Hebbian rule. If now there is a simultaneous correlated excitation in the neighboring neurons, such as would result from similar images projected on the two retinas, an eye dominance pattern will establish itself automatically.

Figure 5.20 presents the growth of such a simulated dominance pattern under appropriate "retinal" stimulation. Since the original connectivity was at random, this is a process of self-organization. Even at this low resolution of thirty-two times thirty-two pixels, the zebra stripe pattern is seen to emerge. Other simulations (Kohonen 1984) show that a crude one-on-one initial map is not even necessary, as long as each retinal neuron has adaptable (Hebbian) synapses to every neuron in the map. The retina only has to be excited repeatedly with large numbers of different images, which have the natural features in common, that there is a gradual change in illumination from one retinal neuron to the next. Then an exact map forms through self-organization. Kohonen demonstrates the same mechanism for self-organization of audio frequency maps

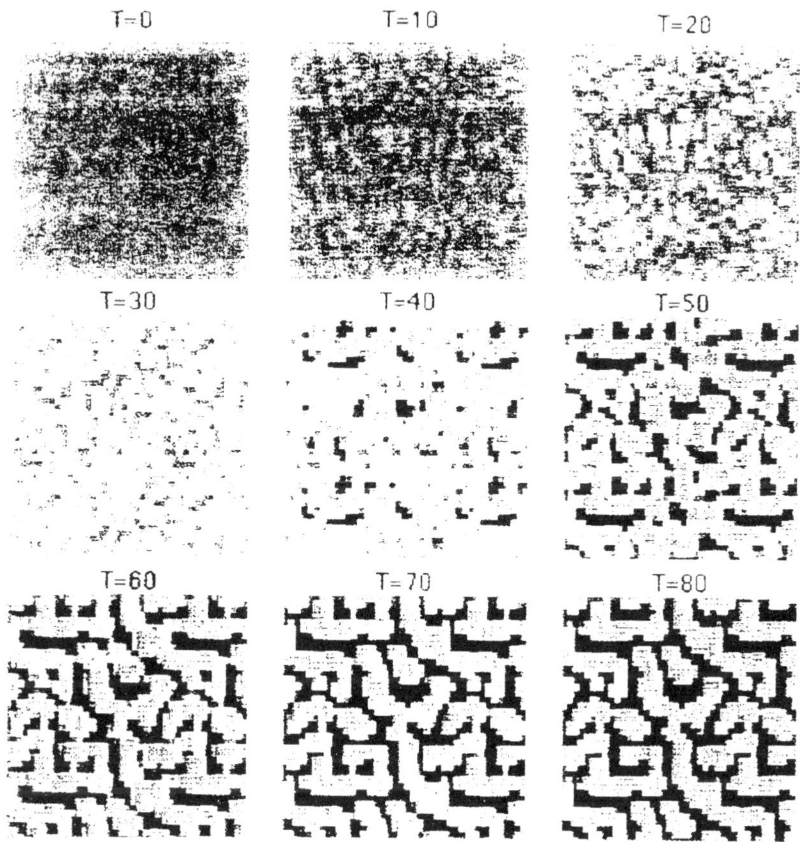

Fig. 5.20
Computer simulated growth of dominance pattern
[From: Miller, K.D., Keller. J.B., Stryker, M.P. (1989), with permission]

found on the auditory cortex. He even continues to show how a buildup of hierarchically organized maps of phonemes can emerge, and he gives a mathematical theory of all such processes. He also has his own approach to the already discussed zebra stripes of eye dominance.

When self-ordering happens in the physical world, this generally is a process of minimization of energy. Crystals, for instance, are self-ordered lattices of atoms, arranged so that the total of potential energy is a minimum. Another such process is the formation of magnetic domains inside ferromagnetic media. Figure 5.21 shows the magnetic domains on an Yttrium-Iron Garnet film. This one has such a striking similarity to the eye dominance patterns in figure 5.18 that it is hard to avoid the conclusion that formally similar physical laws are underlying both phenomena. Crystals and magnetic domains are explained by thermodynamic equilibria. So ultimately a kind of *thermodynamic equilibrium* must form between the retinal excitations and the possible synaptic changes between randomly connected neurons, at least in the visual cortex. The adaptation of the neocortex by the external world–it is a thermodynamic process? But in hindsight this should not be astonishing, unless we want to except the brain from the laws of physics.

Fig. 5.21

Magnetic domains on Yttrium-Iron Garnet Film
[After E. Kneller, Ferromagnetismus,
Springer, 1962, P. 314]

Not all self-adaptation in the visual system is through external stimuli. The mentioned line detectors are already present and working in the newborn cat, even though with a degree of randomness. It is not known how they are formed. But the eye dominance pattern is absent at birth. If one eye is permanently covered by a black patch, the other will take over the entire cortex. If the line detectors are not exercised during a critical period (three to twelve weeks in cats), they will never learn to function properly, resulting in permanent visual impairment. Similar critical periods are known in other cortical systems. If the perception of phonemes is not learned during the period of four months to two years, no language will ever be perfectly acquired. That explains the difficulty that Asians have in distinguishing "*r*" and "*l*." These sounds have not been imprinted during the critical period; they are absent in the Chinese and Japanese languages. Similar difficulties are experienced by westerners who want to learn Thai.

Returning to the visual system, a complication could be seen in the fact that only both right halves of the retinas are represented in the right visual cortex and both left halves in the left visual cortex. What happens if the image of an object or a line crosses the midline? Apparently this is only a logical complication. Neural networks do not follow a large-scale design–they simply adapt. The first time when signals originating in the separate halves come together is when they appear fully processed around the superior sulcus of the temporal lobes, and we do not notice any vertical dividing line. The communication between the halves is through the right-left cortico-cortical fibers of the *corpus callosum* (CC) in figure 5.5. If the *corpus callosum* is cut in an attempt to cure epilepsy, no visual image on the right halves of both retinas can be perceived on the left side of the brain and vice versa. Such split brain cases will be discussed in more detail later.

How does the processing of visual input continue beyond the visual cortex of area 17? What is known is that the retinal map repeats itself several times. *Up to thirteen successive visual maps have been identified in the monkey brain* (Baker et. al. 1981), as cited in (Winson 1985). The successive maps may not be as detailed, and areas may appear scrambled. That is exactly what one observes in successive layers of artificial layered classifiers. Therefore it is safe to assume that each map is used to identify different aspects of the projected external world, classes and classes of classes, so to speak. Actually, the structure of dominance patterns and maps may prevail in other cortical areas (Hubel and Wiesel 1979). Whenever axons from two different maps converge, a dominance pattern arises, contents from the two sources are compared, and the differences classified. Dominance patterns even arise at the endpoints of the axonal bundles that cross from one hemisphere to the other. Hubel and Wiesel call dominance patterns "the fourth profound insight into cortical organization."

Other details have to remain conjectural, yet from the architecture of connections some features of the whole system can be inferred. Figure 5.22 shows the main architectural lines. In the macaque brain, cortico-cortical connections jump from the visual cortex to areas 20 and 21, and then into the superior sulcus of the temporal lobe, where in humans the networks of consciousness are located to end in the brain stem. In the human brain there are

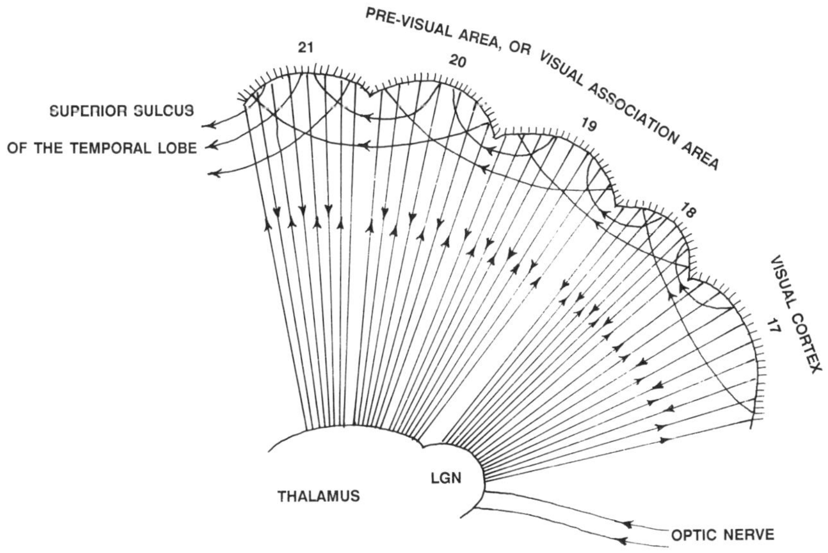

Fig. 5.22
Schematic representation of the basic architecture of the visual system

at least two more stations in this sequence. These are the new areas 19 and 37, and area 18 is vastly enlarged. All those areas are often misnamed as "pre-visual cortex"; actually, they are post-visual. Visual association cortex is a better designation. Here, the visual information is furthermore abstracted and contents relevant to the well-being of the subject are presented in the appropriate compact form to the circuits of the somatic self.

But what is the meaning of the radial thalamo-cortical fibers that are known to go both ways between *several thalamic nuclei* and the visual association cortex? I posit that the classification process may become more powerful if it proceeds in constant interaction with the somatic self, in this way more effectively serving the needs of survival. After the first relevant associative memories are activated at the beginning of the processing chain, some kind of report must be given via the cortico-thalamic fibers. And the signals may trigger responses which, in the absence of better insight, I venture to describe by words like "threatening," "pleasant," "hungry," "relaxed," "startled," "exciting," "incomprehensible" (or "no response"), "desirable," et cetera.

These responses (via the thalamo-cortical fibers) may set switches or modulations for subsequent cortical processing steps, which now will focus on more selected sets of memory associations to the visual contents. Or stated differently, by logical necessity, any cortical classification path must go through a sequence of branching points to arrive at a set of singular endpoints. At each branching point it may be of advantage to touch base on the "feeling tones" of the somatic self in the matter in question. Something like that must be the purpose of the massive radial interconnections. A straightforward detail of such process can be seen in the fact that eye movement and focus are initiated around the *lateral geniculate nucleus.* Changing fixation should be expected to be an integral part of visual processing and controllable from every stage of the processing sequence, not just from the end of the chain.

Near the superior sulcus of the temporal lobe, where the sequence ends, a definite conclusion will have been reached, like "This is an extremely dangerous predator, take to the trees." By now the process has reached consciousness.

I think this kind of distributed interaction between a simple, but genetically determined action-reaction center, and a complex system of self-adapting neural networks is the key to the workings of our mind. This conjectured scenario can be naturally extended to the function of other sensory systems. I will refer to it as the "distributed feedback hypothesis."

The approach taken in this writing is ethological, morphological, and neurological in nature. Therefore it is important to notice that Gerald Edelman, Francis Crick, and Paul Churchland have come to comparable conclusions from purely neurological perspectives. Edelman has performed sophisticated computer simulations of neural networks where neurons and synapses are modeled much closer to their biological reality (Edelman 1987). With his program "Darwin II," he has demonstrated that characters (letters) can be recognized independent of their position or tilt within a given field, if there are two different kinds of networks interacting via circular feedback. His term for such interaction is "re-entrant networks." *Moreover, his networks self-adapt from initially random connections.* He calls pairs of such re-entrant networks *recognition couples*. In *The Remembered Present* (Edelman 1989), a complete theory of global brain function is developed on the basis of re-entrant networks. Part of this theory is re-entrant circuits between neocortex and thalamus.

Francis Crick's "searchlight" hypothesis (Treisman and Gelade 1980), (Crick 1984) also points in the direction of distributed feedback. Crick[24] focuses attention to the fact that the thalamo-cortical fibers are interrupted by the reticular complex. This is a sheet of a special class of neurons on the surface of the dorsal thalamus. At any given time, it seems to select certain groups of the radial fibers going in both directions, and to activate their transmission while suppressing others. The resulting cortico-thalamic activity is compared to a "searchlight" being shone on selected areas of cortical processing. This fits especially well into the picture of the somatic self manipulating the neocortex to its own low-level designs. Crick does not seem to concern himself with the question of "who" is pointing this searchlight. Indeed, this does not become a question if one keeps in mind that this "illuminating" process can be seen as a result of wandering activity in cortico-thalamic circular feedback loops, caused neither by brain stem nor neocortex alone.

Paul Churchland (1995) also stresses the fundamental importance of re-entrant (or "recurrent") networks between the thalamic nuclei and the neocortex, especially noting that damage on both sides of the intra-laminar nucleus (in the thalamus) results in complete comatose absence of consciousness. Damage on one side only results in complete nonfunctionality (sensory and motor) on one side of the body. The conclusions he draws, however, are not quite the same as this book will present in the coming section on consciousness.

The auditory and visual systems are the most important ones in primates. That is why a separate area of convergence of audio and visual pathways can be found, (fig. 5.15). As stated under (c) in the list of basic observations on pathways (p. 128), this area is in the frontal end of the temporal lobe. It probably serves for cross-modal transfer between both modes of perception.

The auditory system is simpler; its sensory input is not as rich as the stereoscopic moving images perceived by the retina. The auditory processing areas, as seen in figure 5.15, are smaller, but the architecture of the visual system is repeated. There are the cochlear nuclei through which the auditory input enters; there is a small auditory radiation that fans out toward the auditory cortex, and there are the radial fibers, going both ways along the cortical processing path. It is safe to assume that the auditory system has the same architecture and functions much like the visual one. This primary map, though, maps the spectrum of acoustic frequencies.

---

[24] At the present time Francis Crick has abandoned the searchlight hypothesis (private communication).

Areas adjacent to the auditory cortex are of special interest. In humans the auditory cortex is virtually surrounded by Wernicke's area[25] (fig. 5.18), which has been known for a long time to be the center necessary for understanding as well as composition of speech. Thus, speech is understood as well as composed in the same area. But what is speech composition but a pre-imagination of the words and sounds before they are actually uttered? Since speech has to be understandable by someone else's Wernicke's area, it seems natural that understanding and generation are linked. In artificial neural networks, memory is indistinguishable from classification. Therefore it is quite conceivable that upon proper excitation from a center somewhere between brain stem and hippocampus, the networks reproduce what they have previously perceived and classified. The same may be happening during the spatial pre-imaginations that become thinking (compare chapter 4).

Association fibers run from Wernicke's area across the fissure of Rolando into the anterior speech, or Broca's area. Being a pre-motor area for mouth and larynx, this essentially is the control center for the speech mechanism. But probably it is more to the point to assume that Wernicke's and Broca's areas form a re-entrant couple, a system of circular feedback.

Already, Broca had found that damage to the area named after him results in a speech disorder (Broca's aphasia) characterized by severe impairment of grammar and the normal enunciation of speech. One might conclude that ideation of speech is predominantly in Wernicke's area, and temporal (grammatical) organization is in Broca's area. This is well in agreement with the general view of frontal lobe function as will be presented in the next section.

We have said that the visual association area, with proper stimulation, could produce spatial pre-imagination which underlies thinking. But things look even better than that. In humans, two large new areas (39 and 40) in the parietal lobe are a meeting place not only of visual and auditory pathways; they also receive inputs from the somatosensory areas 1, 2, and 3. Thus, the somatosensory system joins in an overall synthesis from audio-visual-tactile inputs. Of course, three-dimensionality of space can be explored by the hand as well as by the eye. And the center of Wernicke's area is generally put into the lower part of that same area (40). Finally area 4 is heavily connected to the ominous areas around the superior sulcus of the temporal lobe.

Everything seems to fall into place now. The ultimate synthesis of all sensory input reaches its highest point in area 40. A model of the spatial world must reside in the processing chain leading to it. Thinking, as well as speech, originates here, and space, as well as speech, is understood here. And like in the macaque brain, the ultimate extract of cortical processing is presented to the temporal superior sulcus, ending in the brain stem. Soon it will transpire that consciousness resides here in the temporal lobe.

---

[25] There are varying opinions as to the extent of Wernicke's area. The size pointed out by Eccles is large as compared with others.

Limbic System and the Frontal Lobe

I now return to the last point (d) (p. 128) of the basic observations previously made on cortical neural pathways. All sensory systems send branches to the frontal lobe. What is the significance of this area? Anatomical evidence (Nauta 1971) suggests that the cortico-thalamic fibers from this region, particularly from area 46, connect to nuclei of the brain stem which are involved in sensing and controlling endocrine glands and visceral and cardiovascular action. As in the visual system, the connections go in both directions. In fact the connections go to an entire subcortical network of nuclei and of paleocortex, especially the hippocampus. This network has been named by Broca *the limbic system*[26]. The self-adapting neocortex of the frontal lobe serves this mental system. With this we return to the world of emotions. From sticklebacks to geese, to primates and man, in earlier chapters we have followed an ancient trait that is primarily associated with intraspecies social interaction—territorial aggression, mating rivalry, dominance hierarchy and their taming. But, of course, plain fear of being caught by a predator also is an emotion.

Earlier thinkers like Descartes, William James, and Carl Lange had assumed that our experience of emotion is secondarily derived from increased heartbeat and that visceral response is caused by processed sensory input of, for instance, danger. According to this, we would experience fear because we run from the wolf, not the other way around. But in this century, experimental work with lesioned animals and brain-damaged humans led to the insight that there are special emotional systems in the brain, and that the visceral response is secondary. This, of course, does not surprise anyone with an ethological inclination. Soon the limbic system became primarily understood as *the emotional system*, a view that according to LeDoux might be too restrictive (LeDoux 1986). Of course, all systems interact in one way or another.

In the evolutionary view, emotions, being such ancient determinants of behavior and having strong genetic components, should be found in the old brain, that is in the brain stem and the nearby paleocortex. And, indeed, here are identified parts of the limbic system. The "emotions of being" must reside in close association to the basic circuits of the somatic self and the hormonal milieu of excitation that interacts with it. After having understood in broad outline the evolution of behavior, it should be obvious that emotions are associated with motivation to vigorous action. If forceful action is required or anticipated in order to satisfy a basic need of the self, for instance a ranking encounter or a flight from a predator, the body has to be quickly brought to its maximum active state. The heartbeat has to increase, adrenaline has to be released, oxygen metabolism has to be raised. Often such changes are experienced as highly unpleasant. And they are energy expensive.

Therefore, as we have seen, evolution slowly moved away from unchecked expression of emotion. In the baboon society, open fighting became subdued and the drama internalized. Nevertheless, as R. M. Sapolsky's research shows,

---

[26] Limbic, borderline, meaning the border between brain stem and neocotex. There is no clear consensus on all the constituents of this system, but main parts are: cingular, retrosplenal and insular cortex, which all are buried below the neocortex and have a simpler structure. The pyriform and hippocampal areas (paleocortex) are at the lower edge of the neocortex. The amygdala is at the thick end of the hippocampus, connecting to the brain stem. The septum, hypothatlmus and mediodorsal thalmic nuclei are all in the brain stem or attached to it.

something that may be called "emotional energy" is still rampant in the body and may lead to bad health—in humans to the well-known stress syndrome. In us, raw aggressivity may turn into ambition, and ranking fights into a subtle drive for the possession of a BMW. But the emotional energies still circulate and may cause high blood pressure and other problems.

Of course, pleasure is an emotion, too. Its ancient nature reveals itself as motivator, for instance in the "joy of exercise." A "pleasure center" is found in the septum which is part of the limbic system (Olds and Milner 1954). Rats, with permanent electrodes implanted in the area and provided with means to self-stimulate, took to self-stimulation of up to seven thousand times an hour, essentially as fast as they could press the lever. They forego food and sex in order to be able to experience this pleasure. In some human schizophrenics, electrodes were implanted in the same region (Heath 1963). Stimulation resulted in agreeable feelings with sexual overtones, "some self-stimulated like Olds' rats" (Eccles).

Stimulation in the amygdala can evoke strong reactions of aggression or fear in animals and humans. Delgrado (1969) reports the case of a girl with damage in the amygdala from encephalitis who suffered from unpredictable fits of violent rage. Electrodes were implanted to study the case. Upon radio excitation of the right amygdala, the girl fell into rage and attacked the wall. Local coagulation seems to have improved her lot.

We would rather prefer to give in to pleasure than undergo involuntary aggressive arousal. Yet Darwinian selection does not care, both originally serve the same purpose of survival. Emotions are necessary, but they work better if a certain amount of learned planning, delay, and maybe substitution of one emotion by another can be exercised (compare redirection in chapter 3). *And that is why the emotional system has developed its own area of neocortex—the frontal lobe.*

As the other intelligent cortical systems, through self-adaptation, build a model of the external world, so the pre-frontal cortex[27] builds a model of the world of possible emotional encounters, and in accordance with this model it processes sensory and endogenous (brain stem) input. Thus, our somatic self can be advised whether a particular sensory input is really worth an emotional response. I once had a demonstration of a rigid emotion without such cortical adaptation. Every morning at sunrise a male cardinal (*cardinalis cardinalis*) awoke me by pecking angrily at his mirror image in the window. Subsequently he also began to do this to the rear view mirror on my parked car. This beautiful bird lacked frontal cortex to adapt his emotions to this unnatural situation. The genome cannot know about mirrors.

Walle Nauta writes:

> The frontal lobe is characterized so distinctly by its multiple
> associations with the limbic system, and in particular by its direct

---

[27] Pre-frontal cortex generally means frontal lobe with the exception of the "frontal pole" itself.

connections with the hypothalamus, that it would seem justified to view the frontal cortex as the major—although not the only—neocortical representative of the limbic system. The reciprocity in the anatomical relationship suggests that the frontal cortex both monitors and modulates limbic mechanisms.

And further down he describes the frontal lobe ". . . as perhaps the only realm of the neocortex where neural pathways converge with conduction systems representing the external environment as reported by *all* exteroceptive modalities." (Nauta 1971)

Patients with surgically ablated or damaged frontal cortex undergo mental changes that are hard to classify. In games they cannot keep obeying the rules. They have difficulties staying in the same job for long, and they generally lack motivation. All this is indicative of an inability to plan ahead in the emotional sphere. "My job is unpleasant, but what lies ahead if I would quit?" would be a conscious and rational question. But obviously any continued stable action requires a kind of motivational planning ahead, even if this should remain unconscious. Hardships now have to be balanced against rewards later, otherwise we could not stick to anything for long. When PET scans of twenty-two convicted murderers were conducted by A. Raine and M. S. Buchsbaum, sixteen or seventeen of them were found to have much-reduced frontal lobe activity (Raine 1993).

All motivation ultimately is emotional, and it is the planning for an economic balance of future hedonic needs that provides stable behavior, or shall I say "happiness." The new peripheral baboon male comes to mind; he keeps back his needs for social interaction and sex for months in order to habituate the troop to his presence and reap the rewards later. This requires a model of the complex social world of baboons, so that the unpleasant consequences of particular actions can be foreseen. In humans, the self may well engage in conscious pre-imagination in the field of emotional tensions, similar to the pre-imaginations of movements in space with the aid of the temporal lobe. But it is likely that most of this remains unconscious and happens in the frontal lobe, and if this processing goes on during sleep (compare the role of dreams in the next chapter), we may wake up refreshed and calmed. When the emotional situation comes up we will react in a better way, without realizing that a kind of automatic workout has happened during the night. The frontal cortex must be the organ that makes the ritualization of aggression possible and facilitate the compassionate bond.

Thus, it is not astonishing that the most social predatory mammals—primates and canines—have the largest frontal cortex. The cat has remained a solitary hunter and can do with a minimal prefrontal area; compare figure 5.23.

J. M. Fuster (1985) calls the prefrontal cortex[28] the organ of behavioral syntax. He writes:

---

[28] This is the name usually applied to the cortex of the frontal lobe.

Behavioral syntax, in short, may be characterized as the set of rules that govern the organization of behavioral acts in the time domain, an organization which, to be implemented, requires the purposeful mediation of cross-temporal contingencies. As could be expected, the prefrontal cortex has attained its greatest development in man. For it is in humans that behavioral syntax reaches the highest levels of complexity and creativity. Language is but one example, albeit supreme, of the possibilities opened up by that development.

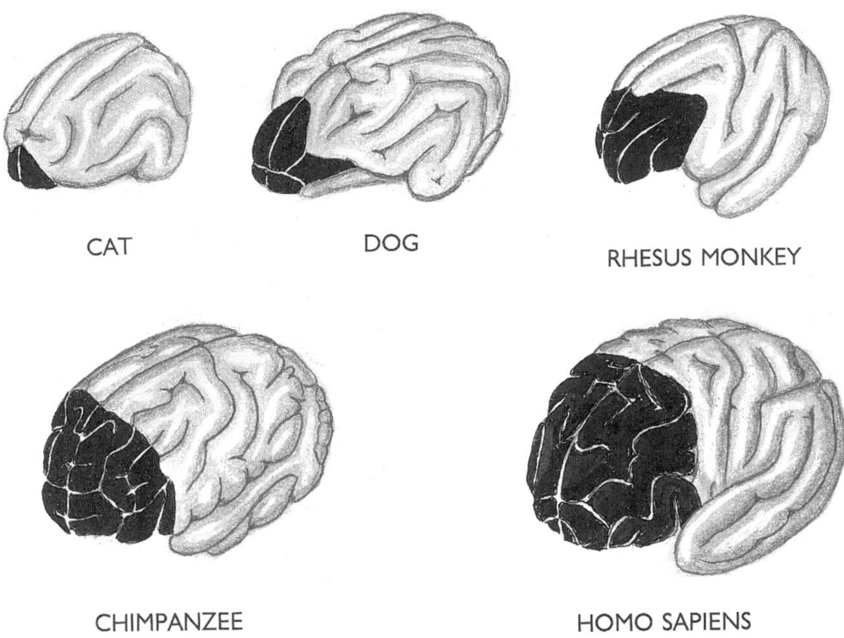

Fig. 5.23
Prefrontal cortices of various mammals
[After: J.M. Fuster (1985)]

Here is another *emergent event*: the skill of temporal organization of behavior can also be used to organize speech. Syntax is born. What is organized now are the expressive movements of larynx, vocal cords, mouth, and chest muscles. And the raw material is ideas, maybe "pre-words" of some kind, that have been formed in the parietal cortex, particularly Wernicke's area. Applied grammar is temporal organization and modification of uttered words to form an understandable sentence, that is, to form an expression that has purposeful wholeness. In this, Broca's area adds the final processing steps. Areas far anterior to it do much of the preprocessing.

Damage near the frontal pole does not result in catastrophic aphasia, but causes inability to construct "complex proposition, especially if they include dependent clauses" (Fuster 1985, citing Lhermitte et al. 1972). In other words, the further anterior an area is to the motor cortex, the more long-term and complex is the planning of motor actions. And that seems to be an aspect of function of the rest of the prefrontal cortex as well—a pre-motor planning area that receives all necessary sensory and cognitive input and composes a properly sequenced course of action. To do all this it has to be part of an intelligent system that is well-adapted to the world but answers to brain stem needs in the domain of emotions. All public speakers know that a good speech rises up from the emotional underground. If you are afraid to show your emotions in public, you cannot speak. Probably this system is much more sophisticated than any of the previously discussed intelligent systems. Its activity, however, mostly remains unconscious, and not much is known about the mechanics of this kind of self-adaptation.

I would like to conjecture that by a process of internalization, the pre-frontal cortex as well has taken on the function of organizing our thoughts in sequences which amount to synthetic theories. Thus, as the most intimate cortical area of our somatic self, it also has become the promoter of the most advanced human faculty. Comparison of prefrontal cortices (fig. 5.23) between species illustrates the point. From cat to man it has grown more than any other area on a percentile as well as an absolute scale. This growth is only rivaled by the temporal lobe which gave rise to consciousness, as the next section will show.

The Networks of Consciousness

"Consciousness is a higher level or emergent property of the brain in the utterly harmless sense of 'higher level' or 'emergent' in which solidity is a higher level emergent property of H_2O molecules when they are in a lattice structure (ice), and liquidity is similarly a higher level emergent property of H_2O molecules when they are, roughly speaking, rolling around on each other (water)" (Searle 1992).

It is quite remarkable how far explanation of brain function can be carried without having to seriously deal with consciousness. On the other hand, if consciousness should be a rather late fruit on the phylogenetic tree, its absence in this volume up to now would be understandable. Might it be that in evolutionary perspective, consciousness is just a rather insignificant side effect? What is the evidence for a seat of consciousness, if any?

The temporal lobe, particularly around the middle region of the superior sulcus, was recognized to be one of the "highest" cortical areas. What happens under electro-stimulation of that region? Penfield sometimes observed strange dreamlike experiences (Penfield 1955). One patient (D. S.) heard an orchestra play so vividly that she thought a record player had been turned on in the operating room. She was able to hum the tune. She did not remember, though, if and where she might have heard that piece before. A young woman exclaimed, "Oh, I had the same very, very familiar memory in an office somewhere. I could see the desks. I was there and someone was calling to me, a man leaning on a desk with a pencil in his hand." In another patient, upon stimulation inside the fissure of Sylvius said, "That bittersweet taste on my tongue," and then, "Oh God! I am leaving my body." Another patient, W. S., said, "I am going into a seizure," and then he became automatic for a short while.

The following cases are from W. Penfield and P. Perot (1963). The patient, E. Wh., saw herself at childbirth and felt as if reliving the experience. R. W. exclaimed, "Oh, gee, gosh, robbers are coming at me with guns!" (He had not been in any such real event). Later he said, "Yes, the robbers, they are coming after me." A. Bra. saw Rembrandt's *Night Watch* and *Self Portrait* and, as in a picture, said, "Christ coming down from the sky," which frightened her very much. M. G. commented, "It's a dream. There are a lot of people. I don't remember the rest." J. V. said, "Oh, there it goes, everybody is yelling. Something dreadful is going to happen." Later he added, "I was holding to a bar, and the bar seemed as though it was walking away from me. I saw someone coming toward me as though he were going to hit me." R. M. said, "No, God said I am going to die." E. B. said, "Someone is telling me in my left ear, 'Sylvere, Sylvere' (her name). It could have been the voice of my brother."

The stimulation in this case was on the right temporal lobe, and the left ear is contralaterally connected to the visual cortex here. N. C. was accurately humming the song she was hearing, "It is the *War March of the Priests.*" M. M. (p. 649) reports a lengthy dream story that includes smelling strong coffee—the only olfactory hallucination Penfield had ever noticed. J. T. said, "Yes, doctor, yes doctor! Now I hear people laughing—my friends in South Africa." And upon the question as to who they are he answered, "Yes they are two cousins, Bessie and Ann W."

What is happening here? Memories, images, sounds are activated in a way that made Penfield think that there is a continuous record of the "stream of consciousness," like a magnetic tape, stored in the brain and played back upon electrical stimulation of the temporal lobe. *But one is more struck by the dreamlike appearance of the reports*. A large majority is mainly auditory, music, or speaking. This is not astonishing since stimulation is near the primary auditory cortex. If one compares the locations of the points of stimulation with the types of experiences, one notices that predominantly visual images were more often produced from points toward the pre-visual cortex, and predominantly auditory ones were ordered more often from the regions adjacent to the auditory cortex. One stimulation (Penfield and Perot 1963) from an area toward the visual cortex reported a crude visual sensation, "a sun or a moon." Other reports mention simply feelings of familiarity, déjà vu, strangeness, fear, or pleasure. What is reported are not so much complete stories as *basic elements or building blocks* of stories and *general feeling tones.*

Therefore my impression of the stimulation is not that memories are played back as if stored on tape, as Penfield would want to have it, despite his admission to cases *where the "remembered" incidents had never happened.* Rather, paradoxically as it may seem, the medial temporal lobe seems to specialize in an activity that is best described as that of a *storyteller or a dreamer.*

But we do not have to rely on such impressions alone. Research with split-brain patients has come to the same conclusion. M. S. Gazzaniga (1983, 1987, 1988) has performed many experiments with split brain patients. In these patients the *corpus callosum* had been cut in an attempt to cure epilepsy. In this way all cortico-cortical communication between the two hemispheres is interrupted. If one presents images to the right half-fields of both retinas, those images are only perceived in the right half of the brain. (The visual fields are contralateral if the inverting property of the lens is added. The left visual field is projected upon the right halves of the retinas and reaches the right half of the brain).

Two such patients, P. S. and V. P., whose operations had been performed at an early age, have developed two minds (Gazzaniga 1988), (Gazzaniga et al. 1984). Both can speak, but they may have different opinions. But generally the ability to speak is present in the left hemisphere only. If a question is asked of the right hemisphere, spoken responses are not possible, even though drawing, pointing, and writing in response may be performed with the left (contralateral) hand. In this case the subject remains unaware (or unconscious) of what the question was, and of his/her response. Consciousness, therefore, must reside in the left hemisphere.[29] Despite the callosal lesion, there is some transfer of information which however remains unconscious. For instance, a numeral or picture may be flashed to the right brain only and the subject asked to match it by selecting from an array of cards in full view. If the test is not too difficult, he generally will select the correct one, but will not know why. It is like in "blindsight" (compare section on Intelligent Systems).

One of the patients was shown a snow scene to the unconscious right hemisphere and a chicken claw to the conscious left. Then he was asked to select two pictures from a set which would match the scene. He selected a shovel and a chicken. When asked why, he said, "Oh, that's simple. The chicken claw goes with the chicken, and you need a shovel to clean out the chicken shed." *His storyteller had quickly composed a story to explain why the shovel had been (unconsciously) selected.* He did not know consciously that the right side had selected shovel as the obvious tool to shovel the snow. But a cue somehow came through from the right into the left hemisphere.

On another occasion, when presented with the written command "laugh" to the right hemisphere, the patient laughed. When asked why, she said, "You guys come up and test us every month. What a way to make a living." Or when "walk" was flashed to the right hemisphere, the subject began to leave the van. When asked why he said, "I am going into the house to get a Coke." Or when a film

---

[29] Except in rare cases where right and left functions are interchanged.

was shown of someone throwing another person into a fire the response of V. P. was, "I don't really know what I saw, I think just a white flash. Maybe some trees, red trees like in the fall. I don't know why, but I feel kind of scared. I feel jumpy. I don't like this room, or maybe its you getting me nervous." And later to one of Gazzaniga's colleagues he said, "I know I like Dr. Gazzaniga, but right now I am scared of him for some reason."

From such experiments Gazzaniga concludes that a cue of some kind, something that I like to call an "emotional feeling tone," *is subcortically transferred between the hemispheres, and an "interpreter" in the left hemisphere is on duty to make a suitable story of any given uncertain data.* The same is seen at work when a posthypnotic suggestion is carried out and has to be rationally explained (Hilgard 1977). Also, such emotional feeling tone occasionally is enough, with some storytelling imagination added, to identify objects flashed to the right brain only. *In Vignette 3 it is shown how good artificial networks can be in filling in from memory what is missing in the input.* This was called "memory recall by association." In that context it is not astonishing at all that stories can be made up from readily available elements whose sole connection to reality may consist in the correct interpretation of one or two facts which cause an association.

The anatomical interpretation of spurious information transfer between the hemispheres is obvious. Both hemispheres are still heavily connected via reentrant radial fibers to the unsplit brain stem. Therefore the experiments verify the assumption made earlier on the simple nature of signals issued from the brain stem: these are not specific feelings and emotions.

This interpreter and storyteller is placed by Gazzaniga "in the distribution of the left middle cerebral artery." (Gazzaniga 1988) That grossly is the already considered area in the temporal lobe.

Further supporting evidence for the importance of this area comes from its removal or disease (herpes simplex encephalitis). Destruction of this area results in the amnesiac syndrome. Lawrence Weiskrantz (1988) calls the condition a "consciousness disorder." The patient cannot remember anything that happened seconds ago. As a result, he is severely disoriented and in need of constant supervision, no new memory whatsoever can be formed, yet they function perfectly well in any other way.

Larry Weiskrantz writes:

> The most striking clinical feature is that these patients, despite retaining normal intelligence and perceptual and cognitive skills, report not remembering anything at all from minute to minute, it really is a crippling disorder. You can be talking to such patients, then leave the room and come back in a minute; the patient does not recognize you at all and admits of no memory of having seen you. So devastating is this clinical condition that for a long time it was thought that these patients

were incapable of learning anything at all. But gradually it has become clear that in fact they are capable of learning a large variety of new things, and of doing so quite efficiently. For example, they can be subjected to Pavlovian conditioning of a sort that each of us would surely report remembering, which they seem to acquire at normal rates. They show verbal learning, they show perceptual learning, they show motor skill learning, and they retain these over long periods of time, sometimes for weeks or months. But the only person who does not acknowledge that anything has been learnt or display any recognition of this fact is the patient himself. So there is a very clear dissociation between the capacity to learn and to retain, on the one hand, and the patient's knowledge of that fact, on the other. (Weiskrantz 1987)

Memories that have been established prior to two years before the damage remain intact. Thus, amnesiacs demonstrate that intelligence can be present without consciousness, further supporting the view of consciousness as just an "add on storytelling facility."

Together with Penfield's electro-stimulation, all the referred observations point to the vicinity of Brodmann areas 21 and 22 as the seat of the storyteller and interpreter whose confabulations are consciously experienced. However, in accordance with the previously mentioned distributed feedback hypothesis, it would be wrong to place the networks of consciousness exclusively into the neocortex. Reentrant networks between thalamic nuclei and neocortex are everywhere. Indeed, consciousness is not just impaired by damage to the critical area of the left temporal lobe; it entirely disappears upon damage to the corresponding thalamic nuclei (Churchland 1995).

To some of my readers, subjectively, this definition of consciousness may sound rather limited. We are used to thinking more of ourselves. To this, Michael Gazzaniga writes: "Does our species' interpretive mechanism have a finite repertoire of algorithms upon which to draw when inferring the meaning of the data it is considering? I am convinced mine does . . ." In other words, *the storyteller-dreamer-interpreter again is a bag of tricks using a fixed number of elements of stories and interpretations to force upon the infinity of situations, yet this limited set is accurate enough to provide a great advantage in the struggle for survival.* I am reminded of the Chinese "Book of Changes," the *I Ching*. There, all possible human life situations are pressed into just sixty-four scenarios. Nevertheless the avid user of the *I Ching* is astonished at how well aspects of his own life are mirrored in most of them.

Of course, the view of the mind's function being mostly unconscious with only a conscious appendix, or island, is reinforced by most traditional psychoanalytic theory. The model of the brain, as expanded here, arrives at the same conclusion from entirely different facts. Thus, both approaches gain confidence.

PHILOSOPHICAL INSERT 3:
The So-Called Problem of Consciousness

I do not wish to engage in the philosophical discussion on the nature of consciousness which is carried out between Thomas Nagel, Paul Churchland, John Searle, Daniel Dennett, and others. The problem does not seem to be a serious one in a system of understanding based on evolutionary epistemology. I hope to have convinced my readers by now that this is the only theory of knowledge worth its salt.

Some important conclusions regarding modern physics have already been derived from evolutionary epistemology in the first chapter, and in the Philosophical Insert 2 (p.94) I believe that the problem of consciousness, if there is one, can be solved by an almost identical argument to the one that solved the problem of modern physics earlier. First, let me return to quantum physics.

Following Gerhard Vollmer (1985), it was said that our evolved mind cannot be expected to grasp realities of the microcosmos because throughout natural history it has built an adaptive interrelation to the realities of the meso-cosmos only. The success of quantum mechanics is solely based on the unexplained and unexpected feature of mathematics to be able to quantitatively model certain aspects of the microcosmos, and not on any inherent qualities of direct human intuition. The wave-particle dualism forever has defied intuitive explanation and has been put to rest, but never explained, by the concept of "complementarity" by Niels Bohr.

It has been generally agreed upon that one and the same real entity is encountered, that sometimes appears as a wave, and in a different setting it appears as a particle. We have equations that accurately describe and predict the phenomena encountered, but otherwise this world is closed to the human mind. It becomes meaningless, except for physicists working in that field who are intimately at home with the mathematics involved, who build mental crutches so that they can juggle the equations around until they fit the evidence, so that the possibilities of nature can be better exploited. But they do not build real concepts. They build mathematical constructs, the origin of whose powers we do not understand.

In dealing with the subjective or *merely auto-connected* (Churchland's word) inner experience of consciousness, we encounter a similar situation. *As our evolutionary origin prevents our mind from entering the microcosmos, so it may prevent our mind from entering itself. There is no external stepping stone to step on and observe ourselves, except when we limit such observation to the mechanics of neural networks.* Actually, already simple logic seems to be sufficient to explain this. The evolutionary argument merely adds that our mind is natural, not supernatural, and hence, not omniscient.

Paul Churchland (Churchland 1995) cites occasions in the history of science where, at the time, any future breakthrough was unimaginable until it was seen to happen. And so, he says, a future breakthrough in understanding consciousness may occur. Yes, possibly, but unlikely; evolutionary epistemology already uncovers one example where a natural limit of our mind's reach is encountered. Why shouldn't there be more? I agree that quantum mechanics breaks through this one limit. But beyond lies a level that is purely manipulative, not insightful. It is nice to be able to make transistors; lasers; computers; compact discs, and CAT, PET, and MRI scans, but they do not add to the understanding of the microcosmos, they only manipulate it. Actually, without understanding consciousness either, we already manipulate it every day. Advertising psychologists successfully shape the consciousness of millions of human population.

A conclusion may be drawn that consciousness and its neural correlate are, in Niels Bohr's sense, *complementary* aspects of one and the same thing. When I make a conscious decision, it is irrelevant whether I think the neural networks, which are me, make the decision, or whether it is my inner experience that does it. Both are the same thing.

Chapter Six

From Neuroscience to Psychology

Sigmund Freud started his career in neurology. As a young man he hoped to put psychology on a neurological foundation. But, of course, this was futile a hundred years ago. Without that foundation, psychology has remained a soft science ever since, prone to misconceptions and ad hoc beliefs. Karl Popper has denied it proper scientific status. Yet it attests to the creativity of the human spirit, and to our power of interpersonal intuition via the similarity bond that, nevertheless, psychologies have been conceived which are quite consistent with the neural mechanics known today. Outstanding among them are the works of Jean Piaget and Carl G. Jung. These psychologies complement the objectively known large-scale features of brain function. They add the many details that are necessary to relate our own inner experiences to our basic neurological makeup. Only after such relation is well-established can we be assured that the evolutionary synthesis is on the right track.

Jean Piaget and Self-Adaptation of the Neocortex

The brain of a newborn human infant already contains all the neurons it will ever have. But this newborn is quite helpless. There are few synaptic connections in the neocortex. Figure 6.1 compares the cortical tissue of a newborn with that at ages three months and two years. Mediated by the brain stem, at first there only are a few instinct movements working. The most important are finding the nipple and sucking. Crying is another one. The hand grasps when the palm is touched. When held upright with the feet on the floor, it will respond with rudimentary walking movements, but this response will rapidly disappear. The baby will begin to smile early on, and the mother will instantly fall for this smile, as intended by the great designer—Darwinian selection. A firm mother-child bond is the only guarantee for this helpless bundle to survive. But the smile is a deception. It promises something the child will not be able to deliver for a long time to come. For now it is not only a pure egoist but also a true solipsist, a state of mind difficult to understand by an adult, and definitely not identical to the imagined solipsism of a philosopher with that leaning. It resembles the state of a primitive animal with very little neocortex, except that it is more helpless, because evolution has caused it to rely more heavily on the neocortex, which is not yet functional.

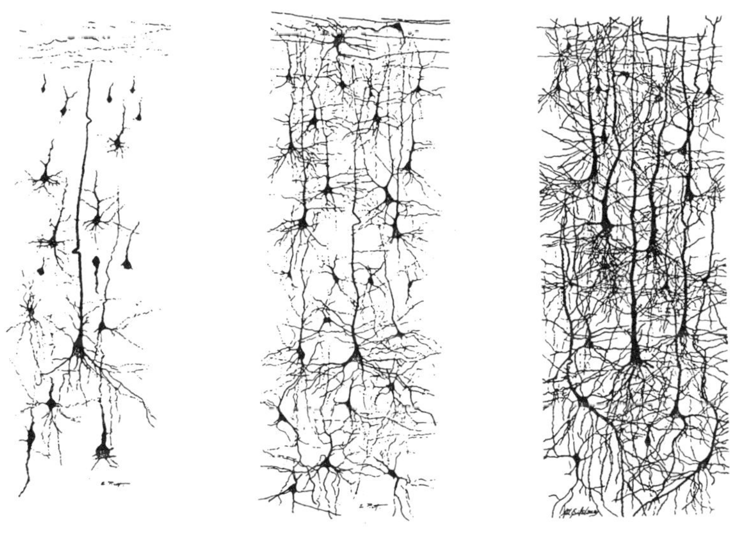

Fig. 6.1
Cortical Tissue of newborn (left), 1 year old, and 2 year old (right)
[From J. L. Conel, the Postnatal Development of the Cerebral Cortex, Copyright © 1939-1967 by the President and Fellows of Harvard College. Reprinted by permission of Harvard University Press]

"The true solipsist," writes Jean Piaget, "feels at one with the universe, and so very identical to it that he does not even feel the need for two terms. The true solipsist projects all his states of mind onto things. The true solipsist is entirely alone in the world, that is, he has no notion of anything exterior to himself. In other words, the true solipsist has no idea of self. There is no self; there is the world. It is in this sense that it is reasonable to call a baby a solipsist; the feelings and desires of a baby know no limits since they are a part of everything he sees, touches, and perceives." (Piaget 1927)

Indeed, it is the raw desires of the somatic self in action! But that action also has a goal: in some animals the most urgent facility to develop right after birth may be to make the legs run. In humans it is the neocortex to adapt. And the baby begins to do just that with gusto and determination. The method is quite simple. Try everything at your disposal (move limbs, body, mouth, head, eyes, fingers) and whatever has *interesting results*, repeat until tired. We remember how self-adapting artificial neural networks learn: simply by repetition. In this way it is discovered that thumbs and toes and the corners of the pillow can be sucked, that arching the body up and letting it fall makes the things that hang over the bed move, or makes interesting noises. It might be fun in itself. We remember the joy of exercise. A metal cigarette case (Piaget's examples) is grabbed, looked at, held up high, and put into the mouth. "The newborn's whole universe is divided into things to suck, things to grab, things to look at, things to listen to, and the like."

All that means nothing less but learning to master the three-dimensional space. The cortical adaptation begins with the mechanisms of visual image processing which organize themselves under the objective necessities of the moving retinal map of the world (compare figures 5.18 and 5.20). Nevertheless the criterion whether any new behavioral acquisition is "good" and worth retaining, ultimately derives from whether it is pleasurable. That means whether a pleasure center in the brain stem is activated. And that is entirely determined by the innate neural circuits of the somatic self. Over millions of years the somatic self must have acquired quite a bag of tricks in order to make the developing animal exercise its full potential. Besides pleasure, our vocabulary knows attributes like exciting, stimulating, provoking, stirring, arousing, awakening, inspiring, and interesting. There is no reason why, unbeknown to him, the babies' inventory of motivations should contain a lesser variety of stimulation.

Following such motivation, the human infant constantly tries to cause something exciting to happen. But it has no notion of what we call "causality." If an interesting event once accidentally follows a certain movement, this movement is tried again. Needless to say, this proves to be the best prescription to discover true causal links in due time. The baby's activities are purely endogenously driven. No parental prodding is accepted. One activity is readily given up for a more interesting one.

Piaget writes:

> Take the example of a nine-month-old infant trying to reach an object hanging from the top of his bassinet. He does things aimed at getting it, but, seeing the top of the bassinet shake, forgets his original goal and concentrates on making the whole bassinet move. Then he goes back to his original intention, pulls on a cord, but then notices his hands as he does so. At this, he loses interest in anything but his hands

and shakes them. This movement gives him the idea of taking his pillow and shaking it, too, in the air. Then he happens to notice the fringe on the edge of the pillow and starts to suck it. This activity reminds him of the position he takes to go to sleep, which he immediately assumes, et cetera. Here is an example of infant thought made up of a series of schematized cycles of behavior linked and mutually activating each other by association.[30] Each cycle has a definite goal, and the means to reach it, and in this sense the process of thinking is partially directed. But there is no hierarchy governing these cycles in relation to each other. They simply succeed each other at the beck and call of circumstances and assimilation. (Piaget 1927, p. 204)

As needed for efficient adaptation, indeed, "the memory of the developing child is unsurpassed" (Piaget) and none of the accidental activities are lost in forming a building block for a working relationship with the world.

It would be interesting to follow Piaget in his atlas of development during the first two years, the so-called "sensory-motor period." I am convinced that one day it will be possible to correlate each of his six stages of development to the progress of organization of connectivity along the sensory pathways. On the average, at the end of the first two years the human mind is essentially formed to the point where language can begin. But since this writing cannot be, among so many other things, also a full account of developmental psychology, I will try to extract only the most relevant results. B. J. Wadsworth (1979) has provided a table of features of the six stages of development. On studying it, one is intrigued to find a certain resemblance to the phylogenetic evolution of knowledge as it has been outlined earlier in this book. The old rule of Karl Ernst von Baer, that ontogeny traces phylogeny, seems to be confirmed again in cognitive development. With that view I have added a sixth column to Wadsworth's table, the "phylogenetic equivalent" of a given developmental (ontogenetic) stage of Piaget (table I). It uses concepts and catchwords from chapters 2 and 4. This table will now be explained in more detail.

Stage 1. During the first month of life, nothing more is observed than the mentioned initial reflex responses. The eyes do not follow an object. Responses and activities in any area, for instance, the "mouth space," are completely independent from any other "space," such as visual, tactile, and postural. To state this differently, cross-modal transfer is not mastered. This stage corresponds to simple pre-mammals with a minimal cerebral cortex. The baby's uncommitted areas of neocortex are still nonfunctional.

Stage 2. (one to four months) Now the reflexes begin to be modifiable. Thumb sucking is mastered through an acquired hand-mouth coordination. Moving objects are followed with the eyes, indicating the onset of organization of the visual system. The head is moved toward a sound, indicating that auditory space begins to be coordinated with visual space. But objects are not perceived

---

[30] We are reminded of Lorenz's sequences of motor habits in simple mammals—one sequence is the key stimulus for the next one, et cetera.

SPACE	CAUSALITY	PHYLOGENETIC STAGE
Egocentric	Egocentric	Pre-mammalien
Changes in perspective seen as changes in objects	No differentiation of movement of self and of objects	Simple motor learning
Space externalized; no spatial relationships in objects	Self seen as cause of all events	True learning with innate reward mechanism
Perceptual constancy of size and shape of objects	Elementary externalization of causality	Individual recognition; spatial thinking begins
Aware of relationships between objects in space and self	Self seen as object among objects; and self as object of actions	Self awareness in advanced primates
Aware of movements not perceived; representation of spatial relations	Representative causality; causes and effects inferred	Thinking exceeds ape level

Table 1

Piagets Stages of Child Development and Pylogenetic Stages of Cognitive Evolution
[Adapted from: Gruber and Voneche (1977)]

STAGE	GENERAL	OBJECT CONCEPT
1 Reflex 0-1 Month	Reflex activity	No differentiation of self from other objects
2 First differentiations 1-4 Months	Hand-mouth coordination; differentiation via sucking, grasping	Vanished objects not pursued; no differentiation between movement of self and of objects
3 Reproduction 4-8 Months	Eye-hand coordination; reproduction of interesting events	Anticipates position of moving objects
4 Coordination of schemata 8-12 Months	Coordination of schemeta; application of known means to new problems; anticipation	Object permanence; searches for vanished objects; reverses bottle to get nipple
5 Experimentation 12-18 Months	Discovery of new means through experimentation	Considers sequential displacements, while searching for vanished objects
6 Representation 18-24 Months	Representation; invention of new means via internal combinations	Images of absent objects; representation of displacements

(Table 1 continued)

to be permanent; no searching is seen when something leaves the visual field. This stage is comparable to the level of the lowest mammals that have motor learning.

Stage 3. (four to eight months) Permanency of objects is now perceived and eye-hand coordination permits grasping and manipulating all objects within reach. This is done intentionally and repeatedly. The baby searches on the floor for objects that fell, tries to make noises with objects, and actively explores everything that is within reach. But the source of activity is always perceived to come from the infant itself; external causes are not understood. The phylogenetic equivalent of this stage is true learning by an innate reward mechanism, as can be found in all mammals from a certain stage onward.

Stage 4. (eight to twelve months) This stage marks the beginning of true intelligence, because means to attain ends are recognized. For instance, a pillow can be moved out of the way to reach a toy. Some events are anticipated. The visual system develops to the point where shape and size of objects are recognized as constant under varying distances and angles of vision. Outside objects or persons are now recognized as possible sources of activity; i.e., cause-event sequences can happen independently of the self. This corresponds to cognitive abilities mastered by advanced non-primates.

Stage 5. (twelve to eighteen months) The conquest of space and causality continues, and is supplemented by self-awareness of the kind that recognizes one's own body as an object among objects and one that can be acted upon. This adds a dimension of feedback in the manipulation and exploitation of space. Such self-awareness also is found on the nonhuman primate level.

Stage 6. (eighteen to twenty-four months) This is the level that can be maximally reached by nonhuman primates and the level from where the child will propel itself into the world of humans by mastering language. Space and causality have been conquered to such an extent that complex action sequences and their results can be pre-imagined before any action is taken. In quality, the child has now surpassed the pre-imagination found among chimpanzees, bonobos, and gorillas.

At the end of the second year, the perceptual part of the neocortex has been fully adapted to the external four-dimensional space-time world, and one can say that a "model" of this world has formed. However, the word "model" is only used for lack of a better one. It is not to be thought as an image, like a small-scale reproduction of a village or the coordinate space of mathematics. Nothing of this kind can be expected to reside in the synaptic connections. It is rather a "bag of tricks" again, an organized multiplicity of responses to sets of stimuli, visual patterns, and actions. But this time the bag is in the neocortex and is better organized, namely according to the inherent order and causality found in the external world. A strong visual "photographic" memory also exists. We only need to see a few characteristic details in order to recall the whole image. That is why it is often so difficult to find an object that is right before our nose. Rather than looking attentively for details, we like to consult our visual memory.

This diversion into developmental psychology yields more than just a proof that a process of cortical adaptation takes place. Piaget's theory of cognition also provides some insight into how this organization of appropriate cortical responses gets under way. The key lies in his three fundamental concepts: *schema, assimilation (into a schema), accommodation (of a schema), and equilibration (between schemas).*

A schema resembles a concept. We may consider it an accumulation of memories (eidetic or otherwise) of objects of a certain similarity, or a list of such objects having some common characteristics. Obviously such lists are more economic than listing separately the objects and each object's characteristics. As in artificial neural networks, all insight begins with classification. That is why Piaget's schemas are so enlightening for understanding of our brain in terms of neural networks.

If a new object is encountered—for instance, a cow—where previously only a schema for various dogs existed (Wadsworth's example), the initial classification may become "large dog." This would be an *assimilation* of "cow" into an existing schema. Yet, on closer inspection, the new animal is found to have an udder and horns; it is also exceptionally big. It will not quite fit into the dog schema.

Logically there are now two possibilities to deal with the novelty. *Accommodate* (expand) the dog schema to incorporate the new deviations from dog, or define a new schema of only one specimen—"cow." Suppose the observer is meticulous enough to reject the first avenue. With more experience, he will later encounter other cows (black and white, black, brown, large, small, short fur, curly fur, young, old, et cetera). How many of the cows will be put into new schemas, and how many accommodated in the old one? Then he might encounter a buffalo, a wildebeest, or a moose. Again, is it *assimilation and accommodation* or new schema?

It would be possible to define a new schema for each animal. But this would not advance the process of acquiring knowledge. To know is more than merely to record. On the other hand, if all the four-legged furry animals would be accommodated into the initial dog schema, nothing would be gained either. One would have one big undifferentiated list of everything. But there is a reasonable middle way: Find an *equilibrium* between size and detail of schemata. This process is *equilibration*. And so the process of bringing order into the practically infinite number of phenomena is governed by *assimilation, accommodation, and equilibration*. The process of equilibration will include a review of contents of the schemas. A number, N, of objects has to be distributed among $k < N$ schemas with the maximum number of common characteristics in each of them.

Piaget's developmental theory has a natural inner logic. This has been demonstrated by Gary L. Drescher (1991). In his book *Made-Up Minds*, he describes a computer simulation of Piaget's first stages. He calls this "constructivist artificial intelligence." A very fast computer is needed. Drescher translates schema, assimilation, accommodation, and equilibration into program files and subroutines to act on perceived inputs which originate from a fixed "world" when attention is paid to some part of it. The world only exists in computer files, and by "attention" he means a set of "exploratory activities" to elicit sensory feedback in the form of visual or tactile patterns. The whole program is called "schema mechanism." A schema consists of "items" within a "context," an action, and items that have resulted from such actions. The dynamics of assimilation, accommodation, and equilibration are translated into a set of programmed rules. The schema mechanism remains restricted to a simplified visual-tactile world.

Initially, the "newborn" program was supplied with 10 primitive "bare" schemas, 175 primitive items and 10 primitive actions, phylogenetically acquired knowledge, so to speak. After running for one day, 7,371 new schemas had been built, 185 new items had been conceived, and a repertoire of 343 new composite actions had been acquired. At this point the computer's memory overflowed.

Examination of the acquired schemas, actions, and items revealed that an insightful cognitive development had taken place up to a level comparable to Piaget's third stage. This approach might demonstrate the natural viability of Piaget's theory, but it certainly demonstrates the inefficiency of computers in dealing with these things. And the computer used (CM2 of Thinking Machines Inc.) was about the best available for this purpose.

In the human brain the actual process of cognition, assimilation, or accommodation has to be instant (in real time) since responsive action may have to be taken immediately. Equilibration, on the other hand, the reshuffling of contents, may occur later. Could it be that dreams are a byproduct of such activities?

Piaget's theory has the ring of truth because it reminds us of the convergence of connections in neural networks. A schema is like a feed forward classifier adapted to the common characteristics of its members. One output line alone may state yes/no as to assimilation of a given object into an existing schema. Or it may output the degree of a match. Is there a hierarchy of schemata—first dogs and cows, then mammals, then animals, and then living things? Maybe yes, after ultimate equilibration, but at first it must be more like older and bigger schemas against more recent ones. I can see agglomerations of interconnected neurons (incorporating a schema) competing with other agglomerations

During sleep I see the sleep center (in the brain stem!) causing simulations of sensory input from memory, stimulating competing schemas, thus opening the door to reshuffling of contents until contradictions are minimized and common characteristics in each schema maximized. We remember the remarkable reports of great scientists and mathematicians to whom a solution to a problem was given in sleep, or after a relaxing interruption of their conscious attempts at the solution (Poincaré 1921), (Koestler 1964, chapter 7), and (Hadamard 1949). Their schemas had equilibrated in sleep with the result of knowledge gained. I am also reminded of the eye dominance patterns, the grain structure of polycrystalline solids, and the magnetic domains in a ferromagnet. After a disturbance, grains or domains change shape, grow, shrink, and recrystallize until a new equilibrium is reached that has the lowest possible energy content. Energy may be compared with the number of contradictions, originating from misclassifications. Might it be that equilibration of the neural networks of the cerebral cortex also could be cast in terms of thermodynamics?

Dreams and Self-Adaptation of the Neocortex

On a previous occasion I have mentioned that mammals have invented a new method of brain utilization that has led to their rapid success. To this there is good objective support. Only mammals exhibit a brain activity during sleep that is characterized by rapid eye movements (REM). Humans report dreams when woken during or shortly after REM sleep. And *all* true mammals that have been tested, have been found to have REM sleep. Moreover, one can fix the point on the phylogenetic tree where REM sleep probably began. There is one ancient quasi mammalian branch, the *monotremes*, or egg-laying mammals, who do not exhibit REM sleep. There are six surviving species of monotremes in Australia and New Zealand. Among them is the echidna or spiny anteater (*Tachyglossus aculeatus*) in figure 6.2, and the duck billed platypus (*Orynthorhychus anatinus*). The echidna has a brain organization that is markedly different from the placentals and marsupials. The brain is larger than to be expected in a rat-like mammal of comparable size, and it has a huge convoluted forebrain. Taking into consideration the animal's simple behavior, one may conclude that the spiny anteater does not utilize its brain as well as true mammals.

This story began in 1972 when T. Allison and coworkers (Allison et al. 1972) studied the sleep of the spiny anteater and did not find REM episodes. Now it had to be explained in which way dreaming can improve brain function. Francis Crick proposed in 1983 that dreaming has to do with partial "unlearning," a process that had stabilized learning in early artificial neural networks (Crick and Mitchison 1983). There is a degree of ambiguity in the distribution of synaptic weights for a particular memory. When many more memories are added, the ambiguities will have to gradually disappear. This is helped by a process of limited levelling out of previous synaptic changes, or unlearning. In this way a network can learn the maximum possible number of patterns.

This idea, however, needed to be made more sophisticated in order to apply to the much more complex neural systems of the brain. A great step forward was taken by sleep researcher Jonathan Winson (Winson 1985, 1990). He developed a theory of the evolution of electrical activity during sleep. According to this, REM sleep would occur first about one hundred and sixty million years ago in the mammalian branch that leads to all of today's placentals (which include primates) and marsupials. The egg-laying mammals branch off earlier (around one hundred and eighty million years ago) and display the slow (theta) wave in the waking state only. The rest of the mammals, together with REM sleep and its accompanying electrical pattern, also exhibit a theta wave during sleep.

Fig. 6.2
Spiny Anteater or Echidna (*Tachglossus aculeatus*). The only mammal that is known not to dream

Winson pointed out that the *hippocampus*, among other functions, acts as a switching center between various cortical modes of functioning (Winson 1985). This control function, to a large extent, is chemical in nature. Several nuclei of the brain stem send unusually thin (and slow working) axons to the *hippocampus* and other limbic areas of the neocortex. Release of a neurotransmitter into the cleft of these synapses results in a change in response of the target neuron with respect to regular synaptic inputs. Basically there are broadside inputs from the neocortex and outputs back into it. Between these inputs and outputs the *hippocampus* acts like a million pole, triple throw switch that is switched by chemical synaptic input from the brain stem. Thus, it can throw the cortical activity into different modes of function.

In the mode of "theta wave exploration" the animal is very alert to sensory input, and vigorous sensory processing and memorizing takes place. In the "aware but relaxed" state, nothing much happens. In "theta and REM sleep," all muscles are immobilized, but processing activity is strong. In cats, the immobilization of muscles has been surgically destroyed. Such cats get up and act out their dreams during REM sleep. Human subjects report dreams when awaken after REM periods, of which there are four or five during each night. Sleepwalking and talking in one's sleep, probably have to do with a failure of that blocking mechanism.

Jonathan Winson calls the nightly cortical activity "off-line processing," and associates it especially with the prefrontal cortex. He conjectures that the survival value of this activity is to assimilate the previous day's experiences and on this basis to plan informed responses for the future. In other words, learning from previous activities and experiences is reorganized and solidified at night.

.It is natural to assume that with increasing corticalization, this kind of "off-line processing," which remains unconscious, can gain sophistication. In humans, the subject of one such activity may be "internal housekeeping," such as the equilibration of Piaget's schemata. With the discovery of off-line processing, entirely new vistas open up on the subject of the unconscious, so dear to analytic psychology. *There is a rich, formative activity going on in our brain that integrates everything previously encountered.* This goes on untiring and automatic every night, and an unconscious personality (Winson's term) is forming without us knowing it.

What may be the survival value of the unconscious personality? Since it is formed in an autonomous manner during sleep, it must include the autonomous needs and wishes of the somatic self, which in fact, initiates sleep. *Those needs come from a world primarily unrelated to current external realities.* They are initially not defined within the context of the sensory-cortical world. Responses to sensory input, if they are to conform to the somatic-hedonic needs, can only be prompt if the raw needs have been previously processed and included in the neocortical processing chains. In this way, whenever called up by association, an optimal action model is ready for immediate use. The location of this model likely is in the prefrontal cortex. This view would be consistent with the results of damage to the frontal lobe, as reported in chapter 5.

"Unlearning," in Crick's sense, and "equilibration," in Piaget's sense, really look like similar processes. We remember that in artificial neural networks, the slow forgetting of older contents facilitates a more balanced (equilibrated) distribution of synaptic weights, so that more items can be stored in a given number of synapses.

The Central Interpreter-Storyteller at Work

All the mental and autonomous systems have the built-in characteristic of wanting to be active. It is activity that sustains living systems. Muscles and synapses deteriorate when not used. And our storyteller is no exception. From gossiping to mytho-poetry and scientific inquiry, it is he (she?, it?) who relishes in making connections between one thing and everything in order to "explain" why. And from earliest human beginnings, stories must have been told around the fire.

Among the tribes who would most closely resemble early Stone Age peoples are the African Bushmen. Despised by blacks as well as whites, today they are close to extinction. Bushmen probably are the most avid storytellers of all human tribes. Figure 6.3 shows a photograph of such a storytelling gathering. By looking at the old man's expression, one can almost hear the excitement of his revelations and sense his own enjoyments of weaving the story. The stories explain things like why a certain beautiful antelope is difficult to kill (it has a magical cleverness that is explained at length), why the stars are hunters, and what the first people were like. Their stories do not yet amount to a mythology, and will be discussed in greater length in Part II.

Fig. 6.3
A bushman 'Old Father' weaves a story
[N. R. Farbman, Life Magazine © Time Inc.]

From the beginning, things needed explanations. Yet, originally, explaining was quite unrelated to finding truth. The concept of objective truth is only a recent one and still lacks strength even among most civilized peoples. It really is like I have concluded earlier: "... *tainted with inaccuracies and illusions, it (the explanation) is just the best fit of available elements to the present situation.*"

The nature of Gazzaniga's central explainer seems to be the Rosetta stone for a new understanding of human psychology and epistemology. Isn't it at the core of all the aimless human productions and shifting cultural achievements seen throughout history? It also reflects what we know about the abilities of neural network classifiers, which force the infinity of items presented into categories, and the associative memory that makes connections between the items, thus fitting together pieces of a puzzle to a picture that is more than the sum of its parts, yet prone to error. Biologically, this enterprise has a twofold teleonomic purpose. The first is to take a shot at reconstructing reality that is *accurate enough to be of survival value.* The second is to conjure up a scenario— an explanation or a story that has the power to *motivate* a response to a situation as perceived—*that causes emotional arousal* which in one way or another connects to the hedonic brain stem "will." Therefore Gazzaniga's central interpreter-storyteller, more appropriately, should be called the "emotional puzzle assembler."

Almost by necessity, this fundamental discovery leads to another one: All the "insights" obtained by the puzzle assembler are, of course, remembered for future recall and incorporation as pieces of a jigsaw puzzle in still other and later explanations. In this way a huge mental system is built up, a system of conscious world explanations that is the most powerful ingredient in the extraordinary success of our species on this planet.[31] *This system of remembered and mutually referenced conscious explanations I will call the "ego system."* The name derives from Jung's definition of the ego as the center of the field of consciousness. Everything in the ego system, at one time, has been at the center of our conscious attention.

From the nature of the emotional puzzle assembler, the following key conclusions can be drawn for the evolutionary synthesis: From the beginning, the growth of insights, convictions, standard responses and the like, have a basic ingredient of the random and accidental. The projected "wants" are more consistent, but some can become exaggerated at the expense of others. Objective truth, nevertheless, does enter to the extent that the model of the external world is built up with more and more accuracy as more and more sensory input is processed, schematized, accommodated, and equilibrated. But this model is rigidly limited in scope. Only what is relevant to foster better informed action for survival is considered, and what is pleasant or relevant in the enterprise of building emotionally motivating interpretations.

Thus, an expanding world explanation is built up by each individual, which is highly subjective, or tailored to the internal vegetative-emotional needs as well

---

[31] It is interesting that a psychologically-oriented "artificial intelligencer," Roger C. Schank, who apparently is not familiar with the neurological storyteller, nevertheless has linked methods of memory access to our passion for storytelling. The title of his book is *Tell Me a Story, a New Look at Real and Artificial Memory* (Schank 1990).

as to facts and misconceptions about the external world. But beyond the individual sphere, the explainer and storyteller also communicate with explainers of other members of the tribe or of the culture. And in this way, through a kind of intra-tribal equilibration of schemas, a common belief is formed. Young members of the tribe, through the ancient similarity bond, will come to accept the commonly held views. Every isolated cultural community will develop their own special world explanations and norms of behavior which will be taken for absolute, since no better insight is available. When two such accidental cultures meet, there will be disagreement to the point that one's own integrity is felt to be endangered. Add to this the instincts of tribal chauvinism and aggression, and you have a perfect example of a bloody tribal war with possible religious overtones.

Thus, human history became a tragic comedy of errors and confusions of our emotional puzzle assembler, held hostage by ancient instinctual patterns that have become atavistic.

Yet prehistoric Stone Age hunters, today represented by the few surviving Bushmen and Australian aborigines, do seem to have been able to live in paradisean simplicity and harmony among themselves and with nature. This may have been the relatively happy condition of humanity throughout the one million years of prehistoric human evolution. The real difficulties have come with a new level of emergence—animal husbandry, agriculture, and material property. I will discuss these matters in the second part.

But despite all the subsequent developments, the original storyteller and explainer remained at work from the time of the weavers of creation myths to the intellectual exploits of Newton, Kant, and Einstein. In yet another emergent step, the written word enabled an unbroken record of man's achievements in an explanation that now reaches back over more than three thousand years. Thus, it became possible to review the explainer's fads and fashions over many generations, and to distill out of them some recurring themes which owed their constancy to the fact that they had a stronger correlation to reality. In this way, finally, some understanding of reality was accomplished that today promises to be objectively true, and thus, universally binding.

The insight that human society is a collection of neural network puzzle assemblers with emotional needs has the most fundamental consequences for epistemology and the assessment of human intellectual pursuits.

Being gregarious animals who use writing, there is no limit to the circular exchange of explanatory stories within a subculture and between generations. This produces all kinds of varieties of "group think," particularly when no clear factual information is available, or when frustrated hedonic needs can be alleviated by fantasy. People living in misery tend to invent a godfather who listens or who will soon return to help. Groups of intellectuals prove their professional superiority by diligent "scholarship," studying the opinions of all the other members of their tribe and expanding on them, so that they themselves can be studied by others and, in turn, expanded upon.

The best examples of the emotional puzzle assembler cum storyteller at work come from philosophy and religion. There is not one philosophy, but a worldwide superculture of many academic tribal groups with each anxiously defending its own imagined territory. Where there is no fixed reference to measure against, such as the unchanging nature in natural science, fashions and opinions change and fluctuate. After all, the transmission of opinion never is complete, and every philosopher wants to add a new opinion which is different and more impressive than all the others.

This, of course, derives from the old ranking rivalry. It becomes real rivalry when the most profound "insight" is honored with an endowed chair at a prestigious university. This brings real power. In this way Hegel had managed to stack the majority of philosophy chairs in Germany with his followers. For a while his tribe had won over all the others. And as Allan Bloom (1987) has pointed out, while the Nazis were defeated by American GIs, the American philosophy departments were conquered by followers of Nazi-colored Heidegger.

But suppose that, every now and then, some members of a circular storytelling society make a good observation of some aspect of real nature, or even a probing experiment, and tell about it. Word gets around. Such a story is different because it cannot be simply changed. It can be checked out. As more events like this are accumulated in the common repertoire, the stories begin to approximate reality. Since this can happen across generations, we have entered a new system of knowledge gathering that functions like evolutionary knowledge gathering, except that it is much faster because it is not genetic, but cultural. *Like its evolutionary counterpart, it adapts to the reality of nature.* It may even increase exponentially with time because previous real knowledge will tend to help in the acquisition of new, related insights. Ultimately there will be today's scientific explosion of knowledge. Natural science may still be driven by ranking rivalries of the scientists and the joy of exercise of thought. But their results are the only real truth we ever can uncover. Everything else is illusion, self-deception, deception, mytho-poetry, or just plain weakness, and no future can be built on this.

If measured against this scale, in looking at the current style and opinion prevailing in the educational establishment of the United States of America, one cannot avoid falling into a deep depression.

I will finish this section with a graphical representation. Figure 6.4 explains the developed preliminary functional scheme of the human mind, or more appropriately of the human spirit since this includes more than just intelligent insight, action, and reaction. Through sensory input from the physical world (Popper's World One), the cultural world (World Three), and the influence from other human beings (social world), the described self-organization of cortical intelligence takes place, including its inherent memories. A special most fundamental class of long-term memories is listed separately. It contains the culturally induced convictions and skills which are preconditions of the ability to function socially and to preserve technological tradition. Of course, these also have once entered through the senses.

The long-term memories and intelligent systems supply the pieces of the puzzle of a given situation. The networks of the explainer fit them together as well as possible. That process is influenced by the instinctual-emotional world of body, brain stem, and limbic system. Thus, the story weaved has emotionally motivating elements. It will cause action for survival and social expression. From here there will be feedback on the contents of the cultural world, the social world, and the physical environment will be changed.

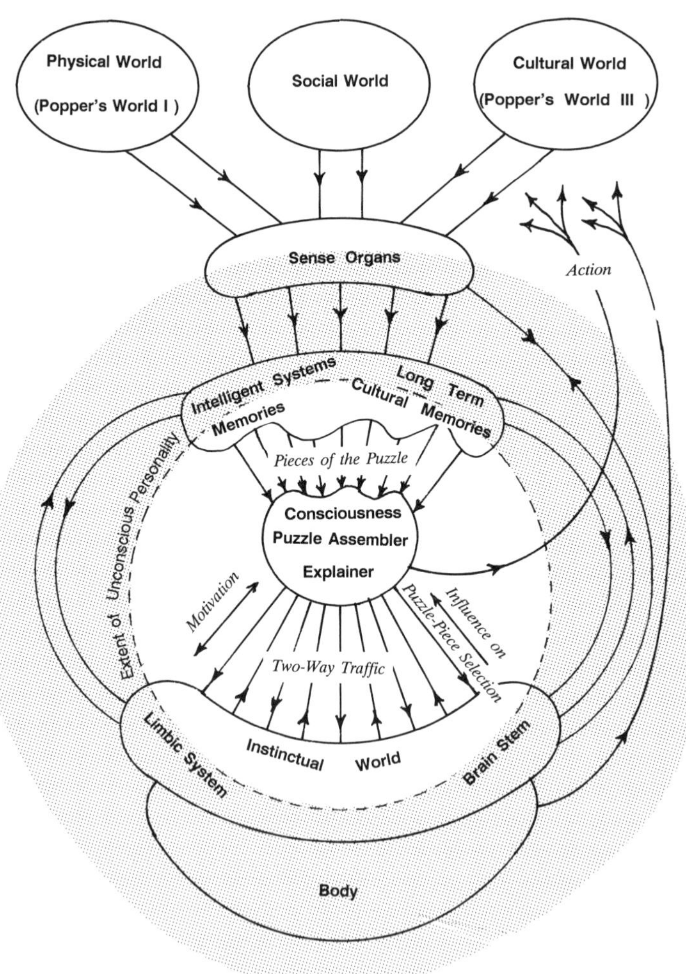

Fig. 6.4
Schema of the Human Psyche

The shaded area outside of the circle marked as "Extent of Unconscious Personality" is not accessible to consciousness. The inside is accessible, even though only a small part of it may be conscious at one given time: contents of memories, a realization of a drive driving us, or processed sensory or somatosensory input. Neither the sensory processing itself, nor the workings of the inner organs are accessible to consciousness. Compared to the totality of brain function, the circle of consciousness has an area that is disproportionally smaller than figure 6.4 suggests. It is safe to assume that all the marvelous achievements up to the primate level function mostly outside that sphere. And we, too, are utterly baffled about what our unconsciousness can do for us unattended.

When memories of conscious explanations slip into the unconscious, they will be treated like any other hypothesis about the real world. They will undergo equilibration with each other and with existing schemata. But since the frontal lobe is involved, no action models will be allowed to consolidate that run against the grain of hedonic-somatic needs. *Therefore, under natural circumstances, the accidental productions of the conscious central explainer can never consolidate into an all-encompassing autonomous system.* The "natural" person will shun the cocksureness of a successful member of modern technology and commerce-driven society. Such a member, in return, will judge the natural person as uncertain and without a developed "personality."

In the natural scheme of the human mind there is no space for "higher principles" or "spiritual goals" of the species. Still, the conscious self and the unconscious self, which experience all this subjectively from the inside, will report a rich inner life, full of joyous delights of discovery, apprehensive decision-making, experience of emotions of the compassionate bond, and the lure of the mysteries of the yet unexplained.

The explainer, like all human faculties, likes to exercise itself. Already, the pre-imaginations of chimpanzees are such exercises. On the human level we can draw from a much more elaborate repertoire: actions of others we found "interesting," that is, emotionally activating or satisfying; previous events of social interaction, and natural events and words which as yet have no clear meaning. Take this all together and the exercises of the explainer become fantasies. And since the elements are pieced together to a new picture of potential or actual reality, we might as well call them "creative fantasies."

Since we live under the spell of the compassionate bond, the explainer also likes to share fantasies with others. They now become stories for the sole benefit, enjoyment, and edification of others. The others relive the fantasies through the similarity bond. And since the stories of the explainer are tailored to emotionally motivate, common sympathetic excitement rises in the circle around the fire. (Compare fig. 6.3, p.164). *Thus, the ancient determinants of imitation, joy of exercise, and the compassionate bond, joined by neural network intelligence and speech, produce the most human of all human endeavors—companionship of the mind.*

But stories of the explainer are just stories. If their contents are not purely accidental ("spirits protect the antelope"), they are reinforced by blind tribal-cultural circular feedback. And since they are emotion raising, they call for action. Thus, unfortunately, they are taken for absolute in the absence of advanced reflection, as, for instance, exercised in this writing. *Therefore the human spirit begins its existence hovering in empty space, with no guiding stars in sight, rebounding upon itself, and thinking that it is master of it all.*

It's little wonder that he invents his gods in this image. All this, however, will change as soon as natural science and objective knowledge enter the stage. Their Promethean nature brings danger as well as blessing and hope.

C. G. Jung's Psychology and the Neurological Model

Carl Gustav Jung (1875-1961) has developed a remarkable system of concepts to explain the dynamic of the psyche. Even though he has had great influence on literature, and a growing number of Jungian analysts practice worldwide, academic acceptance of his work has suffered for two reasons. The first was due to his early friendship with Freud and the subsequent rejection of Freudian orthodoxy, which led to bitter accusations from the Freudian camp. The second was due to Jung's personal dualistic view, which made many of his writings appear to have a mystical character.

However, his basic psychological system can survive a change in underlying philosophy. With this in mind, the model of global brain function that I present is well in tune with Jungian views. There are members of the analytic community who have come to the same conclusion. Foremost among them is Anthony Stevens (Stevens 1982, 1993, 1996), who has given a comprehensive interpretation of the Jungian method and system in terms of the instinctual underpinnings studied by ethology. Based on work by Paul MacLean (MacLean 1973), and J. P. Henry and P. M. Stephens (Henry and Stephens 1977), he even makes hypotheses on the neurological correlate which are close to the view held in this book. However, the recent advances in understanding self-adapting neural networks have an impact on the matter. If the neocortex essentially is self-adapting, and the brain stem contents essentially are innate, the antagonism must be between the functions of these organs, not between right and left hemispheres as some psychologists assume (Jaines 1976). Any left-right antagonism would be secondary, and derived from it as a result of the emergence of consciousness in the left hemisphere.

It was already mentioned that the conscious central explainer closely resembles Jung's concept of *ego*. The two other main psychic systems are the *personal unconscious* and the *collective unconscious* (Jacobi 1942).

The personal unconscious, according to Jung, contains memories of previously conscious events and subliminally perceived events. The contents cannot always be retrieved. Some are repressed by ingrained attitudes. The schematic of figure 6.4 can be interpreted in Jungian terminology. Previously conscious events are, of course, memories of the central explainer's explanations. They are represented by the areas on top, designated as "memories." Note that the flow of traffic between the central explainer and this area leads in both directions. After something has been explained, it can recede back to the shaded area, the personal unconscious. But also sensory input moves downward through this field. We remember that sensory processing chains and memory are one and the same. Therefore, the personal unconscious can also contain subliminal events that had never reached the puzzle assembler. Those events may have been processed unconsciously and equilibrated with other contents of the unconscious personality, including contents rising directly from the instinctual sphere of limbic system and brain stem. This is indicated by the arrows on both sides, inside the shaded unconscious area.

Jung's collective unconscious would be represented by the shaded areas at the bottom. The central explainer is influenced from this region as it weaves the stories and gives them emotional coloration and appeal. This coloration is necessary because the stories in turn have to reinforce emotions and call on various instincts in order to elicit forceful action. That is represented by the arrows leading downward from the central explainer. Earlier I have compared a healthy personality to a conductor who conducts the orchestra of instincts to make beautiful music.

Among the things that can go wrong is "inflation" of the collective unconscious. That happens when the central explainer is swamped by raw instinctual contents rising up along the arrows. Then we act irrationally. But normally there is just a "compensating" influence from our biological determinants trying to insure that our biological needs are included in a balanced way, in the overall system of potential reactions to external stimuli and in willed activity.

The workings of that compensation can be subjectively experienced as archetypal images in dreams. Archetypes themselves are "primordial determinants" of the psyche. According to Jung, their nature within psychology is not known and not knowable. Therefore they are "psychoid" rather than parts of the psyche. But they really are species-specific, phylogenetically acquired determinants of behavior, expressions of the "will" of chapter 3. Archetypal *images* appear in dreams as agents for the formation of an unconscious personality that in this way tries to integrate the inner world of instincts with the externally derived world of ego. They are bizarre and alogical because they are prelingual.

Jonathan Winson (Winson 1985) critically examined psychoanalytic theory since Freud's time in light of the neuroscience of sleep. He comes to reject Freudian concepts like repression of dream material, redirection, and the dream censor. Basically he agrees with C. G. Jung's dream interpretation; for Jung, dreams were simply attempts of the unconscious to correct or compensate for conscious attitudes which may not be in tune with the goals of the unconscious personality and, we might add, the somatic self. Winson (Winson 1985, p. 228) lists the following quotation from Jung, which seems to agree with his view:

> The dream gives a true picture of the subjective state while the conscious mind denies that this state exists or recognizes it only grudgingly . . . when we listen to the dictates of the conscious mind we are always in doubt . . . The dream comes in as the expression of an involuntary psychic process not controlled by the conscious outlook . . . It represents the subjective state as it really is. (Jung 1933, p. 4)

The brain's system of remembered conscious explanations previously was called the "ego system." By necessity, an antagonism exists between the ego system and the instinctual needs because they are of different origin and serve different masters. The ego system originates in the self-adaptation of the neocortex to the present-day external world, whereas the instinctual needs originate in the natural history of mammalian life. In figure 6.4 this antagonism is between the systems on top and on the bottom. I will show in Part 2 that in today's technological culture the ego system has become hypertrophic. This causes the antagonism to grow and become an extreme disturbance in the human psyche, resulting in unhappiness, neurosis, crime, and cultural decline.

In Jungian psychology, the attempts to close this imbalance is a lifelong drama called the "individuation process." This would be synonymous with a continued growth in sophistication of the unconscious personality. Neurologically this process should be recognized as an attempt at preservation of the brain as a single functional unit. Jung considered the discovery of the individuation process his most significant achievement. In *Psychology and Alchemy* he shows that medieval alchemical literature amply demonstrates that the opus of the alchemist was a mental technique, comparable to but different from Eastern meditation, which aimed at bringing the conscious ego attitude into harmony with the archetypal world of the unconscious personality. This archetypal world, however, for Jung and for the alchemists, was spiritual. Many eminent members of the medieval clergy had more or less secretly practiced the art as a compensation for the increasing exteriorization of official Christian religion. Yet the fact that this art also was known in China and among Arabs demonstrates its origin in the universal dynamic of the human psyche.

The existence of this natural antagonism between the brain stem needs and the neocortical system, particularly when amplified by the modern hypertrophy of ego powers, is the neurological key to Jungian psychology. I will call this the "neurological dualism."

Also, I venture to speculate that all philosophies and religions that build on a spirit-matter dualism have an origin in that neurological condition, including

Jung's interpretation of his own observations on patients. I advise the reader to try *Psychology and Alchemy* after finishing (and assimilating) the second part of this volume. It will be as if a dense fog has been lifted.

To conclude this brief review of Jungian psychology, and for use in the second part, let me now introduce the most commonly observed archetypal images and their probable teleonomic purpose (Jacobi 1942), (Stevens 1982, 1993, 1996).

One of the most common is the *animus/anima pair*. In men, the *anima* is the female image that prods us into the proper attitude toward women and all that comes with it: reproduction, nest building, and care of the young. Similarly, women may dream of a male figure that represents instinctual aspects in their relation to men. Whether archetypal dreams are remembered or not, the influence of the archetypes in forming the unconscious personality continues all the same. Anima and animus generally appear as human figures, and they are quite different due to the very different roles men and women have to play in reproduction.

In the young man, anima often is a seductive, beautiful woman luring him into the biological trap of sex and reproduction. But the anima figure also is an *inspiratrix*, a muse and counsel. That is because the relationship has to last so that the children are provided for. We remember that in *Homo* there is a radically extended period of maturation. Also, the male has to be inspired to produce his maximum possible creative contribution to survival. Generally the anima is experienced as positive, pleasant, and sometimes possessing supernatural powers. To artists, she is their muse.

The *animus* of the female, on the other hand, often is fearsome and dark. Biological thinking provides the obvious answer of why this has to be: for the woman, the act of procreation pulls her into a sphere of difficulties she might not be ready for. "Beware of men unless you are able and willing to bear the consequences."

The positive feminine archetype is the *Great Mother*. She is the inner role model for mothering children and men. In a sense, only after this has been sufficiently activated, is the woman ready for procreation, and then *animus* looses his macabre coloration.

The *Wise Old Man* can be encountered in dreams of men and women alike. This archetype must have derived from the alpha male of the band, or the father. We remember the importance of ranking hierarchies, "someone on top must know all the answers." On the level of the human mind this figure, indeed, can offer creative solutions that are wise in the sense of biological wisdom, that is, solutions which are a good compromise between present and future hedonic needs and life's necessities. Owing to the nature of "wise old man," the frontal cortex must play a role in his origin.

The *Trickster,* or *Puer eternus* ("the eternal boy"), is a strange figure that leads to all kinds of mischief and deception. I take it as the archetype that encourages the adventure of conscious manipulation of the world. We remember playfulness in animals as a source of innovation. "Forget all rules, try something unexpected." Among Stone Age hunters like today's Bushmen, we will find this archetype strongly expressed. Those peoples have just only begun to explore the possibilities that lay in continuous consciousness and its ego system.

Related to the *puer eternus* is the *hero archetype*. In many men this is the dominant archetype for a good part of their life. It must be the manifestation of the basic aggressive tendencies, of the "macho" man. We remember the chimpanzee males, who like to patrol the outer limits of the group's territory in search of "heroic" adventures. Like the *anima*, the hero archetype can have a spectrum of developmental stages, from the raw aggression among street gangs to the creative scientist standing up against ingrained misconceptions of his time.

Another often mentioned archetype is the *self*. This one is most difficult to understand, because Jung, in his dualism, equates it with the spiritual influence of God. *It urges us on to the path of individuation.* In the neurological context, this can be understood as originating from a systemic need to equilibrate the two subsystems of the brain: the worldly neocortex and the instinctual brain stem. Being an influence from outside of the familiar conscious system of explanations (the ego system), in dreams the brain stem's prodding can be experienced as uncanny or divine.

Archetypes and instincts are complementary aspects of the same thing. When the phylogenetically acquired needs modify the cortical networks of the unconscious personality, we dream of archetypal images. Probably as a result of this, instinctual-archetypal images enter into mythology and religion, and thus indirectly influence behavior. That is why mythology abounds with archetypal images.

The concepts of archetype and instinct both suffer from a certain inherent lack of sharpness. According to Jung, no list of archetypes or their images can be compiled. Depending on circumstances, one image may represent several archetypes or only parts of one. No clear limits between them exist, only the major ones are permanent, and even there different aspects may become important on different occasions. The same can be said of instincts. The major ones can be isolated by ethological studies. But again they may have different appearance in different context, or manifest themselves in combinations. Konrad Lorenz has invoked the image of a "parliament of instincts," which in a flash conveys the idea. The same description can be used for archetypal images: "the parliament of archetypes."

Why are dreams bizarre and unclear? Because they are prelingual. Sometimes dream figures speak, but no logical sense can be found. Mostly they are like animated pantomimes. They are faint shadows of what is happening when our unconscious personality undergoes equilibration.

Most dreams do not seem to deal with any archetypal image, but are simple equilibrations of perceptions acquired in the last few days. According to Jung, archetypal dreams generally are in color, whereas ordinary dreams are black and white.

When an archetype influences the unconscious personality, and this modifies behavior, we can say the archetype has been expressed. This is in analogy of genes being expressed in a phenotype. And as environmental influences can prevent gene expression, they also can prevent archetypal expression. We already met the neurotic chimpanzee mother who stole and devoured children, and who passed that habit on to her daughter. A young woman will have more difficulties to express her great mother archetype in the absence of a role model, preferably her own mother. On the other hand, this archetype may become hypertrophic, and the dominant content on which self-esteem is built. The result is the overbearing mother who knows what is best for everyone and who mentally destroys her children.

The Don Juan only expresses the seductive aspect of his anima and cannot proceed to the muse, thus eternally degrading women to walking vaginas, and rendering himself forever unfulfilled. And the woman who cannot express her great mother archetype will be stuck with the negative animus. Thus, she will either remain eternally afraid of men, or in a perversion of nature's intent, accept them only as bearers of a phallus: "I do not like them, but why can't I just take what gives some pleasure?" This would be the attitude of the conscious central explainer faced with contradictory signals from the unconscious personality. The inability to express the great mother archetype seems to be a main problem among some of today's feminists. Among today's males, Don Juanism is rampant. Both maladjustments are a result of living in a dehumanized, unnatural culture of a commercialized asphalt jungle, and both are mutually reinforcing.

The central explainer, of course, brings together everything that comes before him, whether it originated from the brain stem via the frontal cortex or from the external world via sensory processing, and thus consciously the appearance of sanity is maintained. No neurotic can ever recognize his dilemma if left to his own devices.

Progress in the formation of a well-adapted unconscious personality is synonymous with an increase of what I have called "Knowledge II," even though at this point this remains unconscious knowledge. As such, it nevertheless makes our behavior better adapted to the total human situation.

Do Jungian psychoanalysis and therapy have a future? Probably it will have a great role to play in a future healing of our sick culture because it seems to reflect the objective reality of the human psyche. As will be shown in the remainder of this book, the human psyche is subjected to great suffering in modern society. These are problems of adaptation to the new emergent level of technologically amplified ego power. Jungian analysis is exactly the correct antidote and can be expected to increasingly influence us in the future. Once the human situation is realistically understood, it should be capable of breathing new life into our declining art, literature, and social health. As a medical therapy, however, it will have to make the transition from a healing art to a healing science based on the objective evolutionary and neurological understanding of the human psyche.

PART TWO

BEYOND HUMAN NATURE

"Zwei Seelen wohnen, ach! in meiner Brust,
die eine will sich von der andern trennen;
Die eine hält in derber Liebeslust,
Sich an die Welt mit klammernden Organen;
Die andre hebt gewaltsam sich vom Dust
zu den Gefilden hoher Ahnen."

(Johann Wolfgang von Goethe, 1749-1832), Faust

"Two souls, alas, are dwelling in my breast,
And either would be severed from its brother;
The one holds fast with joyous earthly lust
Onto the world of man with organs clinging;
The other soars compassioned from the dust,
To realms of lofty forebears winging."

(Translated by Walter Arndt)

Peface to Part II

With the first part I hope to have uncovered what our real nature is, to have outlined the boundaries set to our understanding, to have shown how we are a result of the rational dynamic of the universe, and to have shown how we mesh with it.

I tend to believe that a truly attentive reader, even one of the physicalist denomination, may not have escaped a shock of disillusionment. When it struck home how accidental the world views produced by the central explainer really are, I myself had to go through a kind of conversion experience. When I concluded that our whole menu of cherished opinions and convictions mainly is a result of accidental circumstances, such as place of birth, experience during formative years and the people we interact with, I was deeply shocked and humiliated. I found myself to regard my friends and enemies with more compassion than before. After all, one can only feel sorry for other innocent victims of circumstances like myself. Later on, even monstrosities like Stalin or Hitler began to seem more tragic than appalling. Was I overreacting? Was I overlooking something? *At any rate, the matter of applying the new insights needed very thorough consideration.* I remembered that since ancient times the theme of all good drama is tragedy. Homer, Sophocles, Shakespeare, and Goethe did superbly dramatize the human dilemma, but they did not pretend to know how Homo sapiens had come to be in such eternal troubles.

As for myself, I felt that I had found a *key to this in the emergence of the conscious ego system, and in the neurological dualism that is caused by it*—a burden that comes with the transition to a new emergent level. Also, having unrolled before my inner eye the incredible story from the Big Bang to the slimy, and then to the sublime, I couldn't believe that all this was to end on this planet in misery and disaster. The next emergent level was bound to come, the "Age of Aquarius" of Jung's dreams. I decided to search for signs of its appearance in history and in the world of today, and I was rewarded. As a byproduct, when the main lines of human history can be explained in a coherent fashion with help from the new hypotheses, confidence in their correctness can only increase. The result of these studies is the second part of this book.

There have been attempts to interpret history in terms of Freudian analysis applied to personalities who have changed history's course (psychohistory). To this, my attempt is not comparable. First of all, Freudian views are insufficient to explain the reality of human nature. Secondly, history may well be influenced by "historic" personalities, but it is the collective mental condition—the currently fashionable storytelling and its clash with our true instinctual needs—that determines the long-term course. Understanding the constraints of the human mind and its evolution, however, permit to infer the collective mental condition of a time, thus giving a new dimension of depth to the understanding of the past. Once the past is successfully interpreted in this way, it may become possible to speculate about the future more knowledgeably.

Evolution by natural selection is not gentle, but most animals probably do not feel the suffering that comes with it. When man became conscious, however, a new dimension was entered, and for each of us it turned into an intensely felt struggle—an epic struggle between individuals, tribes, nations, and beliefs. Equipped with an unfathomable unconscious personality, a bag of physical needs, another bag of emotional needs, and a superficial central explainer of it all, we now enter the world stage. By necessity, trouble is brewing. *Armed with the new insights, let us explore history, diagnose the malaise, and look for signs of healing.*

Chapter Seven

Neurological Dualism in Cultural Evolution

The 1.5 percent genetic difference that separates us from chimpanzees causes two main improvements: more intelligence and a voice box. Together, both project us into a new emergent level: culture, transmission of culture across generations, and hence, cultural evolution. Yet the venture into a new system level of human existence did have its cost. When animal nature was left behind, more and more things ceased to function naturally. The neurological dualism widened sharply, a condition that expelled us from paradise. Ever since, we have to use our newfound ego power in order to fend for ourselves. Now we even are forced to responsibly run this entire planet.

Cultural Versus Genetic Evolution

Cultural tradition begins with local bird dialects. In *Homo sapiens* it reigns supreme. Not only is local language passed on from generation to generation; also, religious beliefs, skills of agriculture and animal breeding, artistic and technical skills, and tools and property are all inherited. As compared with genetic inheritance, this is a new mode of transmission of characteristics. Ownership of an ox and a plough is a characteristic of great survival value. In this way we have infinitely more power and opportunity than our basic biological makeup alone would provide. Yet this is a lesser mode; if there would be one generation only, that through some violent event would lose all written (or similar) records and all material goods and manufacturing facilities, we would be thrown back to live in grass huts and collect roots and berries.

But as long as our personal development is under the spell of the new mode, it is not astonishing that our naive central explainer thinks that we and our forefathers are essentially above nature. Awareness of the Darwinian connection to the animal world meets emotional resistance.

Historically, culture was perceived as a phenomenon that was social, rather than biological. Auguste Comte, a French philosopher before Darwin, had coined the word "sociologie" in the belief that laws of social interaction could be found similar to the laws of physics. In fact, the term "social physics" was in circulation in the nineteenth century. It is fair to say that even today nonbiological thought still is dominant in sociology. Where it came closest to consider biology, culture was seen to be a supra-organism. This goes back to Herbert Spencer's philosophy (1820-1903), which saw development everywhere. As Darwinism took hold, it became natural to see competition and selection between cultures of struggling groups as struggle for cultural survival. Thus, Darwinian selection was extended by analogy to cultural evolution. Yet generally this process was taken to be independent of any biological determination.

In the absence of a clear Darwinian picture of the human mind, such views, indeed, are well defensible. If in the process of cultural development, the existence of variation, selection, and retention can be shown, the scheme of a quasi Darwinian selection can be formally applied to a "cultural evolution" (Campbell 1965). Moreover, cultural evolution can be historically and archaeologically traced. Its speed is found to largely exceed the speed of biological evolution. Therefore, within the two thousand or four thousand years of reasonably known cultural evolution, it may be possible to take human nature as almost constant. Our genetic makeup then would be still close to the hunter-gatherers of twenty thousand years ago.

Nevertheless, this "biological constant," be it a true constant or not, in some way has to be a determinant within the process of cultural evolution. The situation seems similar to the material constraints in genetic evolution. There are ingredients in human nature which result in "cultural universals," so well-known to cultural anthropologists. For instance, there is no known culture that does not have some kind of an institution of marriage; everywhere one finds certain forms of greeting, and hospitality plays an important role, as does food sharing. The craving for social status is predominant everywhere. Family defense and cohesion are universal, and everywhere there are certain beliefs in things outside of physical reality. There are taboos, laws about property, behavioral etiquette, methods of settling disputes and, of course, much storytelling.

C. H. Waddington (1960) in The *Ethical Animal* and earlier works was among the first modern biologists to point out in which way cultural evolution and biology mesh. And that is how Waddington saw it:

> . . . the child will acquire certain specific food habits, becoming accustomed to and accepting a particular diet. This is a process

analogous to the development of specific formulated ethical beliefs. In order to find a basis for criticizing these food habits we have first to enquire [sic] what is the function of eating. We find that it is to make possible the growth of the body. Inspecting the growth of human beings on a wide basis we discover that it manifests a general character which we describe as health. We can then ask of any particular food habit or diet how effective it is in bringing about healthy growth. The criterion we are applying here is one of general accordance with the nature of the world as we observe it. If any individual approaches a nutritionist and says that he prefers to grow in an abnormal and unhealthy manner, the nutritionist can do no more than tell him that if he does so he will be out of step with nature. The criteria, of biological wisdom in the case of ethics, or healthy growth in the case of eating, which can be derived in this way, are immanent in nature as we find it, not superposed on it from outside. (Waddington 1960, p. 30, italics added)

Indeed, in view of what I have called the "compassionate bond," biological wisdom may be a more powerful inducement to ethical behavior than religious commandments. Also, we will soon discover that the most enduring commandments are in harmony with biological wisdom, if not a result of such.

But presently, generalizing beyond ethical beliefs, we can say that good culture is a learned habit that permits optimal phenotypal development of our genetic potential. Ideally, culture also could be expected to provide maximum possible happiness for all.

It should not be difficult to add the logic of cultural evolution to the overall process of evolution. This will be explained with the aid of figure 7.1. On the left side is our biological subsystem. The *genetic code* guides ontogenesis of a *somatic system* as well as a system of species specific (or *instinctual*) behavior. Yet both also are under the influence of *environmental constraints*. Insufficient nourishment will cause small and weak individuals. A cold climate will cause the development of an increased layer of insulating fat and an ability to enhance blood circulation when needed. Insufficient parental care will hamper perfection of social instincts. The lack of stimulus at the right time may hurt the perfection of insightful behavior. Thus, genetic and environmental influences together will be involved in the formation of the *biological phenotype*. If culture could be disregarded, natural selection would act on this biological phenotype alone.

With emergence of culture, things become more involved. There is a cultural inheritance, which is in analogy to genetic inheritance. Richard Dawkins has called a (hypothetical) unit of cultural inheritance a "meme," but this analogy should not be carried to the extreme. In part, this inheritance resides in Popper's World Three, in written records or other meaningful artifacts, and in part it is stored in individual memories that lead to oral storytelling and other educational efforts. In the figure this is described by the word *tradition*, the inner imprints of this on the phenotype are summarized as *beliefs, customs, and technologies.* Of course, there are many more, such as language and mental skills like writing, politicking, and artistry. All these cannot be independent of determinants from the biological phenotype.

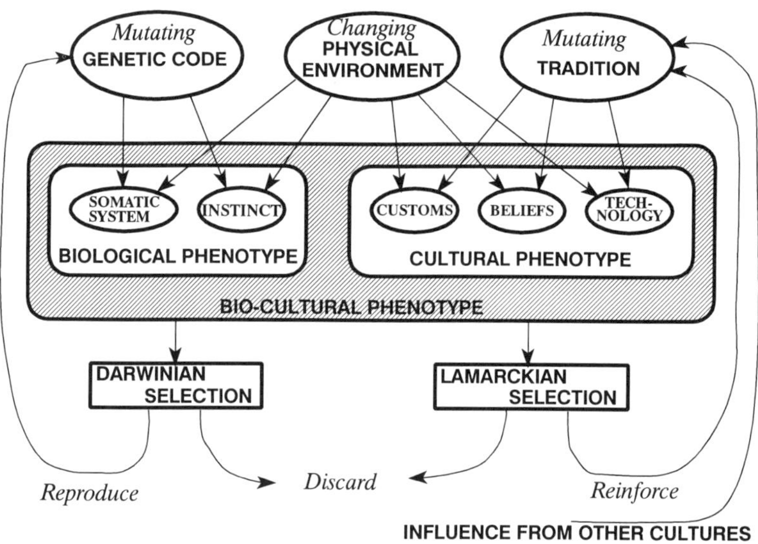

Fig. 7.1

Schematic representation of bio-cultural evolution

Customs and technologies will rather have to satisfy hedonic-somatic needs than run counter to them. Beliefs perhaps are more independent. If, however, they result in actions destructive of hedonic-somatic needs, one might suspect that they will not get very far. Yet under conditions of a successful culture, in the past there has been found so much excess survival strength that contra-biological beliefs and customs could be tolerated, such as human sacrifice and asceticism. But the physical environment also will influence beliefs, customs, and technologies. Jungle dwellers, desert dwellers, Eskimos, and inhabitants of a modern city, all can be expected to have different beliefs, customs, and technologies. All inner imprints of culture form a *cultural phenotype*, which, as the schema in figure 7.1 shows, is to some extent analogous to the biological phenotype. Tradition and genetic code act in similar ways on both of the phenotypes, and the environment influences both.

The important thing is that when it comes to natural selection, the achievements of both phenotypes will play a role in reproductive success. Thus, it becomes necessary to see both together as a unit of selection: the *biocultural phenotype.*

The biocultural phenotype, *as a unit,* will influence cultural tradition. Each generation to some extent will find improvements, possibly in response to environmental challenges. Thus, tradition changes before it is passed on. Also, there will be errors and omissions in transmission—cultural mutations, so to speak. Particularly in modern "developed" societies, the young generation is found to actively resist some of the traditions. All other individuals also take an influence on the common tradition. Other groups' cultures may be imitated. At any rate, the culture changes. The change is further accelerated, because acquired cultural characteristics can be inherited, whereas acquired biological characteristics cannot be inherited genetically. Cultural evolution therefore is Lamarckian[32], mutational, and diffusive. This Lamarckian selection may include competition between ethnic groups which may be won through aggressiveness. Thus, aggressive militarism may very well be culturally bred. History seems to provide a shock full of examples for this.

Darwinian evolution had a component of self-organization. Cultural evolution has an even stronger component of that. Self-organization of commerce, for instance, is a cultural phenomenon. Economists have a hard time to understand its dynamic. Only recently was it recognized that here is another example of a complex, nonlinear system, organizing at the edge of chaos. Here, any interference by design seems to diminish the efficiency, as all state controlled economies have demonstrated.

Stories from times of war and early exploration show that individuals from different cultures, thrown together by accident, immediately will begin to organize collective actions for survival and well-being. Nevertheless, if the individuals for some reason had lost their different cultures, a beginning from scratch seems to be difficult. If a number of urban street gangs were transplanted to an uninhabited island, how long would it take them to develop a decent community? Quite possibly, they would almost exterminate each other before some new order could emerge. William Golding, in his Nobel prize-winning novel *The Lord of the Flies,* has conjured up a situation where a group of British schoolboys are stranded on an island. They soon lose their superficial education and form violent gangs which develop cruel tribal wars until they rescued by the navy.

Aggressiveness is a good illustration of how biological and cultural characteristics are often difficult to separate. So far, intergroup aggression was taken to be biological. Chimpanzee wars were given as an example. Now, however, aggressive militarism appears to be culturally bred also, or we might better say "cultivated."

Culture refines and amplifies biologically given tendencies. We can speak of erotic culture as compared to raw sex, of cultured architecture as compared with the overnight nest built in a tree, and of culinary culture as compared with the

---

[32] Lamarck was a contemporary of Darwin. He held the view that change over the generations happened because acquired characteristics, such as strong muscles due to exercise, would be inherited.

basic satisfaction of hunger. Cultures seem to increase pleasure, including the pleasure generals derive from a well-planned and well-executed war, even though this pleasure might not be shared by their drafted foot soldiers. Thus, all of us develop under genetic as well as cultural control. The question of how much of each is involved does not always lead to any helpful clarification.

The genetic makeup was considered "almost constant" on the time scale of cultural evolution. To what extent may this assumption be correct? We remember that sexual reproduction insures a gene pool of considerable variability without mutations. This is of great survival value when need arises to rapidly adapt to changes. A detailed explanation of genetic variability on the molecular level is rather complicated. But the main points can be understood as follows: Each cell (except egg and sperm) contains two sets of the twenty-three chromosomes, one set originating from each parent. We remember that each chromosome is one long, coiled up string of the DNA double helix. Thus, each gene (or protein template) appears twice. In a fraction of the pairs the genes from each parent are different or "heterozygous." In the expression of such genes there is competition. One parental gene can be "dominant," the other "recessive," meaning that the dominant one will be expressed. Only if both are recessive, will that particular trait be expressed. Since recessive traits are less often expressed, many accumulated mutations can "hide" among them. A given characteristic generally is determined by the concerted action of more than one such gene pair. There can be even large numbers of pairs involved, some favoring, others opposing the characteristic.

Natural as well as artificial and cultural selection now can favor the ever-present recessive traits. In this way extremes can be bred in a short time, such as bald, meaty, and docile domesticated pigs from wild boars. But such artificial breeds have a hard time surviving without human care. In that process no mutations may be involved and the old "wild" genes are still present, be it in form of newly-recessive genes or being distributed among many individuals, each individual carrying only a small number of wild genes, but each different ones. Thus, among all individuals taken together, the full set of wild genes is preserved. This is proven by the wild boar population on the Bahamas. They all descended from domesticated pigs that were introduced to the islands a few hundred years ago. Left without human care, under severe selection pressure, they quickly reverted to the more rugged wild form. Also, stray dog populations (in India) quickly revert from the many large and small varieties of artificial breeds to the original, medium size canine.

With this in mind, one cannot rule out similar human genetic selection under cultural influence. But when it comes to basic mental functions such as intelligence, we should be aware that before one could even begin to determine genetic influence, one should have to first clearly define what intelligence is, and then take out the cultural part. None of this seems to be possible, as Stephen J. Gould has developed in his book *The Mismeasure of Man* (Gould 1981).

Yet in more basic brain stem activity, particularly aggression, is it likely that genetic change due to sociocultural influences could be real? Konrad Lorenz (Lorenz 1966) cites Sydney Margolin (Margolin 1960), who has studied American Prairie Indians in reservations (mostly Utes). These tribes had a long history of warfare. At the time they were found to be extremely prone to traffic accidents, supposedly due to their aggressive driving habits. They also were prone to neurotic disorders which Margolin ascribed to frustrated aggression. Unfortunately, Margolin's paper seems to represent an isolated study of its kind. Japanese also have suffered from centuries of wars between local warlords on overcrowded islands. The character of the samurai caste, with its ruthlessness, overdeveloped sense of honor, and readiness to self-sacrifice, could be taken as a result of selection for aggressiveness.

But in both of the above examples we should not forget the difficulty of distinguishing between cultural and genetic transmission. The studied Ute Indians, after all, may have been neurotic because of retention of a cultural tradition, which under changed circumstances has become atavistic.[33] The samurai may be a product of similar retention. One might speculate that traditions of aggressivity die harder than others because there is stronger allegiance to one's own group against cultural diffusion from outside.

Therefore, the question of genetic versus cultural short-term adaptation remains undecided. Most current anthropologists tend to disregard the genetic influence, but I suspect more out of political correctness than conviction.

I have previously decided not to use too many sociobiological explanations. But the reader should be aware that a consistent and successful mathematical model of genetic as well as cultural evolution is available. It carries the selectionist approach to a high degree of refinement. After E. O. Wilson's (Wilson 1975) *Sociobiology: A New Synthesis*, he has cooperated with C. Lumsden on biocultural coevolution (Lumsden and Wilson 1981). This approach has been further modified and extended by Robert Boyd and Peter J. Richerson (Boyd and Richerson 1985). They call their model "dual inheritance theory," meaning genetic and cultural inheritance interacting, not unlike the scheme proposed in figure 7.1. Many interesting details about the mechanics of such evolution are found and often illustrated with observation. However, it seems to be telling that the model fails to explain altruistic behavior between unrelated individuals and with respect to large groups.

Under "the moral imperative" in chapter 4, I already have tried to explain altruism as an emergent event, rather than from genetic theory. Therefore it may be of interest to see what the ultimate distillation of genetic theory has to say on that subject.

Boyd and Richerson begin their chapter 7 on the evolution of cooperation with the puzzle of the Japanese kamikaze pilots of World War II. There were many more volunteers than could be accommodated due to lack of airplanes. One

---

[33] In medicine, atavism means reversion to a primitive type. Occasionally women are found with extra nipples or men are found with hair covering the entire face.

problem was that experienced pilots, who were needed to escort the kamikazes, themselves wanted to be kamikazes (Millot 1970). Boyd and Richerson write:

> And how did men come to have such self-destructive beliefs? Millot argues that the complex of beliefs that gave rise to the Kamikaze tactic can be traced back to the Samurai military code of feudal Japan which called for heroic self-sacrifice and put death before dishonor. When the Japanese military modernized in the nineteenth century, the officer corps was drawn from the Samurai class. These men brought their values and transmitted them to subsequent generations of officers who in turn inculcated these values in their men.
>
> We find this kind of historical explanation unsatisfactory for two reasons. First, it is incomplete. It tells us why a particular generation of Japanese came to believe in heroic self-sacrifice for the common good, it does not tell us how these beliefs came to predominate in the warrior class of feudal Japan. Second, it is not general enough. The beliefs that led the Kamikazes to die for their country are just an especially stark example of a much more general tendency of humans to behave altruistically toward members of various groups of which they are members. (Boyd and Richerson 1985, pp. 204-205)

Yet after a long and ingenious exposition, which yields many interesting details of potential processes of evolution of altruism, the *model fails to explain the phenomenon.*

Existence of altruistic behavior is a testing stone for the sociobiological approach. In its heavy reliance on selectionism—that is on the biological mechanics of variation, selection, and retention alone—the success of sociobiological theory carries it across borders between emergent levels. Thus, it becomes like particle physics trying to model, for instance, the protein chain that becomes an electrically gated sodium channel in a cell membrane. In principle this should be possible. In principle it also should be possible to explain the extreme behavior of kamikazes through the Darwinian triad. However, there are better ways. They rely on the nature of intra-group cohesion, which can be described by ethological concepts and by comparative psychology. This might not satisfy those who want to build a consistent, if abstract, theory. But it explains things in ways meaningful to less ambitious minds.

Following that vein, first I would like to point out that cooperation and altruism are concepts which alone do not seem quite accurate in describing the kamikaze craze. The driving force behind the phenomenon was primarily intertribal aggression and tribal chauvinism. Both are phylogenetically acquired, species specific behaviors which are not a prerogative of the Japanese, but which can be culturally amplified and refined. In this case the amplification was through inflammatory propaganda. Excitation of the masses to belligerence and violence is easy. There is hardly any monarch, dictator, or even elected leader in known history who has tried this and did not succeed. But in this case there is another strong cultural refinement.

When the awareness is heightened that aggression and self-sacrifice for the common good *are linked*, another phylogenetically acquired behavior can be co-activated: the compassionate bond. If that linkup can be accomplished, the result must be irresistible. That was the essence of samurai tradition. Like all feudal overlords, the samurai caste was a separate tribe within a tribe. The war situation had enabled them to impose their tribal values upon the common soldier.

In that connection we are reminded of Konrad Lorenz's view that the phylogenetic origin of social bonds is from redirected intraspecific aggression. That is why aggression and compassion, love and hate, are seldom found alone. Love for one's countrymen and hate of the enemy always went hand in hand, but in Japan this was carried to perfection.

Into the above discussion only long-term, phylogenetically acquired characteristics needed to be introduced. Indeed, short-term, culturally induced genetic change, if it should happen at all, remains unimportant.

One often hears the argument that if the five original races (see the next section) have been genetically isolated long enough to develop external racial characteristics, their biological determinants of behavior might also have diverged. The reasons that not too much weight should be given to this view are the following:

1. The neocortex self-adapts to culture and environment, not to a genetic program.
2. The brain stem is so ancient and uniform in higher primates, that it appears unlikely that it has measurably changed in the last one hundred thousand years in *Homo*. Yet this possibility is difficult to rule out completely.

In conclusion, one may ask: Why is genetic evolution topped by cultural evolution? Is there a special instinct or drive to develop culture? I think not. Both kinds of evolution are driven by the same situational logic. Beginning with a certain level of brain development, cultural transmission begins to complement fitness. This further fosters selection for mental capacity until in Homo, culture can become the most visible determinant.

As viewed from inside, however, the cultural animal enjoys its cultural existence because life is made easier. The stimuli that previously led to stimulation of pleasure centers now may lead to superstimulation. Instead of birdsongs there may be Beethoven's Ninth Symphony; instead of gathered roots and grubs there may be a refined culinary work of art, and instead of coarse sounds of communication in the pursuit of prey there may be a philosophical conversation in a garden.

Yet we will soon discover that everything has a price. Overstimulation and a life of ease have not been foreseen by our genome. Excessive success of our ego-power leads to its overdevelopment. Dissociation of the two brain systems begins, and this leads to a modern pathological condition that I will call the "hypertrophic neurological dualism."

In the Garden of Eden

The many forms of the story of paradise testify to humanity's longings to return to a state that was simple and where things took care of themselves, in other words, where responsibility was not ours. A conventional psychologist may explain this as a wish to return to the safety of childhood. I would rather emphasize the wish to escape the constant need to explain and decide. The latest fruit on the tree of evolution seems to be cumbersome. Once the new path is taken, nothing comes automatically. Using the metaphor of the fruit, I am awed by the depth of the story of the forbidden fruit whose eating will drive out of the Garden of Eden. The Garden of Eden stands for a largely preconscious existence, dominated by automatic survival skills, phylogenetically deposited in the ancient part of our brain, before any neurological dualism could develop. As in animals, in the Garden of Eden the two subsystems of the brain must have been in exquisite harmony.

But in reality was there ever such a state of archaic existence in simplicity, harmony, and happiness? Of course not. Or if we ever were in such a state on a prehuman level, we probably could not have had the awareness to experience it. Yet compared to the complexities and frustrations inherent in modern civilization, life in the recently vanished original Bushman culture must have been close to such paradise. Of this we have reports of early missionaries, many recent anthropologists, and the compassionate description of his encounters by Laurens van der Post.

Before I attempt to give a brief account of Bushman culture, some historical facts are in order. According to Carleton S. Coon[34] (Coon 1966), the paleontological record shows that the Bushmen (or Capoids) are one of the five "original" races in existence at the end of the Pleistocene era, about ten thousand years ago. The others are the Caucasoids, Mongoloids, Congoids, and Australoids. The Capoids occupied all of north Africa down to the Tropic of Cancer, which is more than one-quarter of the African continent. They may have extended to, or communicated with, Caucasians on the Spanish peninsula. The Bushmen became the most visionary cave painters and rock painters. This art was practiced well into British colonial times. Similarities to cave paintings found in Spain and the close proximity of the famous "temple caves" (Campbell) in the Pyrenees, suggest a spread of the art from north to south, or at least a common origin.

The receding ice in northern Europe caused a population explosion among the European Caucasians, who pressed southward. This, and the drying out of the Sahara, displaced the Bushman from north Africa into east and south Africa. A later expansion of Iron Age Congoids (particularly Bantu) further confined them to South Africa, where they were "discovered" in A.D. 1492, shunned, murdered, and isolated by Congoid tribes. It is quite probable that the hunting

---

[34] C. S. Coon frequently is accused of "racist" beliefs. I think this is largely a misunderstanding caused by egalitarian hysteria. It cannot be denied that there are objectively testable genetic peculiarities between human subpopulations, such as absence of certain blood groups, frequency of certain genetic diseases, and various external features. Only the unsubsantiated claim of genetic "inferiority" of some would be considered racist. Cultural evolution, on the other hand, may be more progressive in certain populations as compared with others. Such differences result by necessity from various circumstances, such as a favorable climate or historic accidents.

and gathering Stone Age culture of the Bushman was the last preserved descendant of hunting-gathering cultures of the northern African steppes which had flourished during the Pleistocene and Neolithic periods.

D. Pilbeam (Pilbeam 1972) argues convincingly that during a period after the last Ice Age, the whole of the European and Asiatic steppes was populated by a more or less uniform culture which was similar to the "great Capsian Hunt[35]" of north Africa. Here, the first artistic productions emerged. Through rational planning and better tools, life's misery had been reduced to a previously unknown degree, whereas the new burdens of existence in a high culture had not yet made their appearance. In short, if there ever was one, here must have been the original Garden of Eden (Pilbeam 1972), and the North American plains may be incorporated as parts of this happy state of early humanity. That is how Laurens van der Post describes his first encounter with the "wild" Bushman:

> The young man stuck his spear in the sand and with his right hand raised, palm open and fingers up, walked shyly towards us, saying in a tone I had never heard before, "Good day !, [sic] I have been dead, but now that you have come, I live again."
>
> We had made contact at last! I was so overwhelmed by the fact that for a moment I barely knew what to do. It was the young man who, after the exchange of greetings in his own tongue, a drink of our best water and a smoke of tobacco, put us all at our ease with his command of natural manners.
>
> He was no taller than Dabe[36] but slighter, with fine bones, and of course, much younger. Nor was there any sign of bitterness or hurt in his eyes. His features were regular and sensitive in the classical Bushman model. His eyes were wide and large and looked steadily into mine when I asked him a question. They had the same vivid light in them which occasionally one sees in Europe on the faces of gypsies in Spain. He was naked, with a loin strap made of duiker skin around his middle, and his skin of a fresh apricot color was still stained in places with the blood of an animal recently killed. All in all he had a wonderful wild beauty about him. Even his smell was astringent with the essences of untamed earth and wild animal-being. (Van der Post 1958, p. 225)

Van der Post had found a group of about thirty people who lived in primal harmony with themselves and with the unforgiving nature of the Kalahari desert. They had mastered survival in an admirable fashion. Precious water could be obtained at "sip-wells" and stored in shells of ostrich eggs. At certain low-laying places they drove hollow canes into the sand. After persistent, strong sucking, water appeared. The mouth was used as a pump. Large and often delicious tubers were dug out of the sand for food as well as for their wetness. The women knew

---

[35] After Capsia, a north African archeological site.

[36] Dabe was a "tame" bushman who had grown up at a plantation, but knew the language. He is reported to have had an air of bitterness and hurt about him.

more than one hundred edible plant species (Lee 1979). In food gathering, the children participated with great pride and were lovingly rewarded.

For hunting, the Bushmen used bows and poisoned arrows of an ingenious three-part construction. The poisoned front end came off and remained in the wound. If that was ascertained, the animal was followed until it collapsed. The hunter apologized to the animal and thanked it for being willing to feed his family.

The sparse clothing and more extensive carrying bags were made of animal hide that was masterfully tanned with the juice of a certain tuber. Their camps were mostly temporary, and they had not more possessions than what could be carried along.

On the average, the daily food gathering could be accomplished in a few hours (Lee 1979)]. Enough time and energy was left for what life was really for: enjoyment.

The successful hunter was greeted by the women with excitement and melodious songs. Sometimes in the evening the women would sit together and sing. Here is Van der Post's account of it:

> The melody was charged with all the inexpressible feelings that come to one at the going down of the sun over the great earth of Africa. They called the song "The Grass Song" and with the difficulty of interpretation neither Dabe nor the singers could readily explain it. I can only recall the feeling and render the words inadequately:
>
> *This grass in my hand before it was cut*
> *Cried in the wind for the rain to come*
> *All day my heart cries in the sun*
> *For my hunter to come.*
>
> They would sing this over and over again, the song becoming more charged and meaningful by repetition, as if the heart, too, was enjoined to a constant act of importunity as in the New Testament injunction to prayer, in order to make life and its powers accessible to its deepest entreaties. The song put us all under a spell, so that I was not surprised that often the young men hearing its crescendo of longing could contain themselves no longer. They would drop what they were doing and come out of the bush, their feet pounding the desert like a drum, their hands stretched wide and their chests heaving with emotion, crying as if the sound had been torn alive and bleeding from the centre of their being. "Oh, look, like the eagle, I come!" (Van der Post 1958, p. 246)

They also had musical instruments. Men had a one-stringed instrument derived from a bow. Women played a four-stringed lyre. One particularly gifted musician was a young man named Nxou. During the day he would often drop his work and play. But especially he played at moments where nature's beauty impresses most, a particularly clear starry night, a romantic moonrise, or at the most celebrated daily event in Africa—the rise of the sun. No one but Laurens van der Post can describe the mood of this art which expresses not only harmony with nature, but also the sadness of the awakening individual, who has learned to know the suffering and death that are so much part of this nature.

They were still aware of their cave art. The last Bushman painter was probably massacred by British or Dutch colonists in the nineteenth century. But the knowledge of the paintings had survived. The paintings are in caves and under overhanging cliffs. Van der Post had reproductions of some of them to show. The reaction of the older men and women was tearful recognition of cherished treasures. The younger generation was overjoyed.

Of course, Van der Post was a romantic, and he falls into the well-known trap of seeing "noble savages" where there only are fellow human beings mastering their survival with means strange to us. But with this reservation, his account still can catch the Bushman spirit better than any of the dry anthropological reports.

A most charming tradition was in wooing a bride. The young man made a tiny, artfully designed bow and arrow out of split bone. He then proceeded to shoot that miniature arrow into his chosen girl's derriere. If she pulled it out and destroyed it, he had failed. If she kept it intact, he had won. One cannot help to think of Cupid's arrows in Greek mythology. Who knows, maybe this is a tradition so old, that at one time news of it found its way via ancient Egypt into Greece. When Van der Post wanted to film this, he had to convince some of the Bushmen to stage it.

The most beautiful girl had just married. After some consideration, in view of the Bushmen's love for make-believe games, the husband consented that she could play the role. But Van der Post's film man wanted Nxou to play the suitor. This caused a great disturbance. Finally, Nxou said why he was unable to do it: "Look! That man is my friend. I have known him all my life. Although he says he does not mind, I know his heart will be hurt to see his woman pretending to be mine." This displayed a tenderness of heart not easily found in modern civilization. Van der Post does not say whether it was her less photogenic husband who finally played the role.

The most striking characteristic of Bushmen is their participatory intuition of the animal world. Africa had the richest animal population anywhere. It wasn't enough that each animal and insect was known, but most of them had characters of their own derived from observations which I can only describe as compassionate. They could, so to speak, get under the skin of the animal, and they were masters of mimicking them. Owners of pets know that the compassionate bond can transcend the boundaries between species. Their pantomimes were full of such animal mimicking, and every time it was not only easy to guess which animal was imitated, but there was also a humoresque and sudden realization of what was funny and significant about it. At the same time one discovered a human, all too human, attitude that thus was demonstrated for the edification of all.

Another pantomime represented humans. This one seemed to have been a standard repertoire. It was the acting out of an ancient real or mythological event of a war. An "apricot Helen of a middle-aged Menelaos" had been abducted by a young archer from a different group. A war started:

> When these people performed their pantomime of the war, they divided into two teams facing each other on their knees in the sand, about fifteen yards apart. They would then taunt one another with the challenges and battle cries of another age, the shouts and movements getting louder and more violent until at last they were twisting and turning with thrust, parry, counter-thrust and evasive action as though indeed spears and arrows were raining down upon them. Though they never moved from their knees, the gesture of heads and bodies, the expression on their faces, and the cries of the wounded and dying, enabled me to relive vividly the atmosphere of a battle of their past. (Van der Post 1958, p. 244)

But this was a "war with a difference," as Van der Post states:

> For no sooner was Helen reclaimed and the ravishing Bushman killed than both sides were filled with fear and revulsion against their deed, as if suddenly among the acacias of their vast Kalahari garden the voice of God himself had made the leaves tremble as He reprimanded them for their mutual sin. They instantly sat down to talk to one another and resolved that it must never happen again. Accordingly they divided the desert into two zones, promising never to cross the demarcation line between them. (Van der Post 1958, p. 243)

Is this an educational drama to counteract tribal aggression? It is difficult not to reach that conclusion. *On this early stage of Stone Age hunter, man was quite capable to culturally compensate for the primate heritage that was threatening his happiness.* The main threat, of course, came from the three universal factors: ranking rivalries, mating rivalries, and intertribal aggression. We remember the tough emotional world of baboons. Apparently early man was well capable to escape these predicaments. In the Bushman they are eliminated through cultured mutual respect, the institution of marriage, and the cultivation of art, including direct "psychodrama."

I would not have presented Van der Post's romantic account in the absence of other confirming evidence. Lorna Marshall has visited !Kung[37] for many years, beginning in 1951, six years before Van der Post (Marshall 1976, 1962, 1969). Eibl-Eibesfeldt (Eibl-Eibesfeldt 1972) has written the account of a human

---

[37] The "!" stands for one of the four different clicking sounds produced with the tongue. The Khosian languages are unique in having these sounds.

ethologist on their basic aggressive tendencies which, contrary to the belief of some nonbiological sociologists, are not different from any other tribes of the human species. He, too, noted their remarkable skill to culturally compensate for aggression, so that the first impression, indeed, is that of an egalitarian, happy society.

According to Eibl-Eibesfeldt, anthropologists who have fallen for the illusion of the "noble savage" did not distinguish between tribal belligerence and acts of interpersonal aggression, and have belittled the latter. Nevertheless, even Lee and DeVore (Lee and DeVore 1968) write: "I recognize that life in pygmy camp is far from peaceful, husbands frequently beat up their wives and vice versa. Yet such outlets, as well as explicit concern shown for avoiding aggression, are insurance policies against aggression of an intended, calculated type . . ." This view ignores results of ethological research. *Intra-tribal aggression is not a substitute for war.* On the contrary, the pecking order makes life within a group possible and, if anything, it amplifies extra group belligerence.

Some Stone Age hunters simply had established the successful cultural traditions of avoiding war and defusing intra-tribal aggression as much as possible. It is wasteful to do otherwise.

That's an enviable state of affairs. *What then went wrong between the great Capsian hunt of twelve thousand years ago and modern man?* I will try to answer this in the next two sections.

The Fall From Grace and the Ego System

In a sad way, we are fortunate that one case of a breakdown of the happy existence of primal man happened right before our eyes. It is the fate of early cultures to fade when brought in contact with later ones. The Aztec empire and culture was destroyed by only a handful of Spaniards under Fernando Cortez.

Compared to this, the Bushmen were exceptionally resilient. But ultimately, deprived of a habitat, they began to interact and trade with the Bantu. The Bantu had cattle, agriculture, and iron. To make things worse, the government of Botswana began a program to encourage agriculture and the keeping of livestock. It gave away donkeys to pull plows and organized the purchasing of handcrafts such as ostrich egg necklaces. Soon the Bushmen were found in permanent camps, with a kraal of donkeys and goats. They had money to spend and were wearing mass-produced clothing. Some began to work for the Bantu. Presently there is a "Bushmansland" reservation, and the neo-Bushmen have "agricultural cooperatives."

When the Bushmen were forced to abandon their lifestyle, John E. Yellen found that their values actually had not changed. We remember retention in cultural evolution. They considered themselves still hunters who temporarily did all kinds of other things on the side. Nevertheless he found that the shelters had become more elaborate. One telltale sign was the following: Previously, the shelters all opened toward a common central plaza so that people at any time could share the pleasures of companionship. But now the huts became more scattered, and the entrance turned away from the communal area, as if everyone had something to hide. Indeed they had. *They had discovered personal property.*

The women began to accumulate beautifully colored fabrics which were stored in chests. They obtained beads, iron pots, blankets, et cetera in quantities well beyond their needs. All this could now be purchased from their earnings (Yellen 1990). The happy communal life and the sharing of goods stopped. Figure 7.2 shows such a new family with their amassed useless goods. They still said that sharing was important, but did not do it, and, as if ashamed that they did not, they turned the huts so as not to be seen. Now they quarrel among themselves. Often, the disputes are settled by their Bantu landlords, who refer to them as "our Bushmen."

Fig. 7.2
A Bushman couple with useless treasures
[From: Yellen (1990), courtesy of J. E. Yellen]

The simple opportunity to obtain personal property seems to have destroyed the fine-tuned balance that was necessary in order to maintain their culture. At least J. E. Yellen came to that conclusion. Property, of course, gives social rank.

Previously the nomadic way had supported an egalitarian culture because there could not be any accumulation of property beyond what could be carried along. Besides, if there were any aggressive encounters or disagreements, instead of fighting it out one could simply leave and join another group. To that end elaborate kinship relations existed. This enabled the Bushman to live virtually anywhere. So no strong leaders were needed to settle disputes. Also, to seasonably migrate to where there was more food and water was an obvious advantage.

But permanent camps also had their value. Better shelters could be built. Better household practices and tools could be invented. One other thing speaks for permanent camps: the great respect and sociability was especially extended to old family members. They had a hard time to migrate. The cruel, but necessary, custom was to leave them behind to die. So there were good reasons for moving around as well as for staying in a permanent camp. It did not need much to tilt the situation the way it went.

But then the trap sprang. In contact with modern civilization, if you have a permanent camp, you have property that makes life much easier. If there is disagreement, you have a strong inducement to stay and fight. To ease the tension you need privacy; the huts turn around. To settle the arguments a leader becomes necessary, or worse: a landlord. In short, *Paradise is lost. The ego system becomes isolated. It thinks to find comfort in useless but colorful possessions which remind of the beauty of an existence in wholeness and social harmony. If you do not have self-respect, having property may help.*

Yellen writes: "The changes occurred so abruptly that the pattern of camp design can be said to have been unambiguously transformed from 'close' to 'distant' *within a few years.*"

We have witnessed the moment of the fall from grace. Even the details of shame, covering the body, and expulsion from the first habitat, conform to the Biblical account so excellently felt by early mytho-poets.

What happened to Bushman culture in the span of a few years may have taken others thousands of years. That is, if no advanced culture was there to interfere. But ultimately the expulsion from paradise was bound to happen. The adapted neocortex is a bright and practical image of today's world, whereas the brain stem darkly mirrors the natural order during the last three hundred million years. Polarity is already built into the system. Evolution is blind. There was no one to foresee the consequences of the neurological dualism.

In chapter 6 it was pointed out that under natural conditions the successful productions of the central explainer are remembered and form what was called the "ego system." But the memories cannot consolidate into a powerful autonomous system, because other parts of the brain are participating in the equilibration as well. The biological wisdom of the brain stem, when it enters into the unconscious personality, will prevent a view that is dominated by neocortical, externally adapted contents alone. The conscious ego system must have remained peripheral throughout prehistory.

But in a cultural environment that knows written records and possessions, the situation is radically different. The productions of the central explainer can be stored in writing. They can be passed on to others, and compared with each other and with reality. A mutually reinforcing company of central explainers of many generations is formed, *which is very much independent of any biological wisdom.*

To a certain extent this already begins to happen through oral communication. Add to this some permanent possessions; they also reinforce autonomies of conscious explainers. With possessions that have been acquired through conscious decision, explainers can identify. Possessions are solidified memories of previous conscious transactions. *The whole complex of the central explainer, the internal and external records of its explanations, and its acquired possessions and skills, now become part of the ego system. With permanent external attachments, the peripheral phenomenon of consciousness has consolidated. It has become a self-referenced fabric of opinions and prescriptions that have been successful, and now can begin to dominate life's decisions.*

Introverted Indian philosophy has a well-defined word for the trap of the ego system—"Maya," the veil that is obscuring true reality which they consider to be the inner (World Two) reality only.

What we have observed in today's Bushman were only the faint beginnings of the ego system. In today's Western civilization, its power has grown out of all proportion. It has created our technological civilization, and we have become addicted to its magic. And since all this is far removed from any biological wisdom, we have sled into a difficult situation. Once again, we are awed by the truth expressed in that early myth about the forbidden fruit from the tree of knowledge.

Early on, attempts to keep human existence wholesome became an important goal. This was not because the situation was understood—what was understood was only the bloodiness and misery of it all—but because there remains a systemic need to equilibrate between the contents of the two main subsystems of the brain, as it is expressed by the Jungian archetype of the "self" (see chapter 6). I take this to be the one pressing evolutionary adaptation that became necessary after *Homo sapiens* developed consciousness and became such a success. *This adaptation, though, is not genetic but cultural, and our success in this has remained partial.* Our present situation probably is comparable to the first land animals who just barely wiggled and waggled along on flimsy fins. The fish ventured on land without knowing what was ahead. We obtained consciousness in the same mode. What lay ahead was this: the logic of the situation forced us into the profession of creator and of nature's minister. But evolution still remains blind, and the attempts to master the new situation so far have remained shots in the dark. Now I will try to incorporate the new concepts of ego system and individuation into a more formal schematic of the inner workings of the human psyche and its interaction with the social and the physical world.

PHILOSOPHICAL INSERT 4:
Evolutionary Epistemology Revisited

Evolutionary epistemology was introduced at the end of chapter 4 (Philosophical Insert 2). The concepts of Knowledge I and Knowledge II were presented there. Now the reader should be prepared for a final attack on these matters. Knowledge I was said to have ascended throughout natural history, and was practical knowledge for survival. In neurological context, and on the primate level, we had recognized acquisition of this knowledge as the self-adaptation of the neocortex to the external realities. Later, on the human level, this produced the natural sciences, classical physics, modern physics, and an extremely powerful technology. At the roots of all this we found Kant's a priori categories of pure reason, and we already have recognized their origin in the evolution of cognitive abilities.

Of Knowledge II it was said that it is interested in self-orientation and wants to lift the veil of mystery, and that it is the domain of artists, theologians, and humanists. Since this, again, is a phenomenon of individual growth, not a phylogenetic one, it must be neocortical adaptation. But adaptation to what?

For once there is the brain stem world with its ancient instincts to which the rational neocortical world finds itself hostage in many ways. It is made to sleep and dream, it is forced to plan future responses and activities so that they result in maximum pleasure, and it occasionally experiences strong, even painful, emotions which intrude unexpectedly. Especially in dreams this second reality that is external to the neocortex, often is experienced as uncanny, threatening, unexplainable, spiritual, powerful, and the like. Therefore, the second kind of cortical adaptation must deal with this ancient world of survival instincts and feelings, which on first inspection appear to be totally irrational. But so probably appears the external physical world to the newborn infant, until after two years it will feel at home in it. This second adaptation builds Knowledge II. Ultimately this leads to what is felt to be wisdom.

But there is much more to this story. Wilhelm Dilthey, a philosopher at the turn of the century, had defined what in English language would become the "human sciences." The original German word was *Geisteswissenschaften*–the sciences of the human spirit. He wanted to put these on equal footing with the natural sciences. Historically speaking, he has succeeded, but he also has created much epistemological confusion. His influence on many other thinkers has greatly contributed to the "postmodernist" disregard for the reality uncovered by natural science, because his "sciences" of the spirit were put on equal footing.

In order to succeed, Dilthey had to provide a solid foundation for his human sciences that could compete with the place that experimentation and observation of nature has in natural sciences. He wrote:

> Understanding (in the human sciences) is *a rediscovery of the I in the Thou*: the mind rediscovers itself at ever higher levels of complex

involvement: this identity of the mind in the I and the Thou, in every member of a community, in every system of a culture and finally, in the totality of mind and universal history, makes successful cooperation between different processes in the human studies possible. (Dilthey 1910, drafts for a Critique of Historical Reason)

As an epistemological first principle, he compares the "rediscovery of the I in the Thou" to Kant's a priori categories of pure reason. *My readers will readily associate Dilthey's principle with the workings of the similarity bond.* This was found to be much older than reason, and more powerful. Transmission of inner emotional states, or feelings, through unconscious cues is prelingual and among nonhuman primates much more effective than any transmission of thought through language can be. It is the ability "to get under the skin" of conspecific individuals, as discussed in chapter 3. To compare this first principle with Kant's categories seems to be quite appropriate, except that it relates to a different domain of the human mind—not to reason and Knowledge I, but to Knowledge II. Through the rediscovery of the I in the Thou, we can participate in other individual's struggles to close the neurological dualism, to perfect the inner adaptation (Jung's individuation process), and thus gain a measured amount of wisdom. This process generally is preconscious and we have found its main location in the frontal lobe and, probably, the right hemisphere. Art, music, good literary fiction, dance, and myth, must be typically human advances in this field of interpersonal communication.

Inspired by Dilthey's concept, we now discover that human culture (Popper's World Three) and the similarity bond, together, form a positive feedback loop, and hence an emergent event. By linking up with the many "Thous" that have left records in art, literature, philosophy, social forms of interaction and social structure, we can gain insight into their predicament and their reactions to it, so as to benefit our own inner integration, or the closing of the neurological dualism, and so in turn leave behind new records of our own achievement. This interaction is "prelingual," even if it may use narrative, as in novels and poetry, because what is conveyed is quite different from rational discourse. *Quite unexpectedly, we have found an evolutionary theory of the arts.*

Chapter 2 (The Roots of Cognition) essentially laid the evolutionary foundation for cognitive abilities. In chapters 4 and 5 this resulted in understanding of how thinking and the Kantian categories came into being, which, ultimately created natural science and technology. But linking up with Dilthey's insight, we see now that chapter 3 (The Roots of Will), also has given an evolutionary foundation to quite a different human pursuit: the humanities. What some of us always thought is true; the arts are not just pleasure and entertainment. They serve a very serious purpose: *to teach the intelligent use of emotions,* which is synonymous with inner integration. Today's social decline is a direct result of the failure of that inner adaptation. Quite logically, this also is accompanied by a decline in the arts.

Lately this kind of inner adaptation to the emotional brain stem imperatives has entered into public discussion. Daniel Goleman (Goleman 1995) in his book *Emotional Intelligence* gives an account of research in that field, and of the many successes in various schools across the country to teach healthy emotional control. It is proven that such emotional intelligence also improves cognitive intelligence as measured by the intelligence quotient (IQ). Obviously raging emotions at least distract.

As encouraging as these attempts are in principle, one should exercise caution. Since the time of Sigmund Freud the waiting rooms of psychotherapists are filled with patients who are dysfunctional because in their youth they had been emotionally miseducated. Generally their emotionality has been repressed by punishment, rejection, or indoctrination, rather than positively balanced. We have to be sure that we know what we are doing.

In the following I will bring all the epistemological conclusions into perspective, and on that occasion it will transpire in more detail how in a healthy cultural environment such inner adaptation proceeds naturally. Popper's three worlds of reality (Popper 1972) will be of great help in understanding the whole structure. In order to avoid too many confusing numerals, in the following I will simply refer to them as "the physical world," "the inner world," and "the cultural world."

Figure 7.3 schematizes the major contents of the three worlds and their dynamic interaction. The inner world consists of the neocortical province and the brain stem province. The neocortical province contains knowledge adapted to everything it encounters outside of itself, the physical world, the stores about knowledge of the physical world in the cultural world, and the laws and social norms that are helpful in succeeding in society. Special importance have the possessions. They reconfirm and remind that when knowledgeable actions have been taken, they have resulted in success, and that we safely are at home in, and connected to our surrounding action space. The arrows in both directions represent the positive feedback loops that build and sustain the ego system. They show the direction of transfer of Knowledge I. All this essentially builds on the consciousness in the left hemisphere. The shaded frame indicates the extent of the ego system.

But the neocortical province also is involved with inner adaptation to the demands of instincts (archetypes) rising from the phylogenetic brain stem province. This adaptation is much less conscious, or altogether unconscious. It happens around the limbic system and the frontal lobe, and probably radiates into the unconscious right hemisphere. When this begins to rise into consciousness, we feel the need to find meaning in the busy activities of the ego system and in the external world. Therefore one Knowledge II loop also links to the physical world. The deeper root of natural science, it was said, is not to build technology, but to aid self-orientation vis-à-vis the universe.

Nevertheless, almost all Knowledge II loops link to the cultural stores. In the pursuit of this adaptation, rationality and language are of not much help. We resort to prelingual experiences of art, music, dance, mythological images and the like, to explore inner emotional geography that may help us to find a safe place to make a home. In this enterprise we are greatly aided by the stores of art, literature, and religion to be found in the cultural world. They have been deposited there by previous generations, and their meaning can be "rediscovered" via the similarity bond. But we also connect to contemporary individuals in this way, and can pick up achievements of others in the art of inner adaptation—or the process of individuation. Our own achievements then feed back on others again, or add to the store of culture for future generations.

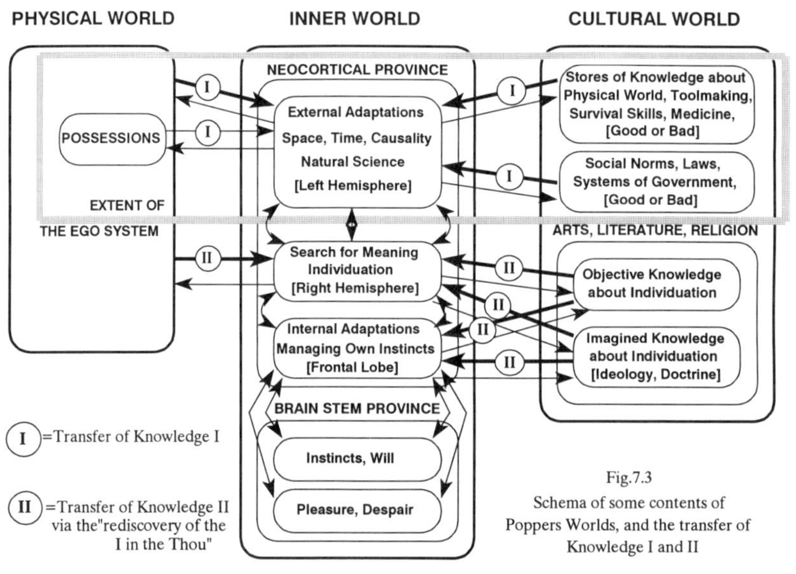

Fig.7.3
Schema of some contents of
Poppers Worlds, and the transfer of
Knowledge I and II

This circular transfer of Knowledge II can result in positive (growing) feedback loops. What grows, however, is not necessarily good for the species. Unfortunately, as compared with physical reality, in this domain there is little to objectively measure against. For this reason the cultural world is full of imagined knowledge, prescriptions, and stories that are powerful enough to precipitate a "discovery of the I in the Thou." But what is discovered may be the "Thou" of an impostor, or worse, someone whose purpose is to mislead for monetary gain. Objective progress in collective personal integration only enters into this world through the influence of strong and exemplary personalities, such as great novelists, artists, musicians and true prophets. They invite us to follow their own path toward a successful inner integration. *But the number of such individuals must be large enough, and their recorded works must speak to the current generation. Otherwise that legitimizing and civilizing system breaks down.*

In the eighteenth century there was a high point in such civilization. But our technology based, ego system centered social system has gone into severe decline because the human truths outside of the ego system have been all but forgotten. Ideologies dominate instead. In this century we encountered Marxism, communism, fascism, religious fundamentalism, cultism, sexual revolution, extreme feminism, postmodernism and, not to forget, plain materialist greed. Erich Fromm has coined the term "techno-fascism" for the dehumanizing power exerted by the blind interests of big industrial corporations in cohort with advertising psychologists.

One clear insight derives from this schema of the human predicament; contrary to the currently prevailing attitudes, there is positive art that fosters inner development, and there is art that works to the contrary. Music can support healthy individuation; there even exists a working music therapy. One interesting study has shown that after listening to Mozart's music, the IQ shows a temporary increase.

In ways not yet clearly understood, musical expression directly communicates with our inner world of emotions. Manfred Clynes (Clynes 1982) has identified a number of prelingual emotional "pre-words," which he calls "essentic forms." They can be expressed with musical sounds as well as with body language, such as ascending and descending finger pressure on a recording device. Anger, hate, grief, love, sex, and joy can all be expressed in this way and correctly understood across cultural barriers by, for instance, Australian aborigines and white college students.

Listening and composing music is accompanied by intense cortical activity in both hemispheres (Petsche et al. 1993). There is no question that music can be educational in the domain of Knowledge II. It also is quite obvious that the rhythm of hard rock single-sidedly fosters adolescent anger and sexual arousal instead of emotional health. Thus, today the "music industry" helps millions to remain forever stuck in a barbarian state of emotional development. An equally strong positive or negative effect can come from art, literature, and religion. At present, the leading proponents of barbarization are in full possession of commercial television. Religion (particularly in America) experiences a regression to either some archaic practices of ecstasy, comparable to shamanism, or falls prey to a rigid fundamentalism that, at best, stops all positive development.

Schools could be most effective in helping inner integration. Mostly, however, they have failed to understand the problem and to notice their high calling. Higher education has degenerated into mere training of operatives for industry and commerce, and of narrow specialists destined to serve a technology-driven scientific enterprise that is blind to human needs. It is virtually unknown that humanity has another vital dimension outside of the ego system, and that inner harmony and relative happiness are achievable goals.

How did we come to be in such a sad state of affairs? This chapter has reaffirmed that human culture is a historical process. To better understand this process, we now have to continue our study of cultural history, which is closely related to the history of religions. The progress of Knowledge II is slow and painful. If, in the domain of Knowledge I, we have progressed from caves to jet planes, in Knowledge II, today we still live in grass huts.

Chapter Eight

From Religion to Reason

This chapter retraces how, throughout cultural evolution and history, reason has slowly but steadily ascended, thus fulfilling a promise in all emergent evolution: knowledge will grow. Today the central explainer and storyteller, with the aid of natural science and written transmission of knowledge between generations, has succeeded (in principle) in explaining human reality as it really is. We are ready to enter into a new level of self-experience and world experience. This would complete the quest of Western civilization to replace mythopoetic intuition with real knowledge.

The Origin of Religion in the Neurological Dualism

The name our culture gave to early attempts toward inner wholeness is "religion." The word comes from the Latin *religare*, which means "to bind again" (Partridge 1983). Word origins often reveal wisdom, because no new word is commonly accepted unless it has the right "feel." The split nature of our psyche is always felt, even if not consciously understood. Thus, religion is supposed to "bind again" our ego system to the original state of animal-nature identity, and harmony, so well-illustrated by Bushman existence.

Religions evolved together with the growing ego system. Most of the oldest cultures were able to avoid a total break by essentially discouraging development of the ego system. I will soon discuss several ancient cultures.

As seen from the point of view of ego consciousness, through religion we pursue *spiritual legitimacy*. This is because our actions, acquisitions, and accomplishments in the real world (Popper's World One) are primarily based on a feedback loop between neocortical realistic adaptation to the world and the success it brings. Within the totality of the psyche, however, this is a single-sidedness which the archetype of the self tries to correct. Or expressed in neurological terms, the brain tries to preserve itself as a functional unit. Since the naive consciousness takes the brain stem influence as coming from outside—from a powerful unknown world called "spiritual"–legitimacy of the actions before this spiritual world is being pursued.

Is the *quest for spiritual legitimacy* identical with Jung's process of individuation? Not quite; in the neurological context, the process of individuation should be viewed as extended self-adaptation of the neocortex to include biological brain stem wisdom. This is an internal personal process that tries to bridge the neurological dualism. It mostly remains unconscious and is indirectly inferred through subtle observation. The *quest for spiritual legitimacy*, on the other hand, as I would like to define it, consists of historically observable collective attempts, failures, redirected activities, and perversions, which result from often misguided collective attempts at individuation, fostered by the archetype of the self, but under the spell of the conscious explainer. *The quest is a collective phenomenon of cultural evolution, and it is studied in the history of religions.*

As the primal wholesomeness becomes lost, the quest may manifest itself in beliefs and practices which are without much relation to inner biological truth. In fact, by pursuing the quest as a faithful member of a cultural or religious group, one may undergo de-individuation. On the other hand, strong individuals can succeed on the path of individuation on their own, by rejecting the communal spirit. Then they become prophets, possibly martyrs. Thus, the quest for spiritual legitimacy is a historical process; individuation is a personal one.

The neurological dualism is a core concept for understanding myth and religion. Joseph Campbell assigns three purposes to myth (Campbell 1968, p. 4):

1. ". . . mythology is to reconcile waking consciousness to the mysterium tremendum et fascinans of this universe as it is."
2. ". . . to render an interpretive total image of the same (the Universe), as known to contemporary consciousness."
3. ". . . is the enforcement of a moral order: the shaping of the individual to the requirements of his geographically and historically conditioned social group, and here an actual break from nature may ensue." (He lists castrations, sub-incisions, scarification, et cetera.)

Of this list the second point can be immediately recognized as the main activity of the central explainer-storyteller. "Contemporary consciousness" practically is a synonym for the fabric of mutually referenced explanations. In prescientific societies, of course, this fabric is unrelated to objective truth.

As far as the third point is concerned, moral order has strong ingredients of biological wisdom: phylogenetically acquired behaviors will find their way into moral convictions. Among them are: the compassionate bond, a degree of monogamy that our species has because of the need for prolonged prenatal care, respect for elders (parents) because of the advantage of imitation and because of ranking order, and group cohesion and extra-group aggression. In a successful culture these innate tendencies are to be modified (cultivated). Ultimately they will be mediated by objective knowledge, but that still is far in the future.

These modifications may have to run against the biological grain. They need to be emotionally motivated. Here, the motivating function of the storyteller comes into play. *By providing contents which act as key stimuli (or releasers) of selected instincts, the storyteller can make the orchestra of instincts play a new tune.* But since the stories (myths) initially have a strong accidental ingredient, incomprehensible or atavistic moral requirements also may arise and persist, such as ritual mutilations. In terms of cultural selection theory, those are "luxuries" which can be retained only if they are not too damaging. They are comparable to the tail of the peacock.

Campbell's first point should rather be the last one. Under prescientific conditions the explanatory fabric soon reaches its limits. Beyond them lies mystery. What cannot be readily explained is frightening: the power of the wind (storm!), thunder and lightning, the growing of plants and humans, the act of procreation, darkness, death, violence, the sun, and the disturbing emotional tensions and frustrations that arise from our primate heritage. But the central explainer will not rest until everything has its place. That is why the gods have to be postulated. Therefore I have to slightly sharpen Campbell's first point: ". . . mythology is to reconcile waking consciousness to the *mysterium tremendum et fascinans* of this universe," not "*as it is,*" but as it is experienced.

Thus, under the influence of the central explainer's fabric of stories, mythology and religion become a crazy medley of powerful key stimuli and super-key-stimuli that make the soul tremble in anticipation of ill-understood events that never materialize.

Shamanism—the Earliest Religion

Bushmen stories are not yet religion. The level that I take as representative of early religion is that of hunters and herders with their *shamans*. Today, such religions are still found among Eskimos, Samoyeds, and Lapps. The shaman probably evolved from the specially gifted cave painter and musician. Such individuals must have been perceived as particularly capable leaders in the quest for spiritual legitimacy. If the dream, indeed, contains guidance to inner harmony, those were the dreamers. A dream could inspire their animal paintings and their music making. And after most of the members of the tribe had lost their dreamlike existence, the ones who still had access to the dream world became valued advisors. C. G. Jung observed the sad moment when even the dreams fail. He once asked an old medicine man in Uganda about his dreams:

. . . He answered with tears in his eyes, 'In old days the laibons[38] had dreams, and knew whether there is war or sickness or whether rain comes and where the herds should be driven! His grandfather, too, had still dreamed. But since the whites were in Africa, he said, no one had dreams any more. Dreams were no longer needed because the English knew everything!' . . . Formerly the medicine man had negotiated with the gods or the power of destiny, and had advised his people. He exerted great influence, just as in ancient Greece the word of the Pythia possessed the highest authority. Now the medicine man's authority was replaced by that of the D.C. The value of life now lay wholly in this world, and it seemed to me only a question of time and of the vitality of the black race before the Negroes would become conscious of the importance of physical power. (Jung 1965, p. 265)

That was in 1925. And the bloody terror that has descended on Africa since proves the point. Another culture has disintegrated upon contact with one sporting a more developed ego system.

Laibons, medicine men, witch doctors, sorcerers, brujos, yogis, astrologers, and shamans are all forms of the same human phenomenon at various stages of development or decline. And with tongue in cheek one might add to the list some of the modern psychoanalysts, as well as "evangelical" pushers of ecstasy.

The worldwide religion of shamanism was a huge and interesting chapter in human spiritual development. But in the context of this writing, it is transitory. Shamanistic cultures have left the state of primal happiness, but are not yet true civilizations with a religion that could contribute to the exploration of the history of reason. I will be brief on this subject and refer the interested reader to Mircea Eliade's *Shamanism* (Eliade 1964) or Joseph Campbell's popular books on mythology.

The shaman of early Stone Age hunters at first, indeed, may have been a true dreamer, a cultivator of archetypes for his people. But as such, he had power, and power corrupts. When the visions and dreams did not come, it was easy to fake them. The explorer Knud Rasmussen (Rasmussen 1930) reports on the life story of Najagneq, an Eskimo shaman, who had shot dead all the family of the girl he wanted to marry. They had objected to the union. When brought to trial by Canadian authorities, his tribal subjects refused to testify out of fear of his magical revenge.

I will assume that during the long history of shamanism, the primal connection has been gradually lost. Yet in its dim way the total brain system remained "aware" that the fabulations of the central explainer are not the proper foundation of life. Now the quest for spiritual legitimacy must have assumed new urgency. Such is the working of the archetype of the self. But since the ego system has become an overwhelming inner reality, naive magical, that is, ego-centered attempts, are made to bridge the chasm.

---

[38] Medicine men

Anything will do as long as it will make believe that our life is securely imbedded in something beyond the daily mechanics. "Spiritual leaders" have an easy task to find followers who believe. Recently I found a newspaper ad from a "shamanic practitioner" doing "soul retrieval." The ancient similarity bond makes us easy victims of suggestion, if not hypnosis. The neurological mechanism of hypnosis at this point is not well understood. However, it must have to do with the superficiality of consciousness and the power of the unconscious similarity bond. Most of our actions and perceptions are automatic, consciousness just being a simple trigger. This trigger occasionally can be replaced by the hypnotist's suggestions. Then things which are there are not seen, and things are seen that are not there.

To observe a stage hypnotist perform is instructive of the human condition. To fall into a hypnotic state is as easy as to fall asleep. One does not even need a hypnotist. Social expectations will do as well. I have observed several rituals among so called "charismatic" Christians. In water baptism there is a mental preparation and general expectation that as soon as one is dunked in the water, a state begins in which the initiate "speaks in tongues." Sometimes the speaking will not stop until the concerned pastor does what is a known technique to end hypnosis. He claps his hands in front of the face of the hypnotized. That gives the act away.

Thus, through imitation, initiation, suggestion, hypnosis, and social expectation, almost any belief can enter into the shared consciousness of a group, and become "religion," as long as it does not grossly contradict biological needs. Occasionally, though, it does, as the mass suicides of the followers of Jim Jones, David Koresh, and Marshall Applewhite have demonstrated.

Despite this negative assessment, in a functional culture such beliefs *can* provide much needed life support. One touching example has been reported by Jung. He once won the confidence of a Pueblo Indian chief and learned about the tribe's religion. They performed rituals on top of the mountain. The chief explained:

> After all, we are the people who live on the roof of the world; we are the sons of Father Sun, and with our religion we daily help our father to go across the sky. We do this not only for ourselves, but for the whole world. If we were to cease practicing our religion, in ten years the sun would no longer rise. Then it would be night forever. (Jung 1965, p. 252)

Here, shamanism ends and mythology begins. This illusion gives them dignity. It makes them feel important, needed, and satisfied that their life is in tune with "reality"—in short, it gives them spiritual legitimacy. But what a touching belief this is! The sun brings light into the darkness. "The sun is God. Everyone can see that," said the Pueblo chief. (Jung 1965, p. 251). And they are proud to help God. To make the matter perfect, according to Jung, the sun also frequently is found to be an archetypal image of the self (the brain stem's will to inner integration).

Even though religion breeds illusions, such illusions can make life easier in many ways. If life is miserable, one can hope for a better afterlife. If no purpose is seen in this bland existence, a purpose can be invented. If one has a hard time escaping the tough emotional-aggressive primate underworld, a set of moral rules can be set which can lift a group to a more agreeable mode of cooperation.

In other words, the central storyteller is not only employed to make life appear better than it is (a clear hedonic goal of the brain stem). But since he also is assembling a best, if incomplete, assessment of the realities of social interaction, rules of conduct can be added to cultural tradition. In essence, the stories are motivational. The rule-givers are assumed to be gods, ancestors, spirits, et cetera, all representatives of the ruling alpha animal that has to be obeyed. Thus, systems of moral tradition evolve.

Religion has survival value because it gives reassurance that we are on the right track. The ego system can continue to grow and the world can be mastered more efficiently.

The Four Ways of Agriculturists

People who practice agriculture find themselves in a predicament quite different from shamanistic cultures. Wheat agriculture was associated with a great leap in the strength of the ego system. This happened around 8,000 B.C. in the Near East, particularly in the valleys of the rivers Euphrates and Tigris, and in the Anatolian plain, which is in today's Turkey, and in the Jordan Valley. Vegetable culture was known before, but unexpectedly, there the tiny grains could not only provide better sustenance, but with the proper methods of cultivation and a fertile soil, they could also create a significant surplus.

All this came in the wake of the receding Ice Age. Plant life expanded and moved into new niches. Wild wheat, *Triticum monococcum*, and goat grass formed a fertile hybrid: emmer wheat, which had larger grains (Kimber and Athwal 1972). This was cultivated. An attentive early reaper of emmer patches must have noticed that some of the oars turned out exceptionally large, with still heavier seeds. This was another accidental hybrid between emmer and a different goat grass. He or she must have intentionally selected this kind of seed for planting, since the new wheat could not spread well by itself; the heavy seeds were not scattered by the wind. The new wheat and the invention of the plough, drawn by the previously domesticated aurochs, became a turning point in cultural history. The surplus could be conveniently stored without spoiling, and would be available all year. Through his inventiveness and planning (Knowledge I), man seemed to have gained independence from nature's toils.

Indeed, this invention was such a great success that according to A. J. Ammerman, L. L. Cavalli-Sforza (1984), and Colin Renfrew (1989), agriculture and the Anatolian agriculturists began to expand all over today's Europe around

6500 B.C. They displaced the local hunting cultures and their languages. Their language was the Proto-Indo-European.[39] Thus, late in the Stone Age, they came to resemble the American pioneers, who had unbounded space to claim and a small, ill-equipped, indigenous population to displace. They first moved northwest, through today's Greece and the Balkans, and then into Germany and France. Then they turned east into today's Russia. Only pockets of indigenous culture and language remained—today's Hungarians, Estonians, Finns, and Basques. All the others from that point on spoke Proto-Indo-European, which later diverged into the modern languages of Europe. According to Renfrew, this movement simply came about because sons of farmers spread out into adjacent spaces. It is estimated that in this way Scandinavia could have been reached in fifteen hundred years. The expanding early Indo-Europeans had a pioneer life and an interesting but little-known mythology, and they did not acquire writing for a long time.

Some linguists and geneticists go even further (Renfrew 1988). Three agricultural subgroups have to be distinguished in the larger Near East area, each with a different, but originally related language. One was in the already mentioned Euphrates and Tigris Valleys, which later spread southeast into the Indus Valley and beyond. The languages originating from this group are known as "Dravidian." This culture was later overrun by Indo-European tribes pressing southward from the Russian steppes and permanently invading India. The other was in today's Palestine (with Jericho as an archeological site), which spread west into north Africa and south into the Arabic peninsula. The "Proto-Afro-Asiatic" language of this group diverged into ancient Egyptian and the Semitic languages (Arabic, Aramaic, Ethiopian, Syrian, and Hebrew).

The population that stayed behind in Mesopotamia, however, developed the first high culture. And that must have come about as follows.

Grain can be bartered. Soon, specialists in all kinds of crafts and trades arose who did not participate in agriculture themselves, but traded their goods and services. Their special skills further improved the lot of all. For instance, the Sumerians of Mesopotamia already had a plough with a seeder attachment. Metalworking, pottery, masonry, and weaving all began to flourish. Most advantageously, the crafts had to be concentrated in centers of commerce—the town was born. One might think a golden age had dawned with ample leisure and enjoyment of life.

But quite to the contrary, instead of enjoying leisure, the surplus was committed as soon as it appeared. Such is the business of the ego system. Agricultural centers are magnets for marauding hordes—an army was needed. An overlord had to command it and settle disputes over land and water. Roads and irrigation systems had to be built. And once the overlord was settled, he wanted to fulfill his primate heritage and launch a war or two on the neighbors. And the peasant had to deliver more grain to feed the army—or else. The most pronounced building in ancient walled cities was the granary with a loading ramp, preferably

---

[39] Later tribes of this new culture went back to the south and conquered India, hence the name "Indo-European." As the ruling class in India they acquired the Sanskrit name "Aryan" meaning "the noble ones"; this is a word now synonymous with racist misuse.

located outside the walls, where the humble peasants came with ox carts to deliver. Oppression began, and overall misery increased despite all advances. And if the crop failed, it was the ultimate disaster.

In a vicious cycle, the failing culture contributed to further spiritual decline. The ego system became isolated and out of control. Brain stem compassion fell back to the uncultured condition, encompassing only the close family. Violence and fear spread. In other words, the neurological dualism rapidly widened. The peasant was not any more his own master, and his master became a slave of his own hypertrophic ego system. What the Bushman society had so well accomplished—cultural compensation of the dark side of primate heritage and plenty of leisure—was mostly lost. Archetypes were no longer cultivated. A dark age began instead of the golden one, and the appearance of a new caste of expensive specialists, the priests, did not improve matters much.

All this happened in Sumer, the land in the middle section of the rivers Euphrates and Tigris, in today's Iraq. There is a cultural continuity from Sumer to Assyria and Babylon, all in the same region. The first written poetic account in the first high culture based on agriculture appeared here—the Gilgamesh epos. We remember that art and literature help in the process of inner integration. True mythology began here because closing of the dualism had become a most urgent matter.

Yet this epos is one of the gloomiest of all mythologies. Gilgamesh is a hero whose main preoccupation is to seek immortality and have it denied again and again, until the abrupt end of the story. Man has to be the eternal servant of the gods and be at their mercy. This is the pessimistic quintessence of Sumerian life experience. A forerunner of the author of the Book of Job wrote:

> "My god, you who are my father who begot me, lift up my face,
> Like an innocent cow, in pity . . .
> How long will you neglect me, leave me unprotected?"
> (Hawkes 1973, p. 202, The Lord of Wisdom text)

> "O Lord, . . . great are my sins. O god whom I do not know, great are my sins . . . O goddess whom I do not know, great are my sins . . . Man knows nothing; whether he is committing sin or doing good, he does not even know . . . O my Lord do not cast thy servant down. My transgressions are seven times seven; remove my transgressions."
> (Eliade 1978, p. 69, penitential psalm)

And these gods had to be continuously placated. Here is an excerpt from the menu of a priest's duties:

> The priest shall mix wine and good oil, and shall make a libation to
> Anu, Antu and all the gods. He shall smear some of it on the door

sockets of the gate of the sanctuary and sacrifice a bull and a ram to Anu, Antu and all the gods. He shall spend the night [there]. The door shall not be shut. He shall offer the meal to all the deities dwelling in the court. In the first watch of the night, on the roof of the topmost stage of the Resh temple . . . [the priest is to recite two prayers, the opening lines of which are given] You shall prepare a golden tray for the deities Anu and Antu of heaven. You shall present water for the hands of Anu and Antu of heaven, and then you shall set the tray, serving the bull meat, ram meat and fowl . . . (Hawkes 1973, p. 196)

Why this preoccupation with gods, with placating them, and with death? In Bushman culture, gods were marginal, and death was just a natural event. Lorna Marshall (1962) has tried to give an account of !Kung religious beliefs. The matter turned out to be difficult because they were changeable. Reading Marshall's work and the old collections of Bushman stories by W. H. I. Bleek and L. C. Lloyd (1911), and Bleek's daughter Dorothea (1924), one gets the impression that a much better description of what they tell is just stories, as a "naive" central explainer and storyteller would produce. They are extremely charming children's stories about all kinds of animals. But the animals do not even have a spirit or soul; they are just imaginatively and lovingly described. There seem to be two gods, one greater in the East where the sun rises, and one lesser in the West. But they may also just be "captains" of other peoples who live there. Their names are avoided, as if in fear, but in a similar way the name of the mother-in-law is avoided. A characteristic story about life after death is simple:

> The hare cried to the moon about his dead mother, and the moon said that she would "return living," which the hare did not believe. For this his lip was split forever, and worse than that, it was declared that: "And they who are men, they shall altogether dying go away, when they die." (Bleek and Lloyd 1911, p. 59)

P. J. Schoeman (1957) reports a Bushman consulting the grave of his father for advice. But this turns out just to be a meditative technique to make a correct decision. Clearly there is not much of a preoccupation with death or life after death. Certainly no trace is to be found of any pitiful dependence on any god.

So why the gods and why the desire for immortality in Sumer? Because after the mind is split into the ego system on one side, and the remote and undeveloped unconscious personality on the other, it appears that life is not lived as it should be. There is no consolation at the end that a good and joyful life has been lived. To have missed the only chance is hard to swallow. Besides, even without any psychic condition, life simply has turned out to be miserable. To believe in life after death becomes a great consolation. And who is to blame for the misery? It must be a power beyond human reach, superhumans—the gods. Moreover, the neglected archetypal images are ready in waiting, wanting to be projected on the mother goddess or fertility goddess, the god of war, the capricious seductress, the heroic god over all the other gods and a long line of others. Thus, here the gods must have come into existence through a response of the central explainer and storyteller to a misery of unknown origin. The ingredients (pieces of the puzzle) that had to be assembled must have been the following:

a. The primate heritage states that misery comes from loosing an aggressive struggle. Your only hope is the instinctual submissive gesture that might placate the enemy.

b. When no tangible enemy is found, one has to be invented. It has to be of human or animal form and powerfully threatening, because that is where threats normally come from. Thus, gods of people who live in misery will be threatening.

c. When misery persists, instinctual submission leads to super-submission to the all-powerful God. But this device still does not work. The perceived aggression does not stop, hence, the endless and senseless compulsive actions, "rituals" that try to placate the victor. This super-submission is a response to super-key-stimuli. We have noted earlier that super-key-stimuli do not exist in nature. But this kind of nature already had been left behind long ago. Now we are in a high culture. Ultimately the super-submission leads to human sacrifice in order to placate the gods. The victims may even participate willingly, thinking to do a good deed.

d. Archetypal images are standard ingredients in explanatory stories. The "wise old man," or the alpha animal, provides a useable frame for a concept of God. However, at first it is an immature, oppressive aspect that prevails. Similarly, problems of sexual attachment in a chaotic social order are blamed on a goddess derived from the negative anima. And so it continues. For every ailment there is a scapegoat, but one that cannot be done away with.

This, briefly, is the human situation well after the fall from grace had occurred. And that there was such a fall, still is dimly experienced as guilt, because this fall is nothing but an internal condition of which, unbeknown to our central explainer, our unconscious personality still has a working knowledge.

Yet, despite all the problems of psychological maladaptation, living by the ego system has its benefits. After several thousand years of agriculture, Mesopotamia gave us the first script, the cuneiform characters. It gave us the first systematic astronomical observations, it invented algebra, and it gave us the first civil law and a system of secular courts.

Among the lawgivers were the kings Ur-Nammu (2100 B.C.) and Hammurabi (1800 B.C.). Around Ur-Nammu's time the legal system became separated from the temples. Later there was a kind of administrator or minister of justice (*ensi*) at the king's court. One of them, Urkagnia, proclaimed a reform based on the principle that "A national king might act not only in the name of his city god, but also in that of the god of justice, the sun god Utu (Hawkes 1973, p. 170). The king was Hammurabi. The monument on which Hammurabi's laws were inscribed in stone shows the king standing "reverently" before Utu (ibid.). The sun god of reason and enlightenment now gives the law, not the chief priest. This is an immense progress. And we should note that the cause lays rather in a further developed ego system, than in one that has been suppressed by religious tradition.

The judges could be merchants, scribes, augurs, archivists, or the ensi himself. Judgeship was a part-time occupation. There were clerks of the court, and a scribe took notes of the court's proceedings. Oaths were administered.

But I am running ahead of my own story. As mentioned before, according to Colin Renfrew (Renfrew 1989), the original agricultural populations of Mesopotamia, Anatolia, and the Jordan valley spread out into new niches. One would expect that each of them developed their independent responses to the pressing need to give spiritual legitimacy to their new agricultural pursuits. Cultural divergence must have been further promoted through interaction with different indigenous populations. If one adds the separately evolved Chinese agriculture to the lot, there must have been four different ways to master the psychological challenge posed by agriculture. Furthermore, there is a worldwide tropical belt where primitive agriculture evolved but remained in an early stage, probably because of the oppressive climate. I will now briefly outline the four different branches on the phylogenetic tree of cultural evolution, and their separate ways toward spiritual legitimacy.

a. *The Indo-Europeans.* Among the four, the peoples who entered the Euro-Asiatic steppes came into the most fortunate circumstances. Colin Renfrew (1989) believes that their unprecedented expansion was peaceful. This probably is not quite accurate.

The indigenous populations which the newcomers encountered were inheritors of the original "Garden of Eden," a well-developed hunting society, close to nature, with shamans and cave painters providing spiritual legitimacy. Whether it was the taking possession of wide spaces, the influence of the indigenous hunters, the good climate, or just the optimism of a pioneering society, the gods of the Europeans did not become punishing gods, or gods providing psychological guidance to the departing soul, they became cosmic nature gods—sky gods. Nature was conquered by man's wit and cunning (the ego system). Nature was stern in its way if you made a mistake, yet yielding if approached correctly, and so were the gods who made the spectacle of nature happen. Here, there was enough elbow room, time, and external success, so as to evolve a strong ego confidence. Too strong, one might say, because a few thousand years later some of the tribes came back south to conquer. Then they were warriors on horseback, and were irresistible. Between 2300 and 1900 B.C.

they ransacked Greece and Asia Minor, took hold of today's Iran, and by 1200 B.C. were firmly established in India.

Wherever they arrived, the new territory was "cosmicised" (Eliade 1978, p. 190) by performing rituals that included the construction of a fire altar to the god Agni (fire). By this, the land was being included into the cosmos of their gods, and thus made rightfully their own. The gods of this cosmos are paternal. The word for sky was *deiwos*, which became *deus* (meaning "God") in Latin. "The (god of the) sky is supremely the father" (ibid.). In Sanskrit his name became "Dyaus-pitar," in Greek, "Zeus Pater," and in Latin, "Ju-piter." Among the storm and wind gods are Donar or Thorr in German and Tanaris in Celtic. He made thunder and lightning. "Master of the wind" was Wejopatis (Lithuanian), who in Iran and India became Vaju, by now the supreme god. Wind was the cosmic spirit who moved all, not just among the Europeans. Let us not forget the absence of all physical understanding among our forefathers. When the storm moves trees, water, and tries to topple people, that is an uncanny power. In dream symbolism the wind has the same meaning, even to someone who is ignorant of mythology. In Egyptian tomb paintings, the souls of the dead have a little sail in their hand, to symbolize that they now are under the influence of the all moving spirit.

It is remarkable that the Indo-European pantheon is predominantly male. The former agriculturists of the Near East had worshipped all the female fertility and mother goddesses. Seeding had sexual connotations—the earth goddess is fertilized, et cetera. But now it is the father who manages affairs through his rational powers. I interpret this to mean that here for the first time the ego system has gained enough objective insight, so as to link real cause and effect in matters of survival in the new niche of agriculture, animal breeding, and land ownership. Land ownership is a satisfying experience, and a strong incentive to attach the ego system to the land and its management. We remember animal territoriality, which also is a predominantly male affair. In contrast, the Sumerian peasant probably had not felt that he owned the land. For him the gods, the priests, and overlords were the true owners. He had to placate the fertility goddess to be successful. The pioneering Indo-European, however, simply did his own thing—not because of racial superiority, but because of fortunate circumstances.

b. *The Indus Valley Civilization and India.* The ruins of the cities of Mohenjo-daro and Harappa testify to a strange culture which on first inspection left no traces because it was destroyed early on by European tribes on the way to conquer the Indian subcontinent. But here was an early high culture with its own pictographic script, likely a result of early Mesopotamian agricultural expansion (Renfrew 1989). To this day the script has not been deciphered. Unearthed bronze figurines and anthropological examinations of skeletal remains show a significant subpopulation of Australoid stock. Therefore it is likely that the preagricultural inhabitants of the Indus Valley region were of this race of which today still pockets exist in India.

Among the figurines and clay seals one finds a god or priest in the typical cross-legged yoga meditation pose. This is the first known appearance of this pose. One is tempted to conclude that here is to be found the origin of the meditative practice, which later on has come to dominate Indian spirituality. This path to spiritual legitimacy proved to be so convincing that even the European (Vedic) cosmic gods of the later invasion were not able to ever displace it. Mircea Eliade writes:

> As we have pointed out, the absence of the Yoga complex from other Indo-European groups confirms the supposition that this technique is a creation of the Asian continent, of the Indian soil. If we are right in connecting the origins of yogic asceticism with the protohistorical religion of the Indus, we may justifiably conclude that in it we have *an archaic form of mystical experience that disappeared everywhere else.* For—to repeat—yoga cannot be classed among the countless varieties of primitive mysticism to which the term shamanism is commonly applied. Yoga is not a technique of ecstasy; on the contrary, it attempts to realize absolute concentration in order to attain enstasiis. In the universal history of mysticism, classic Yoga occupies a place of its own, and one that is difficult to define. *It represents a living fossil, a modality of archaic spirituality that has survived nowhere else."* (Eliade 1958, p. 34, italics added)

It could not have survived anywhere else because this was the specific way of only the Dravidian peoples. How long can a culture live under the illusion that misery comes from oppressive goddesses and gods? At some point this never-ending belief must become outworn by the lack of positive reinforcement. Among the Europeans the gods became more distant as the action oriented ego system became successful in managing affairs. This may be called the "extraverted solution." The Dravidian way was introverted. If some unknown power prevents you (the ego system) to take active control of the external world, try to work on the inner world, which in old times was so much more at hand than it is today.

All kinds of psychic adventures must have been known from preagricultural shamanistic times. Priests knew the experience of intense prayer. Moreover, the climate was hot, and ego-directed physical labor was exhausting. The natural goal of religion still was to escape the inner void of the neurological dualism. The Dravidians discovered that this could be approached in a purely internal way. Here is not the place to discuss yogic techniques. But they certainly have been found to succeed in altering not only the pattern of global brain waves; they can also achieve control over autonomous nervous systems, regulating heartbeat, body temperature, and metabolic rate. "Spiritual" inner experiences are produced, and collectively reproduced. They testify to deep upheavals these techniques cause in the domain of the systems of the unconscious personality.

"The ideal of Yoga," writes Eliade, "the state jivan-makti, is to live in an 'eternal present,' outside of time. 'Liberated in life,' the jivan-makti no longer possesses a personal consciousness—that is, a consciousness nourished on his own history—but a witnessing consciousness, which is pure lucidity and spontaneity." (Eliade 1958, p. 363)

Not "possessing personal consciousness" is akin to eliminating the cumbersome ego system. Living in the "eternal present" excludes active pre-imaginations of ego-directed activities, a sport that must play a major role in the ego's attachment to the world. Instead, the ego system is dissolved by yoga practice.

A present-day anecdote illuminates the yoga way in a flash. Oliver Sacks (1992) reports in the New York Review of Books on the case of a Greg F., who joined the Society for Krishna Consciousness and took residence in their temple. In time he became to be considered an "illuminate." His swami said "he is becoming a saint" because of his permanent beatific "transcendental" smile and withdrawal. He also became blind. When his parents succeeded in removing him from the temple, it was found that he was suffering from a very large brain tumor that had destroyed large parts of his brain, including the optic nerve. His severe amnesiac syndrome had been taken for holy withdrawal from the world. His constant smile reflex and the chanting was all that had remained of his personality, which now, indeed, was free of ego. In the eyes of the brethren this was taken as high spiritual achievement.

The ascetic way is devolutionary. It stops cultural evolution of consciousness. It may give inner peace and freedom to the practicing ascetics, but it deprives the rest of the population of guidelines, rendering it helpless, for instance in dealing with the archaic fertility goddesses of primitive agriculture, which for that reason have survived in India to this day. Nevertheless, in the general population that lived in villages, withdrawn Dravidian spirituality and Aryan (Vedic) gods formed a stable synthesis. This became known as "Hinduism." It resisted all later invasions by Macedonians, Mongols, Moslems, and Christians, and all overlords up to the British viceroys, and it still survives in the relatively happy state of the simple people of the Indian village. C. G. Jung has called this the "dreamlike world of India" (Jung 1965). It is static, eternal, and self-centered.

This brings us to the undeveloped religions of the tropical belt, of which India had a share; these religions flourished there, unhampered by the spiritual elite of self-centered yoga practitioners.

c. *The Tropical Belt.* This is not one of the four branches, but rather the trunk of the phylogenetic tree of cultures. In the tropical belt the dynamics of ego system development forever remained stuck in its earliest stages. Steaming jungles and lukewarm islands are not conducive to vigorous, ego-directed activities. If agriculture developed in such areas, it remained under the spell of the great goddess or of gods of fertility. When psychic development is hampered, the archetypes remain in their raw state. Submission appears to be more important than compassion. The practice of human sacrifices emerges as a form of super-submission. Instead just presenting the jugular vein, as wolves do to announce their submission, it actually is severed. Better yet, the heart is torn out alive. In this way thousands were regularly slaughtered in middle America to placate the gods and insure fertility.

In India a certain creative ferment was initiated by the arrival of the well-developed Vedic gods. Nevertheless, the terrible black goddess Kali has remained in power to this day. Human sacrifices took place in droves until they were stopped by the British Colonial Administration in 1835. Today, thousands of "brides" are still burned alive annually by husbands who are not satisfied with them. In old times the victim's flesh was often torn from his body while still alive, and then ceremonially buried in the fields to insure fertility. Altars had a built-in drain for blood. The victims were mostly male since that was more pleasing to the "great goddess." Joseph Campbell writes:

> . . . among the gentle-eyed Burmese little children purchased for the purpose used to be sacrificed annually at a festival in August, to ensure a hearty crop of rice. A rope having been placed round his neck, the victim was taken to the houses of all the relatives of his purchaser. At each house a finger joint was cut off, and all persons in the house were smeared with the blood. They also licked the joint and rubbed it on the cooking tripod. The victim was then tied to a post in the middle of the village and killed by repeated stabs of a spear, the blood from each stab being caught in a hollow bamboo, to be used afterwards for smearing on the bodies of the purchaser's relatives. (Campbell 1962, p. 163)

To this day the screams of the tortured around the world testify to the tortures the human psyche has to undergo in the process of evolution of consciousness. It is mainly in the temperate regions of Europe and Asia, where previously the great hunt and the first "Garden of Eden" had flourished, that the decisive step in cultural evolution could be taken. Of this, today's Western culture is the fortunate heir.

d. *The Nile Valley and Tribal Religion*. The wheat agriculturists who moved into north Africa and up the Nile valley developed the Proto-Afro-Asiatic language. As compared with Mesopotamia and the Indus valley civilization, Egypt was luckier, since it was protected by a vast desert. The religion of Egypt became even more preoccupied with afterlife. But this was a just afterlife that was viewed much as a continuation of present life. Agriculture was easier along the Nile, and the early gloom and doom of Sumer did not repeat itself. Here, no big cities arose.

Essentially there was a widely stretched out agricultural population, under the Pharaoh, who was considered to be the god Horus himself. And since this god was visible and within reach, he could not be too bad and merciless. The rest of the pantheon also is rather benevolent. Isis can have winged arms that spread out protectively over the dead. Osiris is periodically resurrected. The jackal-headed Anubis guides the dead on their journey to the other world. Hathor is the deity of love, happiness, dance, and music. Often she is represented by a cow, and always she has a sun disc between the horns or on the head. Love and happiness are sun-like, enlightened, and witty.

But the main sun god was Re. On his head is a sun disc encircled by a cobra—the unity of spirit and nature, or shall we say of the bright nature of ego-consciousness joined by the reptilian unconscious? And the Egyptians considered themselves to be "cattle of Re."

The human features expressed in murals and reliefs are warm and personal. That religion has progressed since Sumer is obvious. This well-regulated afterlife is a good solution to the desire for immortality. Under this religion the Egyptians were hardworking and content. Yet it was a static and bureaucratic society. Everything was regulated and all transactions were recorded. The son had a life exactly like the father, and the daughter like the mother. The gods of Egypt became benevolent because life was easier. No excessive threat was experienced that warranted super-submission. The godly alpha animal now could be seen in its role of protector. Yet the desire for an afterlife still is there because life remains hard work. And who wants to die? Besides, there remained a caste with vital interest to maintain their own good life by keeping the doctrines alive.

Man's creative spirit, as well as the strength of doctrines, is illustrated by one important historic interlude: In 1385 B.C., the Pharaoh Amenhotep IV, for the first time in human history, does away with the panoply of gods and introduces monotheism. Significantly, his god is the sun god Ra. To radically break with the past, Amenhotep renames himself "Akhnaton" and even moves the capital. The sun disc, the archetypal image of the self, is put at the center of religion! We remember: this archetype drives us on to individuation. Most depictions of the old gods are destroyed. Akhnaton was a spiritual genius, but too far ahead of the spirit of his time. The population sheepishly followed the new leader, proving to be forever faithful receivers of doctrine, but not understanding much. The episode abruptly ended with Akhnaton's death.

Persistent opinions have it that Moses had lived at Akhnaton's court. But the time does not seem quite correct. Moses, if there ever was such one person, probably lived at least one hundred years later. But he certainly was influenced. From now on, the history of religion begins to converge on monotheism. Even though positive historical documents are hard to find, it seems to be quite certain that with Akhnaton a new religious idea began to spread. At first it must have inspired Zarathustra, and then in succession Judaism, Christianity, and Islam. All of those originated in the vicinity of Egypt, the last three among Afro-Asiatic tribes.

What is characteristic of the religion of the Afro-Asiatic group is the tribal nature of the top god and of the one God of Judaism. The Indo-European gods were cosmic sky gods. In later times they became more and more remote. The yoga way was only self-centered. But Horus was the founding father of a tribe, its possession, so to speak. In the incarnation of pharaoh, he became the direct supreme tribal leader to whom gigantic pyramids had to be built. Horus, the solar sky falcon, already appears on the earliest known work of art in Egypt on the upper Nile (Campbell 1962, p. 49). There seems to have existed a "line of kings devoted to Horus." His first origin likely came from a previous hunting culture, a family or tribal totem, which became the tribal god under whose tutelage lower Egypt was conquered. Thus, all the Nile population became united under this symbol.

"The founding of the united state," writes Eliade, "was the equivalent of a cosmogony. The Pharaoh, incarnate god, established a new world, a civilization infinitely more complex, and far higher, than that of the neolithic villages." (Eliade 1978, p. 86)

The belief in your own all-powerful tribal god must be an obvious assurance of spiritual legitimacy. It could be imagined that ego-directed activities were in reality, performed by "me and my god." Existence under such conditions takes on a flavor of powerful magic. The tribal god can be persuaded to make things happen, and since he is one of the tribe, this tribe as a whole can perfect things that are not ordinary. Typical of this attitude was the Hebrew "covenant" with Jahweh. Originally an unruly and cruel nature force, he could be magically tamed by striking a bargain. "We will behave according to your rules. You will be our only god. In return you will favor us as a tribe before others and behave yourself." Thus, the chosen people came into existence by a magico-legal transaction.

The beginnings of this can already be seen in the super-submission of Sumer. But as the ego system solidifies, and the god is discovered to be capricious, the rational bargain seems to be one natural way to deal with the situation.

The unfortunate thing about tribal gods is that they reinforce the primate heritage of tribal chauvinism. This century's bloody terror in the Near East is a late result of such tribal religions which have survived in the region to this day, to testify to the great power of this ancient illusion.

The covenant on Mount Sinai, of course, had been precipitated by Moses, who possibly had grown up at the Pharaoh's court and had an Egyptian name ("Moses" means "born" or "son"). That is why I let this stage in the history of Judaism follow the discussion of Egypt. But the history of the Hebrew tribes starts much earlier, when there was a first covenant of Abraham which did not yet have the flavor of a bargain.

The Hebrew tribes were not the original agricultural population of Palestine. According to C. Renfrew (1983), this had been one of the three Near East "lobes" of first agricultures, speaking Proto-Afro-Asiatic. But the region was several times ransacked by "barbarian tribes," mountain rustics "who do not know wheat." (deVaux 1978, p. 64), (Eliade 1978, p. 149) The invaders, all speaking Semitic languages, adapted to the local agricultural way of life. The tribe of Abraham was one of the late arrivals from Chaldea in southern Mesopotamia. These ancestors of the Hebrews had been either donkey breeders and caravan merchants or herders. They, too, adapted to agriculture, but firmly held on to their religion. This included blind obedience to their tribal "God of Abraham," as exemplified by the legend of Abraham's willingness to sacrifice his son Isaac for no other reason but the demand of his God. This absolute faith and unquestioned submission of Abraham to his tribal god became the hallmark of Judaism, and later of early Christianity. But in Islam it has remained strongest of all.

e. *Ancient China.* According to Ping-ti Ho (1975), the development in China was independent. But there are similarities with the history of the Near East which testify to the universality of the psychic upheaval caused by agriculture. Proto-Chinese lived in the highlands around the main tributary of the Yellow River, the Wei River. This is called the "Yang-shao culture." It appears to have been centered around today's city of Sian. Here, millet was first cultivated around 5000 B.C. The varieties which have been found at archeological sites, and radiocarbon dated, were *Setaria italica and Panicum miliacum.* In a development that paralleled the events in the Near East, this culture later spread out in all directions. *Setaria italica* is known to be suitable for dry conditions (King 1966). And that was quite necessary, for the climate and soil in the original Chinese agriculture was quite different from the flood plains of the rivers Euphrates, Indus, and Nile.

The area was heavily covered with loess, a fine-grained mixture of clay, quartz sand, and limestone, which had been deposited by wind in the aftermath of the Ice Age. The natural vegetation was that of a dry steppe. The original sites were terraces well above the river valleys, safe from the devastating floods. Irrigation was difficult and was not attempted for millennia to come. The important advantage for Chinese agriculture was that loess can be self-fertilizing due to the chemicals it contains, as long as organic matter is allowed to decay in it. That requires periods of fallow. Much later in the Chou dynasty, a well-developed system of land rotation was recorded in the Book of Documents, (Ho 1975, p. 51). Millet and the later rice are not as easy and abundant crops as wheat, and the soil was difficult. If the traditional system of culture was not followed, crops slowly diminished. Moreover, no animals were available for plowing. All of this was backbreaking manual labor.

Here seem to be the roots of the Chinese preoccupation with changing cycles in nature and with the need to be in harmony with them, which is a main theme in later Taoism. Any mistake resulted in disaster, because the population expanded rapidly. The very existence of this population depended on the "correct" way of doing things. It is probably to record the correct way, that Chinese script evolved right here (Ho 1975, appendix IV).

Compared with Mesopotamia, one notices the absence of the gloom and doom of oppressing goddesses and gods. In fact, most mythologists complain about the lack of the usual pantheon. However, archeological finds seem to indicate that preoccupation with life after death was strong. Food and utensils were included in graves. In later times, little houses for the souls of ancestors were built. The cult of ancestor worship seems to have been present at the very beginning. How is it possible that no gods were blamed for life's misery? Actually, Chinese farmers must have had a much harder life than the wheat farmers of Mesopotamia. Was it the well-known Chinese character of cheerful diligence that already was at work here? For me, the way of the Chinese agriculturists remained a puzzle until I chanced upon the following passage by Ping-ti Ho (1975):

> The Chinese peasant (in Thailand) had a definite place in the temporal continuum of kin. Within the extended kin group—dead, living and yet to be born—he looked to the past as well as to the future: he was not only grateful to his ancestors for what his immediate family had, but was responsible to them for what he did to further the fortune of his family and lineage. His world view was, therefore, historical and kin-centered, and in this context his industriousness and thrift served ends transcending his individual life. *His primary goal was not individual salvation, but lineage survival and advancement.* Protracted labor and extreme thrift were the means to these strongly sanctioned ends. (italics added)

Ping-ti Ho assures us that the above description applies "equally well to the average adult Chinese in modern China." That seems to be the key—not individual salvation, but an important place in the temporal continuum of kin. Apparently one can achieve spiritual legitimacy by cultivating the compassionate bond and the bond of kinship. Under such conditions the individual must feel just as one link in the endless chain of generations whose future well-being during his own lifetime is placed under his stewardship. Ancestor worship extends this chain into the past. The logic goes like this: If my ancestors, likewise, prepared the way for me, they have to be honored, and in honoring them, I honor and give purpose to my own tireless efforts which, in turn, will be appreciated for generations to come. My suffering counts little before this multitude of past and future kin. Later, when titles of nobility were conferred, the degree of honor was not measured as to whether the title was heritable, but how far back the ancestors were to be honored also.

This seems to be a happy way to spiritual legitimacy. "Me and my ancestors and children" seems to be much better than "me and my god." A good old and real animal instinct is cultivated instead of an invented illusion. Development of the ego system is not fostered much by the Chinese way. But the result is an extremely successful culture that multiplies rapidly and happily on limited resources.

In a later development, a sky god, Ti, became the first ancestor of the ruling imperial family. And this family derived its legitimacy from the belief that only its members could converse with their ancestor Ti. Thus, heavenly order and law

(Tao) descended upon the correctly ruled empire. It is not astonishing at all that upon first contact, westerners were considered barbarians and their technical achievements unnecessary. But this heavenly culture collapsed upon closer contact with the West. Beautiful as it may have been, it had a weak ego system. Thus, it was bypassed by the even higher Tao of evolutionary dynamic.

In conclusion, the four ways of agriculturists can be summarized as follows:

At the beginning there was uncertainty and fear. In reaping an uncertain harvest, unknown powers were encountered. Was this kind of pursuit of goals dangerous? Was the failing crop a revenge of someone whose domain (territory) had been invaded? Seeding and reaping was as uncanny as procreation, birth, and death. Hence, the role of a goddess or goddesses which had to be placated.

Gaining certainty, legitimacy, and alleviating fear was accomplished in the four different ways which were built on given constraints of the evolved psyche, climate, soil, and indigenous population.

The *Chinese Way* was built on the confidence derived from an amplified compassionate bond to family and communality with the inclusion of ancestry. A whole lineage could not be wrong in doing what had to be done.

The *European Way* was heroic. The hero archetype derives from male territorial and extra-tribal aggression. It stresses the male companionship of mutual tribal allegiance in war. In the heroic approach fear is overcome by action.

The *Afro-Asiatic Way* relies on a tribal god protector and first founder with whom one can bargain and who will not forsake his tribe. This is tantamount to a "cosmicization" or cultivation of the alpha animal, or the wise old man archetype. The ego directed pursuits are accomplished in his name.

The *Dravidian Way* builds on meditative techniques which open up an illusion of an inner world in which ego system and biological wisdom are one. This happens under the spell of the archetype of the self which undergoes a single-sided inflation. This results in a withdrawal from the world and can only be maintained if underclasses exist who do the work, hence, the class society in India and the multitudes of undeveloped animisms and gods of the Indian peasants.

Early Syntheses

There have been geniuses with unusual insight into the crisis in the evolution of consciousness. Objective insight was not yet possible, but they grasped the dynamics and constraints of the psyche in intuitive, prescientific terms and had prescriptions for a more agreeable life. After all, in the unconscious personality a working knowledge of the human dilemma must exist somehow and somewhere. Similar to the invention of tools through spatiotemporal pre-imaginations, there must exist a pre-imagination in the inner domains of attitudes and their social and existential effects. By coincidence there arose two exact contemporaries, Confucius (551-479 B.C.) in China and Buddha (circa 550-470 B.C.) in India, who left their mark for millennia to come. Both were rather indifferent to beliefs in gods. Interestingly, the somewhat uncertain year of Buddha's death, also was the birth year of Socrates (470-391 B.C.). They all preached human self-perfection. Thus, they became forerunners of modern humanism. They ushered in a period of creative turmoil, called the "axial age."

It would be nice if each of the four original agricultural societies had produced one such overwhelming figure. But the lines of history seldom are straight. For instance, Zoroaster (or Zarathustra) was a prophet whose origin has to be seen in the ferment that was caused by the Indo-Europeans invading Iran. Jesus Christ in turn drew on another boiling cultural transformation, whose reactants came from Zoroastrism, Judaic tribal religion and Hellenism, which itself was another late reaction product of invasion from the north. The phylogenetic tree of cultures, unlike the biological one, has branches that may rejoin again after having branched out earlier.

I will begin with Confucius. He often is considered a philosopher rather than a founder of a religion. But I have to agree with Eliade, who said, ". . . directly or indirectly, Confucius profoundly influenced Chinese religion. In fact, the actual source of his moral and political reform is religious." (Eliade 1982, p. 23) "No man did more to impart enduring shape and character to a civilization than Confucius," writes F. M. Mote (1971). Confucius did all this without ever having intended it, and he died in the belief of having not accomplished anything. Yet, two hundred and fifty years later, the founder of the Han dynasty entrusted the administration of the whole empire to Confucians, an arrangement that was to last for two thousand years. Confucianism had become the longest lasting state religion ever.

What was his secret? According to Mote (1971), there were three pillars. At first, through his example, he instituted the role of private teacher in a system that did not have any schools or professional teachers at all. Secondly he created not only the methods and ideals of education, but also most of the contents of such education. Thirdly he broke down the social barriers and admitted pupils from all social classes, who after proving themselves became "superior men," worthy of the highest positions in the hierarchy of government. Thus, birthright of nobility was to be replaced by the right of the educated. Beginning with the Han dynasty, exams were administered to qualify for public service. Altogether this is a truly gigantic achievement.

But the main pillar was the content of his educational message. Civilizing spiritual growth was to be obtained through study of the liberal arts of the day. Literary works and ritual texts were studied in the context of comparative philology as well as their application to moral attitudes. History and ethical thought were studied, and what may be the right way of governing. Music and athletics were practiced. The main purpose of this study was to develop jen, the most important characteristic of the "superior man." Jen can be translated as, "goodness," "human-heartedness," "love," and "benevolence." The practice of jen means to have concern and respect for the well-being of other human beings. One obtains jen through civilizing one's inner attitudes. This does not only mean to learn a certain quintessential artistic wisdom, but also to observe form and social ritual. Man's civilization is declared the greatest achievement. This can be accomplished by our mind on its own, and it is the highest moral duty to pursue this goal. This is the exact opposite of yoga introversion. It is a closing of the neurological dualism through civilizing education that encourages well-balanced action, whereas yoga introversion is discouraging of such.

We bow in awe before the ancient Chinese culture that had reached this point two thousand years before Erasmus of Rotterdam. Taoism, the belief in the law of eternal cycles in an unchanging nature, and Confucianism, together, succeeded in keeping the ego system in check, so that natural harmony, "correctness" and well-being could be maintained.

All the early syntheses are ahead of their time. It generally is much later that their true value is understood. In the meantime, China gave much lip service to its greatest teacher, but did little to stop the burgeoning corruption at all levels. Western humanism was to become the only dynamic one in tune with the dynamism of emergent evolution. But is it a better one?

Another ancient genius of the human predicament was Gautama Buddha. Buddha means "the awakened," but he was born Prince Siddharta Gautama and lived a princely life until the age of twenty-nine. Then, disgusted with the senseless glamour at court and the willing mindlessness of his concubines, he called for his groom Chandaka, saddled his horse, and left the town by night. A dozen leagues outside, he stopped, cut off his hair, changed clothes with a hunter, and sent Chandaka home with the horse. That at least is what the legend wants us to know. "When he stopped, he also dismissed a troop of gods who had escorted him until then. Henceforth the gods will play no part in Buddha's fabulous biography" (Eliade 1982, p. 75).

After a year he had mastered the philosophy and yogic techniques of the day, as taught by several wise men whom he had followed. He then settled down in a "peaceful place near Gaya" in order to practice the "most severe mortifications." After six years "he understood the uselessness of asceticism as a means of deliverance." Horrified, his disciples left him. He then sat down in a forest under the famous tree, later named "Ficus religiosa," and vowed to remain there until he obtained "awakening." After four nights of temptations by evil spirits followed by enlightenment, he found the "four noble truths" and thus became Buddha. He remained at the site for seven weeks undergoing more temptations, then left for Benares to find his five former disciples, to convert them, and gain more followers who could be sent out to further spread the message.

What is the message? Having emerged from the introverted Dravidian way, to us extroverts they appear abstract and in need of interpretation, yet to the practitioners of yoga meditation they have intense meaning. I will follow Mircea Eliade's interpretation. At the center are the "four truths" (Eliade 1982):

1. The first truth concerns the essence of the human spirit as defined by the word *dukkha*, which means "pain," "suffering," certain forms of happiness, and certain spiritual states in meditation. One is tempted to translate this by "self." The Russian derivative from this Indo-European root is *dusha*, which means "the soul." Buddha states that *dukkha* is not permanent, and in the last analysis is the five "aggregates" or "assemblages," namely (1) the totality of material things, the "sense organs and their objects" or "appearances," (2) the "assemblage" of sensations provoked by the former, (3) the "assemblage" of perceptions and notions resulting from the sensations, (4) "psychic constructions, including both conscious and unconscious psychic activity," and (5) thoughts produced by the sensory faculties and by manas[40] which has its seat in the heart and organizes the sensory experiences. Manas, obviously, is the dynamic brain stem "will" that drives cognition and that can be transmitted through "the rediscovery of the I in the Thou," or the similarity bond. The "assemblages and appearances" describe the conscious ego system.

   In short, the five aggregates come as close to a physiological concept of the human mind before a background of physical realism, as anyone could in prescientific times. They almost sound like the neurological dualism and the central puzzle assembler. It is absolutely astonishing that purely through introspection one can come to such insight.

2. The second truth is about the origin of suffering (*dukkha again*). This is seen in desire, appetite, or thirst (which also cause reincarnation). There are three kinds: desire for sensual pleasure, desire to perpetuate oneself, and desire for self-annihilation. Of course, the brain stem animal still is at work, the pleasure centers lure, and the unresolved neurological dualism is very frustrating. (In the West this desire for self-annihilation caused apocalyptic beliefs).

3. The third noble truth states that one can escape from the pains of dukkha suffering by abolishing the appetites under (2). The result is the state of nirvana, the extinction of thirst. Here, the typical Dravidian-yogic way reappears—asceticism and the extinction of ego system wishes. But it now is tempered by the next noble truth.

4. The fourth noble truth indicates the ways which lead to the condition of nirvana. This is Buddha's "middle way" which steers clear of excessive pursuits of the pleasures of the senses as well as of excessive asceticism. The middle way is eightfold (also known as the "eightfold path"). Its characteristics are: (1) just opinion, (2) right thought, (3) right speech, (4) right activity, (5) right means of existence, (6) right effort, (7) right attentions, and (8) right concentration—all prescriptions for a life of balanced civility, enabling individuation.

[40] Manas (Sanskrit), mind, mentality, the unifying principle involved in sensation, perception, always thought in Indian philosophy as a kinetic entity, will and desire being equally present with thinking. (Runes 1984)

Buddha's synthesis is a quantum leap in human self-understanding with the aid of ego consciousness. It describes all the things that we can do by ourselves, and says that we are free to do them.

It is hard to tell how successful early Buddhism really was in bridging the neurological dualism in a balanced way. But, initially it must have been good progress in Knowledge II. Unfortunately, the set goal still remained in tune with yogic introversion. The thirstless state of nirvana is synonymous with escaping the cumbersome ego system on one hand, and blocking the brain stem desires on the other. Despite having been a quantum leap in self-understanding, it is difficult to see how Buddhism can contribute to the future of a world where objective knowledge and informed action have to prevail.

Of the four ways of the agriculturists, the ancestral, the heroic, the meditative, and that of the divine tribal leader, the last most easily can produce effective codes of ethical behavior. They can simply be ascribed to the demands of the tribal god. Moses' Ten Commandments are such prescriptions. In contrast, Confucianism and Buddhism give the appearance of intellectually constructed ethic for the elite. But since true objective insight is unknown, authoritative commandments may be a stronger medicine, particularly for the masses. This was the strength of early Judaism. Their power has prevailed almost to this day, and the history of the Western world might have been quite different without them.

Unfortunately the person of Moses is shrouded in fog. If one is to believe Freud's "Moses and Monotheism," he is a mythical composition of two different men. The first being an Egyptian from Akhnaton's court, the second a priest of a fierce desert or volcanic God, Jahweh, who received the Ten Commandments. Eliade takes him as one person. Campbell does not want to touch the issue. The Oxford historian Robin Lane Fox is not even sure he existed (Fox 1992). But whoever was involved, a synthesis emerged; not in the objective, nor in the philosophical sense. But it is a powerful synthesis of applied ethics. This synthesis is realistic to the extent that if the rules are followed, the main atavistic aspects of our primate heritage are prevented from being culturally amplified. On the other hand, instincts which are supportive of the new cultural existence are encouraged. The Ten Commandments, in their present Biblical form, are probably editions and re-editions of earlier ones, "but the most important commandments certainly reflect the spirit of primitive Yahwism" (Eliade 1978, p. 180). I will take them as a brief summary of the Mosaic synthesis. It is quite astonishing how the commandments appear to be rooted in a thoroughly realistic assessment of human nature. I will try to illustrate this for each of the ten cases.

The first four seem to be designed to motivate the acceptance of the whole idea:

1. Introduces the tribal god "your Lord" as unique and requests to ignore all others. This sets the emotional stage of the believer's relation to the tribal god. It is personal and authoritative. This god-person is to be associated with

the innate responses to the alpha animal. Thus, one is induced to become a follower.

2. Warns against fashioning images of anything in heaven, in Earth and in water. Primarily this is directed against other gods. But Jahweh himself should not be imaged either, which insures that he can be considered a powerful inner psychic experience, rather than a material thing (idol), and thus subjected to ego-power rationalization. Images foster externalization and profanation.

3. Extends this warning to the name of God. Naming something is comprehending it, becoming familiar with it, again making it an object of ego-explanation. But Jahweh has to remain absolute and threatening, a super-key-stimulus, if his commandments are to be effective.

4. So that the contract with Jahweh is constantly on everyone's mind, one day of every week is to be devoted to particularly rigid observation of rules of conduct and ritual. No work is permitted on that day.

After thus setting of the stage, the remaining six rules spell out what is forbidden or prescribed. In the preconscious world of animals, root aspects of behavior were determined by instinct. There was the similarity bond causing inhibition to injure members of the band. Mating behavior had certain patterns, even though emotional strain was present. Parents were natural role models. Imitation of their behavior was a good developmental strategy. With increasing reliance on the new ego system, however, instincts can be overruled by conscious decision. Therefore, disaster looms. The next three commandments give the clear impression of having been designed to compensate for the main destructive results of the increasingly powerful ego system.

5. Parents are to be honored, otherwise transmission of manual, mental, and social skills, and other cultural achievements is hampered. This obviously is a question of personal and tribal survival strategy. Young individuals are quick to see that their parents are mentally slower, and that they tend to resort to tradition. This may go against the youthful ego exuberance. Nevertheless, if cultural continuity and evolution is to work, the parent's ways have to be learned. Honoring the parents also helps to express the comapassionate bond

6. Forbids to kill. Ego-determined individuals easily outmaneuver their inhibition to kill, and metal knives, swords, spears, and axes make killing easy. The instinctual inhibition would still be functional if the victim was to be bitten to death. In modern times, of course, killing is even easier. Pulling a trigger at a distance or pushing a button may do the trick, and the victims may not even be seen by the murderer.

7. Forbids adultery. Our primate ancestors were not monogamous. Instinctually, Homo sapiens probably has only as much tendency toward monogamy as the extended care for the brood and the lactating mother require. But even this

instinctual pattern can be overruled by ego directed behavior. In a successful culture, though, extended parental care is an absolute necessity. Neglected children are in danger of becoming socially disabled. In turn, such children pass on their disability to the next generation. Besides, monogamous life avoids wasteful emotional stress.

The next two commandments appear to compensate for the absence of amplifiable instincts in matters that have become important in the context of cultural evolution.

8. Forbids to steal. Property is a defining ingredient in agricultural and herder societies. Several instinctual determinants may be supportive of the acquisition of property: territoriality, striving for rank, and care for the brood. Those instincts are put to positive use in a healthy advanced culture. A habit of stealing would cause intolerable social havoc. However, there does not seem to be an instinctual inhibition to steal, except possibly through a weak and indirect appeal to the compassionate bond. This would require active pre-imagination of the disappointment of the bestolen, a rather weak and abstract technique.

9. Forbids lying. The degree of interaction among tribal members for a common goal is a direct measure of cultural success. Such interaction is not possible if deception is rampant. Again, deception among primates does not seem to be biologically inhibited. It has been widely observed by Jane Goodall and others.

10. The last commandment advises to suppress jealousy over property, wife, and the success of other members of the tribe. This is a psychological advice that makes compliance with commandments 6 to 9 easier. It is mainly the ranking rivalry with its dark emotional stresses, as in baboon society, that is being culturally tamed by the tenth commandment.

This brief exposition shows that the authors of the Ten Commandments had a superb feel for the natural constraints under which the human psyche functions. This seems to be possible by introspection and social observation, without realizing that the constraints are biological in nature. They also had properly assessed the need of cultural progress for the survival of the tribe. Commandments 5 to 10 form an ingenious response to the situational logic that arises, when the security of an instinctually based order has been left behind, but objective understanding is not yet capable to take over.

Nevertheless, from our contemporary position, the motivational commandments 1 to 4 are archaic. The god is a tribal god, no individual freedom of decision is granted, the ethic mostly is restrictive instead of positive. Moreover the restrictions do not apply to individuals outside of the tribe. The Old Testament gives ample evidence that killing and enslaving of other tribes was permitted. In this respect the commandments remain within the heritage of primate social order.

But Mosaic law formed a fertile ground for further development within Judaism. I will leave it to the specialists to fight about who and when added what. But in the Jewish tradition before Christ, the imperative of love was well-developed and extended to "foreigners." Rabbi Hillel the elder is reported to have summarized the entire Torah by saying, "Do not do unto others as you would not have done unto you," a teaching to be repeated by Jesus, who also is reported to have said, "Think not that I have come to destroy the law, or the prophets: I am not come to destroy but to fulfill" (Matthew 5:17). Many scholars believe that his teachings were very much in the Jewish tradition of the day.

To this day the strength of Abrahameic faith and of the Mosaic moral system have proven themselves extremely stable in many tribes. Through Islam even a sort of renewal has occurred.

Islam's contribution to the progress of humanity had been a shining example at its time, but short-lived. As an effective, mostly tribal religion, it first had caused rapid flowering of Islamic culture and ego system development. Around A.D. 1000 this was far superior to what Europe had to offer. Hellenistic traditions were eagerly absorbed. Science and philosophy flourished. Medicine was unsurpassed anywhere else in the world. Tolerance of other religions was practiced. Much of the classical texts would have been lost to us had Islamic scholars not preserved them. But with Al Ghazzali's (1059-1111) writing, *The Incoherence of the Philosophers*, which often is translated as *Destructio Philosophorum*, a reaction of mysticism began to sweep away all application of human reason.

Mysticism, of course, is the introverted solution already encountered in the Dravidian branch. It may provide charming inner experiences, occasionally even a measure of wisdom and tolerance, but it prevents the conquest of reality. In 1150 the Caliph Mustanjid of Baghdad ordered all philosophical works burned. This was the year the University of Paris was founded. Muslim Spain maintained a freer spirit for a few more years. In Seville the books were burned forty-four years later. Since that time Islam has condemned itself to a marginal, if not atavistic, role in the evolution of universal culture. It has become a religion of absolute, anti rational submission.

Through the many Jewish communities within the Islamic empire, this mysticism also infected Judaism. Being incompatible with the scientific world view of the West, the dominance of the latter has caused an antagonistic reaction, often described as Islamic and Judaic fundamentalism.

It is only in Christian cultures and in "enlightened" forms of Judaism that a healthier balance was struck between mysticism, dogmatism, and reality, that could serve the further spiritual progress of humanity. Therefore we now have to turn to the events that have led to this watershed in the history of spiritual legitimacy.

The Ego System Comes of Age

In 1976 the Princeton psychologist Julian Jaines presented a thesis under the title "The Origin of Consciousness in the Breakdown of the Bicameral Mind" (Jaines 1976). He had concluded that consciousness is a cultural phenomenon, and that it arose in Homeric times. In particular an "astonishing contrast" can be observed between the *Iliad* which is a glorification of heroism under the fateful dominance of the gods, and the *Odyssey*, where Odysseus is not a hero in the old sense, but he lives by his personal wit and is described as the "great sufferer."[41]

Between 1000 B.C. and 800 B.C. an Odysseus cult arose in Greece. "Odysseus of the many devices is the hero of the new mentality of how to get along in a ruined and god-weakened world" (Jaines 1976, p. 273).

Indeed, a *Götterdämmerung* ("twilight of the gods") of the Indo-European cosmic gods must have been in the making. Wagner got the theme from his friend Nietzsche, who, besides being a philosopher was an eminent classical scholar. At the end of the "Ring" cycle, the gods leave the cosmos, and are possibly annihilated by fire. After several days of performances of the "Ring," one is left with a profound sadness for the loss of this ancient world. It is almost like a repetition of the "fall from grace." Now the world really has become dark. I stand lost and alone. Unwillingly I find myself in the shoes of the superman, and they are much too large to wear.

It is because of the growing ego-power that the cosmic gods of the Proto-Indo-European agriculturists began to lose their impact on daily life. By the time of classical Greece, they had regressed to a quasi-human existence, quarreling among themselves, mating with humans, and admitting the mixed offspring to Olympus. Meanwhile, in the *Odyssey*, a new hero arises. After a series of psychological initiations he becomes, not a god, but *Homo autonomicus*, when he emerges from the waves and proudly announces before the court of the Pheacians and the king's daughter Nausikaa:

> "Behold Ulysses! no ignoble name,
> Earth sounds my wisdom, and high heav'n my fame.
> My native soil is Ithaca the fair, . . ."
> (Homer's Odyssey, translated by Alexander Pope)

This is no less but the male equivalent of the birth of Aphrodite from the waves. But who is born is not a god but man. After this Odysseus is ready to return to his land, to a marital bed, that he himself had built on top of an immovable stump of an ancient tree, and to a wife who had resisted all temptations, not by a god's decree, but by her own cunning and deep earthly love.

I can agree with Julian Jaines if his concept of consciousness can be approximated by the concept of ego system. As far as consciousness is concerned, obviously the Bushman had it. It cannot possibly have arisen in Greece between the Iliad and the Odyssey. But the event so aptly unearthed by Jaines was the

---

[41] The change is so abrupt that some scholars hold the view that the *Odyssey* and the *Iliad* had different authors.

coming of age of the ego system, the permanent inclusion into consciousness of the realities of the physical and social world, which now could be manipulated at will.

In neurological context, Jaines postulates that in the described transmutation, the consciousness of the left hemisphere (left temporal lobe, according to Gazzaniga) lost communication with the right hemisphere, which thus was left under the dominance of the gods residing in the unconsciousness. Considering the archetypal components of god images, this explanation is quite plausible. After the origin of ego-consciousness in the left hemisphere, the right hemisphere quite naturally must have remained under a stronger influence of brain stem, limbic, and (right hemisphere?) frontal lobe activities. The sensory input that is processed by the right hemisphere probably is a decisive component in the formation of the unconscious personality.

But in the context of my exposition, this loss of communication between neocortical and brain stem systems, or in Jaines' term, the "breakdown of the bicameral mind," must have begun much earlier with the evolution of the ego system and must have proceeded more gradually. This is the leading theme pursued in the present chapter. The event observed by Jaines was only a last step in a long sequence begun by the early agriculturists. Also, the left-right dichotomy should be considered a secondary development of the basic cortico-thalamic neurological dualism.

Nevertheless a revolution happened at the time postulated by Jaines. I will try to explain it. The Odysseus cult was not unique at the time. As the last centuries B.C. approach, the fashionable cult heroes increasingly turn from pure divinities, into mixed breeds and then humans. Among the earliest figures of this kind are Orpheus and Dionysus. Orpheus is a human figure and is considered to have been the founder of "Orphic Mysteries," secret cults of initiation that had to do with the immortality of the soul. He is a lyre player who has magic power over animals. His religious message is in "radical contrast to the Olympian religion" (Eliade 1982, p. 183) because the soul of Orphic religion was not just the surviving soul of a shadowy after life—it was divine by its very nature.

This also seems to have been the secret message conveyed to the Eleusinian mystes. Orpheus sometimes is mentioned as an initiator of these mysteries also. The known reports of the effects of initiation are described as a loss of the fear of death. Eliade notices many shamanistic elements in the Orphic legend. When the Olympian gods began to fail, initially there might have been a reactivation of more archaic contents. But what counts is how they were accommodated to the new situation.

While the Orphic preparation to receive the divinity of the soul was vegetarianism, asceticism and purification, the Dionysian way was in drunken ecstasy. Dionysus was said to have been born by a mortal woman, and fathered by Zeus himself. He remains a child in most legends. The initiations took place at night in caves. The showing of the divine phallus was part of the ceremony. The central event seems to have been the divine presence, with music and dance, and it resulted in the belief that an intimate bond had been established with the god. This feeling, the experience of the divine nature of the soul, was then celebrated by the initiates or "bacchants" with wine and merriment, which could include sexual orgies. In this ecstasy man becomes brother of the god. The half-man, Dionysus, "though a child, is made by Zeus to reign over all the gods in the universe" (Eliade 1982, p. 284), as a late Orphic text states.

This divinization of the human soul is the revolutionary element at that time. This became a universal tendency in the Mediterranean world. Other mystery initiation cults were built on the legends of Cybele and Attis in Phrygia, Isis and Osiris in Egypt, Adonis in Phoenicia, and Mithras in Iran (Eliade 1982, p. 379). Particularly the Mithras cult became very powerful in pre- and post-Christian Rome, competing with the Christian belief itself.

The alpha animal god-person had become a friend and relative, as if all mankind now had alpha rank. This state of mind called for new great deeds. Or was it that the deeds were the cause of the new condition? At any rate, the ego system was now divine itself, and no limit existed to human endeavor. Alexander's conquests can be seen as the first historic result. Let me briefly report on this watershed in human history. One could defend that it all began with Socrates (circa 470-399 B.C.). His most bespoken pupil was Plato (428-348 B.C.). In turn, the next famous of the Greek philosophers had been a pupil at Plato's academy: Aristotle (384-322 B.C.).

When Aristotle was refused to become successor to Plato, he went abroad and became the private tutor of the young Alexander, son of Philip II of Macedonia. This country was in the northeastern corner of today's Greece. It included parts of today's Bulgaria and Yugoslavia. It was considered to be somewhat barbarian, but Philip II was a clever ruler. He succeeded in conquering all of Greece except Sparta, and he formed a kind of confederation, the League of Corinth. He also had built up an enormously effective army that used new tactics. The foot soldiers had twenty-foot-long spears that could be pointed forward, out of a line of five soldiers deep, protected by shields against arrows. This was impenetrable, even on horseback.

Alexander's father also must be given credit for choosing Aristotle, who was the most realistic of the classical Greek philosophers. He understood the prevailing lack of objective knowledge and therefore catalogued a huge panoply of facts about nature and animals. Today he is still often called the "father of biology." Alexander was not made into a philosopher through Aristotle's effort, but he must have gained considerable respect for it, because later he always had philosophical advisers. He certainly had learned how to apply detached rational logic in pursuit of his goals. In other words, he was made well aware of his own ego power.

Alexander was determined to become a hero. At the age of twenty, in 336 B.C., he succeeded his father. Having inherited a superb army, he immediately began to expand the empire. But he was a hero with a difference who broke with all traditions. He became a genius of novel war strategies and tactics, as well as a capable statesman. His small Greek-Macedonian army was so superbly organized and in high spirits, that within eleven years he had conquered Syria, Phoenicia, Egypt, the huge and rich Persian empire of Darius, and had crossed the Indus River into India, defeating a large army of the Indian king, Porus. It was as if the newly-freed spirit had swept away cultures stuck in outgrown traditions and corrupt ways. His empire came to range from the Danube in the north to Nile and Indus in the south, from the Libyan desert in the West to the foot of the Himalayas in the East. To establish political unity he married a daughter of Darius and ordered ninety of his prominent officers to marry other noblewomen of Persia. At that time, his goal seems to have been to unite all the people of what was the "world" in those days.

Among his enlightened achievements was the introduction of a unified valuta, mostly in coins minted from the huge and uselessly hoarded treasures of Darius of Persia. This was an economic jump start. Then the whole of the Mediterranean world became one great zone of trade and commerce. Goods began to be exchanged not only within the empire, but also with India and even China. And with this, new ideas were imported and had to be assimilated.

Alexander the Great died in Babylonia at the early age of thirty-two, to enter into folk mythology far beyond the limits of his short-lived empire. When Napoleon invaded Egypt more than two thousand years later, he was taken for a reincarnation of the legendary "Iskander the two-horned." Yet as a person, Alexander probably has to be considered tragic. Having come into the command of a powerful army at a very young age, he must have succumbed to the fateful feedback loop that leads to ego-system hypertrophy. After the first conquests he must have become possessed and did not know when to stop. Historians have considered the possibility that he actually was poisoned by his own officers who had reached the limit of the endurable. I think Alexander, Julius Caesar, Napoleon, Mussolini, and Hitler, all have fallen into this same psychological trap; all ended in a bad way, and all demonstrate to us the destructive power of an inflated and isolated ego system.

Alexander as a historic event is an entirely different story. With his empire a cosmopolitan spirit spreads over the old world. Among his city foundations was Alexandria[42], just west of the Nile delta. This was to become the world center of Hellenistic culture and learning. Later residents of Alexandria were: the first "engineer," Heron of Alexandria (circa 150 B.C.) who founded a school of technology, Euclid (365-276 B.C.) the father of geometry, and the astronomer Ptolemy (109-179 B.C.), whose system of planetary movements remained in successful use until the time of Copernicus.

Mircea Eliade writes:

---

[42] In total, Alexander founded sixteen Alexandrias.

After Alexander the historical profile of the world was radically changed. The earlier political and religious structures—the city-states and their cult institutions, the polis as "center of the world" and reservoir of exemplary models, the anthropology elaborated on the basis of a certainty that there was an irreducible difference between Greeks and "barbarians"—all these structures collapse. In their place the notion of the oikoumene and "cosmopolitan" and "universalistic" trends become increasingly dominant. Despite resistances, the discovery of the fundamental unity of the human race was inevitable. (Eliade 1982, pp. 203-204)

In the aftermath of Alexander's conquests several of his generals founded dynasties in Asia, Egypt and Macedonia, which became "Hellenistic Kingdoms," until conquered by the next world empire—Rome. Hellenistic culture dominated everywhere. It had much to contribute to the ascent of Rome itself.

The unification of the historical world begun by Alexander was accomplished initially by the massive emigration of Hellenes into the eastern regions and by the spread of the Greek language and of Hellenistic culture. Common Greek (koine) was spoken and written from India and Iran to Syria, Palestine, Italy and Egypt. In the cities, whether ancient or recently founded, the Greeks built temples and theaters and established their gymnasium. Schooling of the Greek type was increasingly adopted by the rich and privileged of all the Asiatic countries. From one end to the other of the Hellenistic world, the value and importance of education and "wisdom" were glorified. Education— almost always based on a philosophy—enjoyed an almost religious prestige. Never before in history had education been so sought after, both as a means of social advancement and as an instrument of spiritual perfection. (Eliade 1982, pp. 205-206)

A proper name has been given to this explosion of knowledge: "Hellenistic Enlightenment" (Eliade 1982). Among the enlightened urban intelligentsia the dominant philosophers became Epicurus (341-270 B.C.), and Zeno (333-262 B.C.), the founder of the school of stoicism. Will Durant bemoans the decline of classical Greece, and sees "multitudes of Oriental infiltrations" (Durant 1926, p. 97) and sides with Chrysippus (280-209 B.C.), who found stoicism "hard to distinguish from Oriental fatalism."

But in reality here is an entirely different phenomenon. For these post-Alexandrine urban sophisticates, the gods are dead, man is by himself, and afterlife is uncertain. What other goals can there be but to live out one's life span in the most agreeable way? That is how far the ego-power could swing in the aftermath of that great liberation.

The pleasures of Epicurus are primarily not the pleasures of the flesh, but the pleasures of the mind and of a civilized life. Zeno recommends to make the best of life's pitfalls and to bear the inevitable with a trained detachment.

How sophisticated those thinkers of the Hellenistic enlightenment have been, shall be illustrated by a citation from Carus Lucretius (circa 98-54 B.C.), a Roman poet and follower of Epicurus. He seems to presage Darwin:

> Many monsters too the earth of old tried to produce, things of strange face and limbs . . . some without feet, some without hands, some without mouths, some without eyes . . . Every other monster . . . of this kind earth would produce, but in vain; for nature set a ban on their increase, they could not reach the coveted flower of age, nor find food, nor be united in marriage; . . . and many races of living things must then have died out and been unable to beget and continue their breed. For in the case of all things which you see breathing the breath of life, either craft or courage or speed has from the beginnings of its existence protected and preserved each particular race . . . *Those to whom nature has granted none of these qualities would lie exposed as a prey and booty to others, until nature brought their kind to extinction.* (Durant 1926, p. 830, italics added)

Even though Eliade gives the new education "an almost religious prestige," the Hellenistic enlightenment lacks the legitimizing power of religion. If my hypothesis about the nature of religion is correct, such radical exertion of rational ego-power must call forth an equally radical attempt to reestablish spiritual legitimacy. And that, indeed was the case. While Lucretius ended in suicide, all the marvelous craving for knowledge remained restricted to the urban elite. The farmer, village craftsman, and artisan lived in a different world. Neither the soldiers, nor the bulk of city dwellers were capable to participate in high-strung intellectual adventures. And one should rightly suspect that even the best educated had an archaic soul right under a thin enlightened skin.

To this simpler and healthier soul, man had become like God. This indicated an acute dissociation of the two brain systems, and that was experienced as frightening. Things could not go on like this forever. At the least a strong, if not devastating, response from the higher powers was to be expected. The world would come to an end (the eschatological expectation), or a divine messenger (messiah), if not God himself, would come and set things straight.

Of course, such stories were the result of stirrings from the unconscious personality which indicated a disregard for biological wisdom. Some strong correction was due. Since the agent of such change was not known, the storyteller inserted what always is inserted for an unknown power external to the ego—God.

As a consequence, traditions were recast and new cults sprang up in many places; some old ones degenerated into vulgar farces. But the mystery religions had placed man next to God. And surely a decisive real communication was to take place anytime.

Such was the disturbed and contradictory state of the human mind at the birth of Jesus Christ. He was to become a nucleus of crystallization for a belief that would provide the last and most powerful system of spiritual legitimacy through religion that ever existed. No longer would we have to wait for our soul to enter into its divine existence. Right here on Earth, we were to be impregnated with the divine Holy Spirit, which henceforth would guide our ego system. And later we would enter into the new divine existence even with our own very physical bodies. The inherent Holy Spirit, of course, is a deep biological wisdom that finally begins to come to the surface. It is the first inkling that the dualistic gap between the two subsystems of the brain can be closed, and man can become a truly free and self-responsible agent of actual truth.

Christianity—The Blessed Illusion

Six hundred years before Jesus Christ, Christianity had an overture. A generation before Buddha and Confucius (circa 628-551 B.C.), there lived in Persia the prophet Zarathustra (or Zoroaster). Geographically, Persia (Iran) is so close to the original agricultural sites, that one would expect there to exist one of the longest local religious traditions. In fact when the Indo-Europeans returned south, their cosmic gods, Dravidian influence, and local archaic gods all came together in a melting pot where the most grandiose synthesis of a cosmic religion took place. I have chosen not to include Zarathustra in the section on "early syntheses" because he would explode this concept. Confucius and Buddha give the impression of "rational synthesis," Moses preaches an ethic natural to the species, but Zarathustra is creator of a new powerful cosmic myth.

This myth is characterized by the dualistic view of a creator-god and his adversary, and the need for cosmic salvation in which man takes a decisive role. The idea of a future savior from this situation is proposed, as well as the promise of the resurrection of bodies. There is a divine sacrifice to save the world, a fiery hell for punishment, and heaven as reward. All this was invented here. The image of the Magus, the Faustian man, who participates and wins in the cosmic struggle, the future alchemist, all had their origin here.

Most of this is the result of Zarathustra's spiritual genius, in the form of revelations of his god Ahura Mazdâ, who has created the world by thought. In the legend, Zarathustra is conceived by human parents, but under the influence of a luminous spirit. Like Orpheus, Dionysus, and others, he became a divinized human figure. Historically, he was a simple priest of a traditional Aryan religion. His revelations began at the age of thirty, and he spent much time to try to reform his former colleagues. Initially he had little success to win followers. "In essence," says Eliade, "the reform consists in an *imitatio dei.*" Man is not a "slave or servant of God," but he is advised to follow the good example of Ahura Mazdâ. In this way the divine cosmos becomes an image and model of the inner microcosm of the human spirit, a view that will be repeated in medieval Christian Europe.

This microcosm-macrocosm correlation gives away the psychological basis of Zarathustra's creation: it is a projection of the basic neurological dualism on the blank screen of the unknown cosmos. The dualism between God and the adversary is the experienced dualism between the brain stem and its neocortex. The demanded responsible human involvement in the drama makes this even more obvious. We are asked, and are able, to participate and to resolve the dilemma, because this is a drama inside of our own person.

The Lord, as seen by Zarathustra is accompanied by divine beings (Amesha Spentas): Justice, Good thought, Devotion, Power, Integrity, and Immortality. Some scholars believe that those beings found their way into Judeo-Christianity as the archangels. Ahura Mazdâ is the father of twin spirits, one (Spenta Mainyu) chooses to be a good spirit. The other (Angra Mainyu) chooses evil and death. Thus man, likewise, is free to choose. In a reversal of previous conventions, some of the original Aryan gods (the war gods) choose evil, and they now give their generic name ("daêvas") to the daemons; hence comes the name of the devil.

The isolated ego system, indeed, becomes daemonic when it escapes the restraint of instinctual compassion.

Everyone will be judged at the end. When the Cinvat Bridge has to be crossed, the evil ones will fall into the abyss. Who succeeds to cross where the bridge becomes narrow like a knife's edge, enters the "house of song"—paradise.

Personal individuation, the achieved quest, is compared to paradise.

Zarathustra predicts that by the will of Ahura Mazdâ the world will be "transfigured" and the evil annihilated. Like Christ, initially he seems to have hoped for the immediate arrival of this event and calls himself "the savior" ("Saoshyant") who will precipitate it. But the event does not happen, and:

> This tension will very soon harden into dualism. The world will be divided into good and evil and will end by resembling a projection, on all cosmic and anthropological levels, of the opposition between the virtues and their contraries. (Eliade 1978, p. 313)

And in later Iranian view, the dualism is extended to a material and a spiritual world, the world of thought and the "bony world" (ibid.). Fire altars of the old Aryan tradition are maintained. "The eschatological ordeal by fire and molten metal that he announces had as its objective both the punishment of the wicked and the regeneration of existence" (ibid.). Fire "purifies and spiritualizes the world" (ibid.).

Like in later Christian development, the mystical optimism of the quest to be fulfilled soon ends in frustration. More is needed to achieve it, hence, death by fire and a new beginning are in order. The last days and tribulations are here. Apocalyptic expectations probably are a regular result of frustrated individuation.

The radically new idea of Zoroastrism is that man by freely choosing good, participates in the process of the cosmic transfiguration. Fighting evil is the first duty of every Zoroastrian. Moreover, through the "archangels" Ahura Mazdâ can help the worshippers to find what is good, which is the first appearance of the idea of the Holy Spirit installed in man.

Many scholars have tried to trace the Judeo-Christian views of punishment in hell, of the Savior, of transfiguration, and of the Holy Spirit to Zoroastrian origins. The results have been meager, except maybe for the origin of the angels and of hell in the Old Testament. Yet it is hard to believe that after two centuries of Persian rule and after the great dissemination following Alexander's conquest of Judea in 332 B.C., not more of the analogous beliefs in Judaism and Christianity had actually come from Persia. Judea was a multilingual area and part of the Mediterranean world. The official language imposed by the Achaemenid (Persian) rulers was Aramaic, which had become the language of everyday use. In fact, Jesus spoke in Aramaic to the crowds and used Hebrew with the priests. (It is not known whether he knew enough Greek to communicate with the Roman rulers).

There certainly was no cultural isolation due to language barriers. Lately Zoroastrian links through the sect of the Essenes have been investigated with help from the Dead Sea scrolls. John the Baptist and Jesus have both been shown to have been at least sympathizers with the Essenes (Fujita 1986). But again, any connections to Zoroastrism remain unproven. Also, we probably will never know how many supporters of Akhnaton's monotheism had fled Egypt after their ruler's death in order to spread their belief abroad, in Persia as well as in Judea.

Fortunately for the present discussion, the question whether this branch of religion spread by dissemination or by local reinvention does not matter much. News that answers true existential need always travels with uncanny speed. Neither will I tire the reader with the complex history of the church and the many Hellenistic or pagan elements that entered into it. Nor will I deal with the controversial views concerning the historic person of Jesus.

It is likely that Jesus began very much in the Jewish tradition of the day, but was a charismatic personality, a "prophet," continuing a long tradition known in Judaism. Yet what became canonized as the supranational Christian faith certainly did not all originate with Jesus.[43] In those days there was no interest in historic fact. Whoever was charmed by the utterings of Christ, the legends of his miracles, and of his resurrection felt that here was a "son of man" who at the same time was a god or at least a messenger of a god, who certainly partook in the divine existence. At the time, after Orpheus, Dionysus, and Mithras, this was not a great leap. As he himself had said, the "Kingdom of God" was in everyone and could be experienced within.

---

[43] Robin L. Fox lately has presented a masterly study of the accidental ingredients in Christian and Jewish scripture. It was widespread practice to alter texts in copy and to ascribe one's own writings to famous names, such as Solomon. From early Mosaic law to the Christian gospels, there runs a chain of storytelling that is ever-changing and has little historic truth. Yet he says, "A story may be wonderful evidence of what the narrator and his audience believed and assumed: it may help to hold their view of reality together" (ibid.). And earlier he writes: "Where the truth has been lost, stories filled the gap, and the desire to know fabricated its own tradition" (ibid.). For me, this is "wonderful evidence" that my view of man the storyteller and of the desire for spiritual legitmacy can also be derived from historical detective work. (Fox 1992)

A personal network of extremely uplifting explanations of this new mode of human existence was immediately weaved by everyone and passed on as living divine truth. After all, the central explainer and storyteller was alive and well. Through such personal syntheses, Hellenistic (Platonic) influences, contents of mystery religions, the popular Mithras cult of Iranian origin, and probably a measure of Zoroastrism, all had some influence on the mainstream of the new legitimizing power. This experience was so strong that thousands let themselves be tortured or be fed to the lions instead of revoking their new belief. After all, they already had the "Kingdom of God" in them, would soon be in heaven, and the rest of the world would follow shortly. Death was rather a triumphant entry than an end, much as it had been for Jesus Christ himself. Faith and baptism endowed the believer with Holy Spirit and a personal God instead of a tribal one. That justified unusual ego deeds, because they were in the service of that spirit and hence, good.

This emergent event of Christian religion can be viewed as a liberation from the compulsions of Mosaic law. Faith in a personal God and spirit now gave the power of autonomous ethical behavior. The strength and simplicity of this legitimizing power cannot be overestimated. Like none before it, this religion empowers the ego system to new adventures. It does this by creating a new synthesis in the unconscious personality. But since this system is unconscious and prelingual, it defies rational description. Instead, like all prelingual wisdom, it has to be passed on through parables and irrationalities. No other religion asks to believe in so many impossibilities—resurrection, virgin birth, Pentecostal miracle, and the transfiguration of bread and wine.

The invitation to creative ethical freedom, to me, is best expressed in the story of the adulteress to be stoned: "He that is without sin among you, let him first cast a stone at her" (John 8:7), and on one dares, "go home and sin no more." Here, we are empowered to make peace with our animal nature without discarding the order upon which any further progress in cultural evolution depends. The key is forgiveness and universal love. This is the most effective antidote to animal aggression.

The new empowerment is the reason why science and rational political design from now on can soar. After a short Islamic interlude, almost exclusively they have evolved in Christian cultures.

I have stated before that finding the inner biological truth, like developing the worldly ego-system truth, are slow and gradual processes of cultural evolution. No concept ever sprung to life in a complete and final form. The atoms of Democritus of Abdera were not the atoms of modern physics. The planetary orbits of Johannes Kepler had a connotation of Pythagorean mysticism, which only evaporated well after Isaac Newton. The powerful images conjured up by Jesus stated the inner truth of human nature only in terms that could be experienced by regular people of his time. The new faith only intended to give functional reassurance. Neither Jesus himself nor his early interpreters had any interest or understanding of objective reality. They did not even know that this could become an issue. This concept came only fifteen hundred years later with the *nova scienza* of the Renaissance and was to grow into today's science.

Yet the Renaissance would not have been possible without the legitimizing power of Christian faith. Even though this was built on an illusion like many stories of the central explainer, it launched the greatest and most successful project of civilizing education of all times. As if to demonstrate the legitimizing power of the new faith, the Church of Rome became a sponsor of art, architecture, and scholarship of hitherto unknown proportions. After it had gained political power, this was used more wisely than anyone had done before. The "Holy Roman Empire" and the Church transformed an accumulation of warring European tribes into a high culture that for the first time in world history would break through the barrier of illusion, to seek and to embrace objective truth. As if in preparation for this, it permitted the cream of forty generations to withdraw into monasteries and to devote a life to study. During that period of latency all of the known classical philosophy, mathematics, and science were assimilated into a new Christian inner universe. This enterprise is referred to as "scholasticism" or "scholastic philosophy," and it often is belittled as a period of unoriginal attempts to justify Christian faith by ancient philosophical means.

But medieval scholasticism should be viewed as a state of intense development. Comparable to the state of the pupa of a butterfly. When the time was ripe the capsule burst, and out came the new and wondrous butterfly of the truly independent human spirit.

The earliest witness for the sprouting of a new spirit was Francesco Petrarca (1304-1374) or Petrarch, the "father of humanism," poet laureate of Rome. Still completely under the spell of Christian devotion, there was one event in his life that some historians consider as the birth date of the new humanistic vision that led to the Renaissance. On 26 April 1336 he climbed Mont Ventoux near Avignon[44] (Petrarca 1975). No one climbed mountains in medieval times. But Petrarca made it a point to see how the world looked from above, and he was duly impressed as well as scared to have transgressed. So he felt obliged to write to his father confessor about it. Overwhelmed by the sights of the river Rhone, the sea at Marseilles several days away, clouds below him and other mountains, and after reading a passage from St. Augustine's confessions, he is induced to write: ". . . I closed the book enraged with myself because I was even then admiring earthly things after having been long taught by pagan philosophers that I ought to consider nothing wonderful except the human mind compared to whose greatness nothing is great."

Indeed, an event of great transformation. Note that "earthly things" are seen in opposition to "human mind," which now assumes divine greatness.

Nicolaus of Cues (or Cusanus, 1401-1464) can serve as another illustration of the state of affairs only a hundred years later. His life and thinking were at the

---

[44] Actually, some historians, following Guiseppe Billanovich, believe the letter and date to be fictitious. Petrarca probably did climb the mountain and truthfully describe his emotions, but years after the event.

threshold between the age of faith and Renaissance humanism. Not surprisingly do we see him live a double life. As a cardinal in Rome, a friend of two popes, travelling papal legate, and Bishop of Brixen in Tyrol, he presents himself as an inexhaustible and clever church politician. At the same time as philosopher, humanist, and collector of Greek manuscripts he appears to have transcended all dogmatic limits in favor of a philosophy that is firmly rooted in classical Greece. Yet he has never been officially accused of heresy because his ideas did not contradict what was then accepted church teaching. By his time Platonism and Aristotelianism had been so much absorbed into the mainstream of Christian thought that whatever views were built on the "pagans" Plato or Aristotle could hardly be heretic.

He was so at home with Greek philosophy that in some of his writings the personal God-father of Jesus and Abraham has become "the maximum," a remote philosophical construct. It is easy to visualize the learned alumnus of Heidelberg, Cologne, and Padua, grand representative of the rich Imperial Church, reciting mass in Latin that the public does not understand and thinking of his "maximum" behind all clouds, of which he, himself, thanks to his inherent Holy Spirit, is the supreme and benevolent administrator.

It obviously looks like a contradiction that his other self was mystical. "He was familiar with the traditions of Rhenish and Flemish mysticism and was especially influenced by the writings of Meister Eckhart" (Watts 1982). His work, the "Docta Ignorantia," is supposed to lead back to a mystical unreasoned faith. That seems to have been a strong personal motive behind his tireless life in the service of the church.

In Nicolaus of Cues a condition of the Western mind announces itself which was to last from the Renaissance to modern times: On one side we have a mastery of human reason which well transcends Greek classical rationalism. This reason now has matured in one thousand years of Christian scholasticism where it had enjoyed a dignity that only the belief in the innate Holy Spirit and a personal God can convey. It had become charged with a moral responsibility only faith can convey. When today we say "human reason," we think of humanism, goodness, honesty, struggle against the evils of ignorance, and the blood that has flowed in its defense. But it was born in the monastic schools of Christianity. And it is only since the eighteenth century that reason has fully freed itself from its monastic womb.

On the other side we have a religious belief which has to be mystical, that is completely outside of reason, in order not to contradict the objective world of realities within which God only can exist as a philosophical "maximum" behind all clouds. Of course, Nicolaus of Cues was prescientific. Still his power of reason was strong enough to prefigure the split personality a modern scientist has to develop if he cannot free himself from the bonds of a mystical belief.

The Socratic wisdom of not knowing anything, versus the Christian wisdom of not being able to comprehend God either, must be a true epistemological zero point experience. It must have redoubled search. This was a creative tension that only found relief through the birth of science, the Protestant reformation, world exploration, and ultimately the project of enlightenment.

The most significant point in Cusanus' later thought is that already he himself begins to initiate the creative resolution. As for Petrarca, this came through *the insight that the human mind is autonomous and ready to assume its Renaissance role as being the measure of all things*: "When you consider that the mind is a certain kind of absolute measure, which cannot be greater or smaller, since it is incontractible to quantity, and when you consider that it is a living measure, as it measures through itself, *as a living compass would measure through itself*, then you understand in what way it makes itself a notion, a measure or exemplar, so that it finds itself in all things." (Cusanus 1450, p. 29, italics added)

What is to be measured is stated a few pages back: "Physical and mechanical arts and logical conjectures." Understanding of nature now is possible, and only God has to remain in a mystical distance. In fact, now his trust in the divine power of human reason is so great, that much time is spent to clarify the subtle difference between God's thoughts and human thoughts, so as to avoid the equation "man = God."

Here is the ultimate distillation of the Pauline legend of the Pentecostal miracle.[45] And from here there is only a small step from God the "maximum" to God the hidden law behind nature. That step, however, heralds heresy, and for this thought Giordano Bruno was burned at the stake one hundred and fifty years later. Yet by that time many creative minds already were performing experiments in physics in their own private search for the revelation of that God of nature. And after the disgrace of Galileo's trial the creative minds simply moved north to the heartland of the first Proto-Indo-European agriculturists and their distant cosmic gods, where a new wind of religious tolerance had begun to blow. Through a series of gradual transformations the fierce desert God Jahweh had turned into a power that gave spiritual legitimacy to the exploration of nature.

In the Protestant countries there developed a certain rejuvenation of what was thought to be the true early Christian faith. The scriptures assumed a supernatural flavor which even Isaac Newton (1643-1727) took for an indisputable fact.

The gap between God the mystical being and the observable nature was temporarily closed by an ingenious device: the two-book theory. This flourished particularly in Britain. One book was the Bible, and the other was the "Book of Nature." Both could be read and understood through human reason, and both equally were revelations of God's intents and thoughts. Thus, it was virtuous to pursue both. Newton approached both books with equal zeal and a similar rational method. Most scriptures he considered to be addressed to simple people and were just "Milk for the Babies" (Manuel 1974). But the book of Daniel and

---

[45] The conveyance of the Holy Spirit upon the apostles, as described by St. Paul in Acts, Chapter 2.

John's Apocalypse contained revelations in coded form which could be deciphered in similar ways as nature could.

"If they are never to be understood, to what end did God reveal them?" (Manuel 1974, p. 88, Yehuda manuscript) He prepared a dictionary of the language of prophecy. The listed images and "prophetic hieroglyphs" remind one of Jungian dream interpretation. In fact he frequently quoted the dream book of Artemidorus (Manuel 1974, p. 98). He thought that there was a fixed symbolism and that the early people had expressed themselves in symbols rather than with logic. He believed that to Moses and to the writers of the book of Daniel and the Apocalypse, the whole truth, including the laws of physics had been revealed, and he, Isaac Newton, was chosen by God to define this truth for the people of his times. The great mathematician and physicist he was, this did not prevent him to remain the simple country boy in religious matters, but still motivated by "the greatest system of spiritual legitimacy of all times." The "two-book theory" provided him and many of his contemporaries with an unquestioned guidance in their pursuits. Yet unwittingly his great success bred the key ingredients, so that the matter could be decided in favor of the Book of Nature only. His religious writings largely became forgotten; some remain unpublished manuscripts to this day.

On the continent the two-book theory did not succeed as well. Newton's rival and co-inventor of differential calculus, Gottfried Wilhelm Leibniz (1646-1716), also engaged in religious speculation, yet his god remained very much the remote philosophical maximum of Nicolaus of Cues. Nevertheless, up until the time of Albert Einstein, whatever had become of the God of Abraham and Moses, it exerted great influence. Remote as the unknown mysterious creator had grown, he had a definite character for Einstein: "God does not play dice" and "Subtle is the Lord, but malicious he is not" are some of his famous religious statements. He used this freewheeling idea of his God so often in justifying his intuitions about physics, that Niels Bohr at one point was compelled to reply, "Stop telling God what to do" (Pais 1982, 1991).

In fact, Einstein's creativity appears to have been constrained by this religious feeling. He never was able to accept quantum mechanics in an unbiased way. I think it is not accidental that it was Niels Bohr instead, who immediately grasped the full implications of the quantum concept—Bohr had no relation whatsoever to religion (Pais 1991). By his wife's account, "He was sorry for the role religion played . . . He thought it was not good for human beings to hold on to things which were, as nearly as one could see, not true" (Bohr 1963). Abraham's God could go through many metamorphoses, but he could not be adapted to the dice-throwing chaos of modern physics. On the mythological level this would rather appear to be the devil's domain.

Thus, legitimacy through religion, for the intellectual pursuits of the ego, ceased with the dawn of post-Einsteinian physics. But in the meantime the most grandiose explanatory system of all times had been erected. This was not any more self-referenced as all previous fabrics of the central explainer, but referenced to the realities of nature.

The blessed illusion of Christianity has been the midwife to all this. For that reason I personally hold this faith in great respect. Is it really gone? Certainly millions still consider themselves to be Christians or Jews. Actually, statistics show the percentage of believers growing in the United States. But what is growing are sects based on ecstatic experience, not the faiths founded on long traditions of theological discourse and mental discipline. Only those have been the carriers of the Christian culture that gave birth to the age of reason.

Ecstatic cults, on the other hand, are thriving on unproductive hysteria. Communal ecstasy is just a form of mass hypnosis. The participants are made to imagine all kinds of communications with the Holy Spirit, they "speak in tongues," sing, dance, roll on the floor, claim miracle healing, et cetera. To lead a life based on fantasy is dangerous and borders on insanity. It is comparable to addiction, the super stimulus being the that the void between neocortex and brain stem is being closed, until the next dose of hypnosis. And when such unfortunate ecstatics fall into the hands of deranged or exploitative leaders, bloody tragedy is unavoidable.

Other people who still try to cling to the old ways tend to fall into the religious "rigor mortis" of the "Christian Right," intent on grabbing political power in order to prove themselves right. But this is a purely American phenomenon. In Europe even some enlightened clergy and renegade theologians agree with Karen Armstrong's *The History of God* (Armstrong 1994), who comes to very similar conclusions as I present in this writing, but from a purely historical and theological perspective.

Still, I am convinced that the disoriented street gangs of today, the neo-Nazi skinheads, or the Pol Pots of the jungle would be better served with a convincing dose of Christian illusion than with an attempt of enlightened humanist education. The chasm simply is to deep to be bridged in one step. Randall Balmer, in his television documentary "Mine Eyes Have Seen the Glory" and in the accompanying book (Balmer 1993), gives some good examples of the staying power of the Christian illusion today. Badly educated people in misery, such as in poor black communities and Mexican migrant farm workers, through a revitalized Christian faith have found the strength to escape the social zero point of alcohol and drugs and to build acceptable personal lives. Their primal and ecstatic style of revival reflects what early Christians must have experienced. Belief in the inherent Holy Spirit plays a dominant role in this movement. Still today this can give all the personal ego power necessary to succeed in their struggles of existence. Yet this remains a purely private salvation; it cannot save today's world at large.

Illusion has to end when "righteousness" derived from such experiences grows unchecked by reason. Otherwise a tribe of militant "chosen people" will emerge that is in opposition to the rest of the world, and upon which all their own evils are cast out. This is the old mechanism of "projection of the shadow," discovered by C. G. Jung a long time ago. Soon they would begin to attack democratic political design that has been installed two hundred years ago by people who personally were infinitely more advanced on this path of spiritual legitimacy, on which the "new" Christians are just naive beginners. These loud trumpeters of the Holy Spirit are no St. Pauls, St. Bonifaces, or St. Patricks. There kind seems to have left us. Instead, we have acquired fraudulent "televangelists" fornicating and pedophiliac priests which are tolerated by their establishment, and violent cults suffering from delusions. Indeed, it is not only in the minds of leading scientists that Abraham's God has gone the way of all previous gods. Unnoticed by most people, he has become a *deus otiosus*[46] (Mircea Eliade's term), a weakened remnant of a glorious past. Reluctantly and sadly we are forced to look for a more current power of spiritual legitimacy.

Karen Armstrong, the wisest nun who ever lived, ends her book with these words:

> Human beings cannot endure emptiness and desolation; they will fill the vacuum by creating a new focus of meaning. The idols of fundamentalism are not good substitutes for God; if we are to create a vibrant new faith for the twenty-first century, we should, perhaps, ponder the history of God for some lessons and warnings. (Armstrong 1994)

No, please. No new religion, but the final form of an enlightened humanism: objective humanism. Nothing else can fill the vacuum at this late point in human history.

The Quest Fulfilled?

Throughout the past, the quest for understanding had manifested itself as a quest for spiritual legitimacy. Even C. G. Jung himself, especially in his later years, understood the archetypal world as spiritual—not as biological and instinctual. As we have seen, contrary to other religions, Christianity did not discourage ego system development. It is as if the urge to pursue Knowledge I was so strong that legends and myths had to be invented in order to better justify that pursuit. In that sense religion and the quest to understand serve the same purpose. Only the technique is different.

Christian religion builds a typical emotional puzzle assembler's story; a story that motivates moral behavior and thus, after the dualism has widened, compensates for the lack of natural compassion and social harmony. Following the quest for objective knowledge, on the other hand, one refuses to fall for unfounded stories and tries to find the actual truth. This, of course, was an impossible dream until our days. But now "spiritual" loses its original meaning

---

[46] Latin: *Otium*, leisure. Webster's dictionary defines *otiose* as "unemployed, inefficient, useless."

and becomes "brain stem instinctual," and, if the main proposition of this book is correct, "mechanical."

This insight is so far ahead of the "Zeitgeist" still prevailing today, that acceptance is coming slowly. Even the Jungian analyst Anthony Stevens, who was the first to point out the origin of archetypes in animal instincts and who clearly understands the problems of the prevailing neurological dualism[47] (Stevens 1996), even he shies away from pushing the physicalist explanation. Instead he blames natural science and the scientists for causing the schizoid condition of modern humanity. This seems to be the prevailing stance in his profession. To some extent this is understandable because the primary concern is healing a patient's individual schizoid condition. Nevertheless, Stevens is aware that this condition afflicts the entire modern civilization. In this respect he says "we lack a unifying dream" to guide us.

He is right; only a spiritual revolution of Copernican dimensions can save us. But such revolution is exactly what a conclusion of the evolutionary synthesis in our days could bring about. The Copernican revolution had displaced us from the center of God's universe, and the yet unnamed revolution happening now declares us autonomous individuals who have understood "the inmost force that holds the world and guides its course" (Goethe's Faust, Part I) by discovering that it is only ourselves who are left not only to guide its course, but also to guide ourselves with help from the final understanding of how we work. To me such insight not only is powerful, but also is a natural consequence of that three-thousand-year-old growth of objective knowledge, and of the success of knowledge gathering throughout natural history.

I think that today, worldwide, there are millions of individuals who have reached that state of inner development. Millions of others could in principle be in the same position, but are kept from joining either by mental inertia, lack of self-esteem, or are unable to escape the doctrinaire trap of organized religion. But unfortunately, many more billions remain dwelling in mind-sets of the distant past. Carl Gustav Jung once observed that among his patients he could find people from many different historic times. Some were so advanced they appeared to live in the future. Others were from the Middle Ages, still others from the eighteenth century or from prehistoric Stone Age. And, I may add, the "Evangelicals" existed nineteen hundred years ago.

A similar observation was made by Tatyana Tolstaya (Leo Tolstoy's grandniece). She said, "Even in the middle of Moscow, within a ten minute walk from the Kremlin, live people with the consciousness of the fifteenth or the eleventh century" (Tolstaya 1991). The "National Front" skinheads of Bury St. Edmunds in East Anglia (Great Britain) perform a vicious all-male war dance in a tavern (Buford 1992). It is a communal arousal of tribal aggression. Similar rituals happened among the Papua cannibals and probably among our early Indo-European ancestors. Yet all those belong to the species of *Homo sapiens*. The difference between you, an intelligent and well-mannered reader of this

---

[47] Steven calls it "the schizoid wound at the heart of our culture," ibid. p. 337. Compare also pp. 322-325

demanding text, and the skinheads, lies in them having been deprived of cultural transmission. The less traditionally evolved cultural values we absorb in our youth, the more backwards in time we appear. It is as if Karl Ernst von Baer's thesis of ontogeny tracing phylogeny also applies to cultural evolution. That is why Jung's fine-tuned powers of observation associated his patients with different epochs in history. But they simply were at different stages of cultural education, capable in principle to absorb much more.

Therefore the ultimate spiritual legitimacy, the fulfillment of the quest for objective knowledge, in principle appears to be within reach of all mentally healthy people. Yet the task of actually spreading the basic knowledge is immense. Meanwhile it is safe to assume that, like in times of the "Hellenistic Enlightenment," the new mode has remained a thin veneer on the surface of even the chosen few. In situations like the recent Bosnian war, in the failed cultures of central Africa, or in past Nazi Germany, or in today's city street gangs, a regression has occurred to the raw primate nature of the "Naked Ape" (Desmond Morris).

Is it realistic to expect that the goal ever will be achieved? It is my personal belief that this will happen worldwide within a few generations. In part I base my belief on the observed exponential growth of objective knowledge about ourselves. From a certain point on this is bound to have a growing impact. To make this happen only a few inspiring leaders need to emerge. The phenomenal success in Germany of Eugen Drewermann, the "new Luther," is indicative that the times are ready (Schönborn 1993). The bulk of human population always has been led by the few, and this will hardly change in the future. But mainly the new epoch will be precipitated because we will be forced to wake up by events beyond our control. The events are overpopulation and the weakening of the life support system of our planetary spaceship. We should remember that in the past the human spirit progressed most if put under stress. The Ice Age caused the birth of Homo sapiens proper, and the Hundred Years' War and Black Plague ignited the Renaissance.

Chapter Nine

The Pains of Success—and Beyond

In this concluding chapter it is shown that the evolutionary success of the human ego system has brought us to the brink of self-destruction. Soon the planet will no longer be able to sustain the still-rising billions of human population and its activities. This is a first in all of natural history on Earth. Will the human spirit be able to adapt to this situation and avoid a global catastrophe? In the most optimistic scenario, at minimum a radical change of our habits will become unavoidable. Whatever form the difficult time of transition may take, a lifestyle of harmony with nature should be possible afterward, albeit with a much smaller and stable world population. This will be a culture based on the ultimate and realistic spiritual legitimacy—a new and objective humanism. It will be compared to the adulthood of humanity—the fulfillment of C. G. Jung's dream of the coming "Age of Aquarius"—or the fulfillment of the quest for objective knowledge.

An Unexpected Constraint: The World Ecosystem

One feature of life on Earth is that all species evolved together. Almost from the beginning a species could not change without causing responsive changes somewhere else. Many of the interactions are just competition for resources. For example, increased length of a giraffe's neck may diminish the volume of food a tree-dwelling mammal can gather. If the giraffe now can reach most of the sweet flowers on a particular species of tree, a catastrophe might befall a species of insect which had specialized to collect nectar from this particular flower and pollinate it. Pollination may stop and the tree's population may fall.

But much more consequential are interactions that foster mutual support. Around three billion years ago, certain photosynthetic microorganisms began to proliferate. They were to become the ancestors of plants. Their main metabolic product was oxygen. Thus, Earth's atmosphere became oxygen rich. This enabled the evolution of oxygen metabolism in other organisms, which happened to be an extremely efficient way of energy turnover. All of the animal kingdom relies on oxygen ever since. Animals burn carbon to produce energy and CO_2 gas, which in turn is consumed by the plants who make the oxygen and collect carbon. This system of mutual support insures continued growth of both, animals and plants. Dead organisms contain stored energy. Fungi evolved to fill the ecological niche of reclaiming that energy. In turn, they themselves became new food for someone else. Plants not only created Earth's present atmosphere; in time they also made their own nutrient-rich soil by depositing organic material in inorganic sand.

Therefore the interactions among species, and between species and the physical environment, can be positive, leading to exponential growth of the populations, and thus to an increase in the positive interactions. Or they can be negative. In this case the competing populations will either stabilize, or they may become extinct. At any rate the numbers of competing interactions will not grow, but the mutually beneficial ones always will grow. This simple logic is akin to Darwinian selection. Therefore we have to expect that the organic world in time will abound in positive interactions which will make possible a much larger mutually supporting population of species and of individuals. In this process, as the planet became saturated with life, the physical properties of the entire Earth changed. Besides the atmosphere, the chemical composition of sea water, and even temperature adjusted. As a result, we late observers find that Earth is "miraculously" well-suited to support life. From studies of the conditions on Mars and Venus, and various known facts as well as assumptions about Earth, J. Lovelock concluded that without life, the average surface temperature on Earth today would be 240 to 340 degrees centigrade, and the atmosphere would be almost pure carbon dioxide[48] (Lovelock 1972), (Margulis and Lovelock 1974).

With help from Lynn Margulis he has engaged in some simple but impressive model building of ecosystems. He championed the need to include in comprehensive ecological theories the changes of global physical parameters. In hindsight this appears obvious. The total volume of living organisms should be expected to grow exponentially until limited by what is available as input, such as minerals dissolved in water, gases of the atmosphere, and energy in form of sunlight. That simply means that life will grow until the physical environment is changed. It will not simply be changed by depletion, but amounts of metabolic products will be released into the environment which are comparable to what has been used up. Lovelock introduces the word "geophysiology" for this dynamic.

Most ecological subsystems seem to have a certain stability to perturbation. Natural perturbations are climatic changes, volcanic eruptions, meteor impacts, the continental drift, as well as internal (systemic) oscillations, which may be

---

[48] It should be noted that Lovelock's views are controversial, and he does not help his cause by presenting science in a mythological way.

chaotic. Stability predominates because instable subsystems tend to collapse under such loads and are survived by the more stable ones–again natural selection. As a result, the entire world ecosystem of today gives the appearance of a self-regulating superorganism. In fact much of the same logic of interaction exists within multicellular organisms, where the various cell types with their chemical products and physical activities form a stable interactive system of survival. This works on an entirely different level than the individual cell. Yet organisms die, and the world superorganism is by no means assured of continued survival.

Since the human spirit has promoted our species to a success story of global proportions, our activities have become crucial to the survival of this superorganism. Even without human contribution, computer simulations of simplified ecosystems show sudden collapses. Being a complex nonlinear system, under particular global conditions it can self-organize; under others it can break into chaotic oscillations. Under most conditions, however, any self-organization stops. The system collapses. This demands extreme caution in handling the world's ecosystem. As the CO_2 level rises from the burning of fossil fuels, it is quite possible that at a certain point a sudden change of world climate may occur. It has been proven that the last two Ice Ages came and went not gradually, but quite suddenly, and they were accompanied by rapid changes of the CO_2 level. In fact our activities have already increased that level by an amount larger than the increase that followed the last Ice Age. The dangers of global warming are real, and the consequences of ozone depletion can be grave.

Because of the analogy of the world's geophysiological system to a giant living organism, upon suggestion by the novelist William Golding, Lovelock gave it the name "Gaia"[49] (Lovelock 1979, 1988). Here is an interesting modern mythologization of reality–to look at Earth as a personal vis-à-vis appeals to respect and admiration. Among Lovelock's readers, religiously-oriented people, and "New Age" mysticists eagerly embraced this idea, especially because it came from a respected scientist who had participated in the NASA project of search for life on Mars. Earth's ecosystem, though, should be not worshipped but protected. Gaia is not a theory, it is an image, a symbol.

Lovelock may be a romantic, but other scientists sound similar notes. The ecologist E. G. Nisbet writes: "My own opinion and that of many others is fully in support of the Gaia notion as stated scientifically" (Nisbet 1991, p. 46).

Details of the world ecosystem can be learned from ecological texts such as, Basic Ecology (Odum 1983). Clive Ponting (Ponting 1991) has given a good summary.

The Irresponsible Childhood of Humanity

Prescientific man's naive attitude toward nature's gifts is best illustrated by the history of an ecological microcosm that has existed and vanished on Easter Island. Polynesians settled in this remote island in the fifth century (Jennings

---

[49] Greek Earth goddess, also "Ga" or "Gaea," mother of Titans and gods; from her name the words "geology," "geography," et cetera are derived.

1979). There were no mammals, only lizards, thirty species of plants, and extensive forests. The settlers had imported chickens and rats. Their diet had to consist mainly of chicken, fish, and sweet potatoes. As far as is known, they had no further exchange with the rest of the world until a Dutch ship arrived in 1722. The population peaked at seven thousand around 1550. They had a well-developed knowledge of astronomy, as witnessed by the layout of certain ceremonial stone platforms, and a religion that required erection of giant carved stone statues, more than twenty feet tall. By 1550, deforestation had caused soil erosion, giving the island today's desolate appearance. Crop yields fell. When the wood had run out, no good houses could be built any more. No fishes could be caught with nets because the nets had been made from the paper mulberry tree which had disappeared. The whole culture collapsed. By 1722 there were only three thousand people who lived in caves or primitive reed huts. They engaged in continuous warfare between clans and enriched their meager diet with cannibalism. After 1722, the population further declined until the rest became enslaved and shipped away.

The most important lesson from this sad story is that as the forests began to disappear, instead of conservation efforts, a mass production of the giant statues began. Six hundred of them exist. Certainly, whatever god was served by this tireless effort, he was imagined to be able to save the desperate islanders, who could not even build a boat in order to leave. Then a sudden breakdown of social order happened. Partially-carved statues remained in the quarry. No further tree trunks were available as rollers for transporting them. The naked ape nature became exposed again. Chimpanzee wars began. Each warring clan began to topple the statues of the other. In later times, people did not know any more how the statues were made and transported. They said they "had walked."

My reader, of course, is not puzzled by this ecological history of an isolated world. The spiritual legitimacy of the islanders was not derived from reality, but from a god image, probably a tribal super-leader whose sad stone faces look out over the sea, and one day, like Moses, will take all home to the promised land they had left one thousand years ago. Who would care about trees with such grandiose prospects?

So it all ended in a catastrophe. And there are many regions in the world which ended likewise, even though they were less isolated. The invention of agriculture made things much worse. We remember Uruk in Mesopotamia, the first city sustained by an agricultural hinterland. Today it is buried in desert sand because the rich soil died from overuse and overirrigation. Irrigation, particularly in hot climates, causes salinization. The evaporating water leaves salts behind which render the soil infertile. This, and deforestation, ended all early cultures along the rivers Euphrates, Tigris, and Indus. In the Near East, only along the benevolent Nile did agriculture flourish continuously until the twentieth century. Now the Aswan Dam finally has initiated ecological damage even there. The Egyptians had been blessed forever by yearly floods which brought a deposit of rich soils (silt) from elsewhere. When the flood did not come or was too strong a blessing, the political fortunes of Egypt waned. When they came regularly, Egypt waxed. Today there are no floods, the silt is accumulating behind the dam, and Cairo is being poisoned by chemical fertilizer and sewage in the groundwater. The major sources of the silt are Ethiopia and today's Uganda. Ethiopia once was a fertile green land with a rich culture. Deforestation and overuse turned it into a desert.

The entire Mediterranean region, including north Africa, once was rich agricultural land. Syria was abundant with greens and known for its tall cedars. Then the forests were used up for building galleons and houses and cooking food. The land was overused and overgrazed. The crops were shipped to Rome. Today we have deserts and dried-out stony cliffs called *carso* in Italian.

In Mesoamerica, the Mayan culture died a sudden death from social and agricultural collapse. China today suffers severe flooding and severe soil erosion because forests at the upper Yellow River and its tributaries have been destroyed already in ancient times. The Yellow River (which is yellow because of the silt it carries) has deposited so much silt in the plains that it now has a water level of eight meters above ground (Piel 1992, p. 77). How much higher can the dam be made, and when will it break?

The only regions on Earth where nature has been more forgiving are the temperate zones, especially of Europe, Asia, and North America. Here, glaciation has produced large amounts of ground-up topsoil. Evaporation is much less, and rainfall is more frequent. As compared to the tropical rain forests, the temperate forests bring more nutrients into the soil. Irrigation generally was not needed, and despite deforestation the soil largely remained in working order. The worldwide success of the Europeans was not due to their inherent Aryan superiority, as some would want, but simply due to the good luck of having wandered into a region that was large and whose ecosystem was difficult to destroy. That, however, is not anymore assured after agriculture has become chemical technology and agribusiness.

But even in the lucky regions, populations grew to the limit of what agriculture could produce, and famine followed by sickness was never far away. Clive Ponting concluded:

> An existence under the constant threat of starvation and in the daily reality of an inadequate diet and malnutrition has been the common lot of most of humanity since the development of agriculture. (Ponting 1991, p. 109)

For instance, in the early nineteenth century, inhabitants of the poorer regions of France lived on only 1,800 kcal per day, whereas the Kalahari Bushmen gathered 2,100 kcal per day, working only three days a week (ibid., p.103). Still, at the beginning of the Industrial Revolution, in an average year, if the weather did not destroy the crop, life could be rather acceptable. The picture that Will Durant paints of preindustrial agriculturist dwellings in England's forgiving nature is impressive:

> Each village–almost each household–was a self-contained economic unit, growing its own food, making its own clothing, cutting its timber for building and fuel from the adjacent woods. Each family baked its bread, hunted its venison, salted its meat, made its butter, jellies and cheese. It spun and wove and sewed: it tanned leather and cobbled shoes; it made most of its utensils, implements and tools. So father, mother, and children found work and expression not only in the summer fields but in the long winter evenings; the home was a hub of industry as well as of agriculture. The wife was an honored mistress of many arts, from nursing her husband and rearing a dozen children to making frocks and brewing ale. She kept and dispensed the household medicines; she took care of the garden, the pigs, and the fowl. Marriage was a union of helpmates; the family was an economic as well as a social organism, and had thereby a solid reason and basis for its unity, multiplication and permanence. (Durant and Durant 1965, pp. 45, IX)

Of course, there was a ruling class. There were wars between rulers, mostly fought with the extra sons of the productive peasants. But wars were limited to armies in uniforms. Civilians were left more or less unharmed if they stayed out of the way. This was not the best of all possible arrangements. But a certain balance had established itself that almost deserved the name "civilized."

Peasants everywhere maintain the most unchangeable culture. Their ego system remains in a state of relative subordination to fixed seasonal and daily routines, without which the fruits of Earth cannot be reaped. Their unconscious personality certainly has a store of intuitions about the ways of nature and of the domesticated animals. When change comes, it remains in tune with nature. Thus, despite what I have termed "the fall from grace," and the lack of environmental responsibility, there can be found a continuity spanning several millennia of an existence in relative harmony with nature which ranges from Bushman-like hunter-gatherers to some pretechnological agriculturists. I like to view this period as the long childhood of humanity. This comparison is reinforced by the fact that, like children's groups, the peasants tended to come under the domination of ego driven bullies.

All the past and present catastrophes indicate that the human spirit has yet to find itself at home on this planet. The long childhood of humanity was innocent because we did not know what we were doing. And in innocence we celebrated life whenever any cause for joy could be found. At the apex of such celebrations we find Mozart's music and Renoir's painting. But it all petered out at the latest with the First World War. Since that time, disorder and helplessness rule the arts. Chaos and helplessness also encroach on social order.

Industrial Revolution—Adolescent Prelude to the Future

During the long childhood there has been a slow worldwide progress of arts and simple technologies. Waterwheels were invented for irrigation. Gold, bronze, and iron came into use. As a preview of the things to come, the Middle Ages in Europe saw an unprecedented flowering of mechanical arts. The windmill, clocks, watches, and firearms were invented. All this heralded the beginning of a new mode of spiritual existence which the Europeans began to embrace. It naturally grew in a place where the good fortunes of climate and soil had permitted the longest and most comfortable agriculture to develop that our planet had ever seen. The other ingredient was the rediscovery of Greek art and philosophy which also had been a result of fortunate circumstances. We remember Petrarca (or Petrarch), the "father of humanism." Humanism, of course, is synonymous with the beginning of life by the ultimate spiritual legitimacy of reality.

Once again our species was about to blindly stumble into a new emergent system level. If the previous was childhood, this one should be called "adolescence." Its most outstanding early beacon was to become James Watt's 1765 improvement of Newcomen's steam engine and the subsequent partnership with Matthew Boulton in large-scale production of this artificial source of mechanical energy. As animal and human muscle power began to be supplemented and then virtually replaced by artificial energy, we found ourselves facing a totally new situation, as if transplanted to a wonderland where a mere wish could move a mountain. That is, if you had the means to buy the machinery.

The well-known first result was that the rich got richer and the poor got poorer. This, of course, is primate biology in action. Ranking order still is alive and well. We already saw it in action after the agricultural revolution in Ur and Sumer. Not general abundance, but stronger social stratification into the poor and the mighty was the outcome. The added tools had given more power to the dominant individuals and thus enabled them to better suppress the weak. Biology was culturally amplified.

Now the story repeats itself. As the clever and the powerful gained new advantage, efficient large-scale agricultural practice reduced family farming. An agricultural as well as industrial working class came into existence that had little rights, lived in poverty, and had no prospects of ever improving their lot. Since ranking order is such an ingrained feature of primates, it does not astonish that such conditions at the time were not considered abnormal. The great chemist Humphrey Davy declared in 1802: "The unequal division of property and labor, the difference of rank and condition amongst mankind, are the sources of power in civilized life, its moving causes and even its very soul" (Piel 1992). The influential satirist Bernard Mandeville had already said in 1714: "The poor have nothing to stir them up to be serviceable but their wants, which it is prudence to relieve, but folly to cure . . . to make society happy, it is necessary that great numbers are wretched as well as poor" (Tawney 1926, p. 190).

An old law by King Edward VI forbade associations of the working class. This was renewed by the British Parliament in 1720. Such were the beginnings of the age of "reason." In those days of innocent adventures into a new system level, the foundation was laid for the political troubles of the twentieth century.

In 1776, Adam Smith proposed the theory of a free market economy. Selfish interest would naturally produce socially desirable results. People (today's "consumers") would spend their incomes on desirable goods which manufacturers would produce in order to make a profit. Competition between producers would drive the price down to the least sustainable level, and everyone will be happy. In 1846, free trade was indeed adopted by parliament as England's norm.

Yet there was a dark side to such reasoning. In 1798, T. R. Malthus, in his *Essay on the Principles of Population*, pointed out the obvious, that populations always increase to a level where they would be kept in balance by misery—that is by infant disease, plague, famine, and war. Adam Smith also had sounded a similar note. Only in an expanding state of a national economy are the laborers sufficiently paid, so that their population could increase to meet the higher demand.

Smith wrote, "The wages paid to journeymen and servants of every kind must be such as may enable them, one with another, to continue the race of journeymen and servants, according to increasing, diminishing, or stationary demand the society may happen to require." (Smith 1776)

In other words, there must be an equilibrium of supply and demand in the breeding of humans of the miserable underclass. To this Karl Marx (in 1867) only had to add the self-fulfilling prophesy that the workers inevitably will band together and destroy the capitalist system of free enterprise.

Indeed, there seems to lay a tragic inevitability in the unchecked economic process. This has its roots in human nature. The economic process is driven by competition, which mirrors the natural ranking order within a tribe. Yet the reaction, as it expressed itself in the Russian Revolution, was founded on nonobjective explanations of reality, a result of the ignorance of Marx and Engels about human nature. The revolution tried to replace a naturally evolved, if unfortunate, system of human cultural interaction with a utopian one that human nature was not able to sustain. Competition has proven to be a human driving force that does not have a substitute. To the reader of this book this does not come as a surprise.

In the West, meanwhile, true cultural evolution has outsmarted an ill-conceived utopian revolution. Following a course that has been described as "social engineering" by Karl Popper, a gradual adaptive change of early capitalist harshness has taken place. Local institutions of helping the poor were in existence in Europe since the Middle Ages. But now they proved inadequate to handle the amplified problems of industrialization. It was not so much moral conviction than the will of the ruling class to survive that became the driving force of social improvement. After the French Revolution the ruling feudal class lived in fear of uprisings of working people. In the second half of the nineteenth century Napoleon III won the cooperation of workers and the peasantry through the introduction of state-guaranteed pension funds, countermeasures to unemployment, and producer's cooperatives (Köhler and Zacher 1982). This effort did not grow into a successful system. But it was noted by Otto von Bismarck, who had been ambassador to France in 1861, before becoming chancellor and forger of the German Reich.

It was this French attempt, the fear of the growing socialist parties, and the good influence of the farsighted academic economist Gustav Schmoller which caused Bismarck to embark on a course of social legislation, albeit with a simultaneous suppression of the Social Democratic Party. In November of 1871 he wrote to his minister of commerce: "The only means of stopping the Socialist Movement in its present state of confusion is to put into effect those Socialist demands which seem justified and which can be realized within the framework of the present order of state and society" (Syrup-Neuloi 1953).

Bismarck and Schmoller had read the signs of their time correctly. After Enlightenment, French and American Revolutions, the "wretched" masses no longer would accept to be kept in misery by a dominant few. They had discovered the power of their own ego system. Thus, the newly united Germany became the first state to acquire national health insurance in 1883, Workman's compensation laws in 1885 and a national retirement insurance in 1889. White-collar workers were incorporated in 1911. Unemployment insurance, however, had to wait until 1927. In hindsight Bismarck's reforms probably changed the history of the Western world in the twentieth century. They sufficiently immunized Germany against communism after the collapse of 1918. Think what a Marxist-Leninist Russo-German axis could have done to the civilized world. It would have been the richest and technologically most advanced entity that ever existed.

It does not surprise that among the industrializing countries it was the one with the least resources that came in first. Here, the pressure on the laboring masses was strongest. England and France at the time were building colonial empires. Fortunes could be made, or at least hoped for. The United Sates had an expanding frontier. Later it was to become the richest nation, and in terms of social security, the most backwards to this day. It needed the Great Depression to precipitate Roosevelt's Social Security Act in 1935. Unemployment insurance followed, but universal health insurance is still lacking today (1998). England and France undertook serious social legislation only in 1946, after the war had crippled their economies, and decolonialization had become a reality. Today every industrialized nation has a safety net of social institutions in place which insure a minimum of sustenance for all. Until recently that minimum often has been very comfortable, as for instance in Sweden and Denmark. But lately, due to increasing fiscal poverty, that net has begun to deteriorate everywhere.

Thus, progressive societies in essence had adapted, at least temporarily, to the need of taming the social woes created by industrialization. Note that these "progressives" were exactly the ones who had experienced the Renaissance movement at the most. It was due to this empowering heritage that here general education could further spread, and political stability could grow.

But this certainly is not true worldwide. Of the 5.4 billion world population today, only about 1.2 billion or roughly 20 percent live under such fortunate circumstances. The rest either are trapped in preindustrial conditions, or simply are poor. People previously in the Marxist camp have experienced a complete breakdown of social practices and values, and for them it will be a long process to regain health. Conditions in Africa not only remained preindustrial; there also was a breakdown of old social and economical systems without the sprouting of new ones. It was said before that an evolved culture can be destroyed by just one generation not being able to transmit the ordering values. This today is the prevailing picture in the previous Russian empire, among African tribes, and in many other places.

A most serious effect of the Industrial Revolution was this: as the agricultural revolution had done eight thousand years ago, the Industrial Revolution now also unleashed a new unprecedented population explosion (see fig. 9.1). This happened because of progress in medicine and because agricultural output multiplied through the use of machines and chemical fertilizers, and through added production in the "colonies." The colonized people also were superficially infected with the European ways. This mostly destroyed their indigenous cultural base, but made them multiply rapidly. This time the ever-increasing numbers reached levels which began to exceed what our planet can sustain. As pointed out before, populations grow exponentially with time if there are no constraints. The constraints may convert the exponential curve into a sigmoid (S-curve, fig. V2.3), with a constant saturation level. But as already noted, this constant level is an equilibrium sustained by misery.

Worse than that, instead of simply saturating, the world population may overshoot its ecologically sustainable limit and then experience a catastrophic collapse. This is often observed in animal populations that collapse together with their sustaining ecosystem. It is a result of the multiple positive nonlinear feedback loops that build up the ecosystem, as explained at the beginning of this chapter. Similar collapses and wild oscillations are well-known and understood by designers of all kinds of industrial and robotic control systems that also consist of such interacting loops. This whole area is called "system dynamics research." Some computer simulations of the economic and ecological future of the world predict just such a collapse beginning around 2025, unless we take immediate countermeasures (Meadows et al. 1974), (Meadows et al. 1992).

Yet one does not have to rely on the computer simulations alone. It is enough to follow the yearly reports of the respected "Worldwatch Institute," or the statistics compiled by the "World Resources Institute," in order to see where

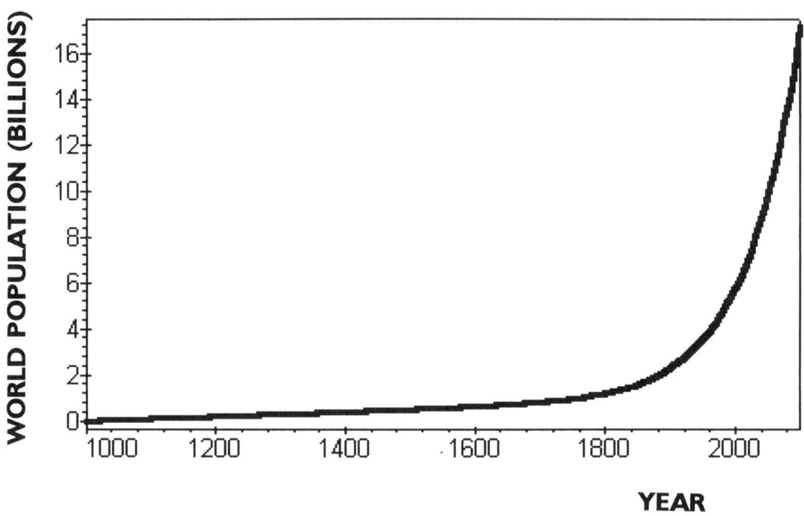

Fig. 9.1
The popualation explosion
[After: Ophuls, W, (1992)]

our technology based society is heading. If worldwide industrialization should be able to continue at the present rate, population may stabilize at ten to twelve billion by the year 2070 because all "developed" countries so far have been seen to stop growing. It is widely believed that this would be the absolute maximum that could be fed under the best of circumstances, that is if no climate changes occur from the increasing CO_2 level or ozone depletion, and if law and order prevail everywhere to insure productive use of all arable land. But this is unlikely to happen because social instability already is growing worldwide, and global warming has begun. Moreover, such a large industrialized population cannot be sustained for long because of soil erosion, pollution of ground water with artificial fertilizers, and the rapid depletion of nonrenewable resources. Water already is becoming scarce in many places. Oil will not flow forever. We are currently pumping at a rate of about 3.1 billion (3.1 10^9) metric tons a year. Even though hard data on world reserves are difficult to obtain, estimates show that if the current trends continue, by the year 2025 only 30 percent of that yearly volume will be available, and by 2085 all the oil essentially will be used up. The maximum coal production would be reached around 2100, and around 2500, coal also would be gone (Piel 1992).

Minerals are also being rapidly depleted. Without certain cheap metals, industrial production as we know it will be severely limited. Assuming continued exponential growth, tin essentially will be gone in sixteen years, lead in eighteen, zinc in thirty-two, copper in thirty-three and nickel in fifty-six years. Simply by maintaining current volume of mining, without further increase, tin will last for twenty-one years, lead for thirty-seven, mercury for forty-two, and copper for sixty-six years (Ophuls 1992). Of course, we will not run out of these materials overnight, but mining will quickly become prohibitively expensive. New finds are made on a limited scale, and new technologies can extract the minerals from lesser quality ore, even from previously processed and discarded material—but this cannot continue forever. Cost and energy consumption set a limit, and the more volume has to be turned over, and the more chemicals used, the more environmental damage is done.

It appears unlikely that the 80 percent of the world's population in poor countries will ever see the level of wealth of the present industrial nations, simply because there are not enough resources left on Earth. Scarcity of nonrenewable resources and environmental destruction may even rob the already wealthy of their good life.

Thus, the industrial age, as we know it, is grinding to a halt. The future will have to begin without the prelude ever having come to the expected shining conclusion. Adolescence has no conclusion, it only points to the next stage.

The present final decades of old-fashioned industrial society are characterized by the predominance of a hypertrophic ego system. The most visible indicators of this are excessive greed and the absence of long-term planning. Even women who previously, with help from their care-giving, reproductive nature, were able to remain keepers of some treasures of biological wisdom, now have joined the bandwagon. The taming of the social woes of technology again is beginning to slip. Education, particularly in liberal arts and classical humanism as it was practiced well into this century, had tended to prevent excesses through the existence of a middle class and a cultural elite. This is fast disappearing, if not already gone.

It is a tragic error to think that technology is "value neutral." During the millions of years of primate evolution, all the power tools at our disposal have been our own bones and muscles. Much later we invented digging sticks, fire, bows, and arrows. These certainly were not value neutral. They were extremely valuable stimulants of cultural evolution and of the ego system. But every stimulant, be it chemical or mechanical, can become destructive if it is strong enough to disturb the balance of an evolved system. Eibl-Eibesfeldt (1972) tells the story of a Papua who, as a reward, was invited to a ride in a small airplane. He asked to take a few rocks along. It turned out that he wanted to drop them on the village of his enemies. Given the unexpected technological stimulus, this hunter-gatherer had immediately invented bombing. Intertribal aggression is only the most visible instinct that becomes reinforced through the super-stimuli of effective weapons. The devastating wars of the past century prove this point.

Tribal chauvinism and the steamship, together, have built the British empire, thus destroying many budding indigenous cultures.

The automobile has destroyed town communities, and now we are trapped to use it, to pollute the air and live in social isolation in sprawling suburbs. Television and computer networks have made social isolation even stronger, halting the personal development of Knowledge II that so much depends on direct social contact. Instead, our mental health is thrown upon the mercy of exploitative Hollywood enterprises. The majority of the inhabitants of the United States divide their waking time mainly between selling it for profit and sitting before the screen.

Today, what should be education in civility and aesthetic judgment, under the influence of technology, has been reduced to mere training of operatives of industry and commerce. Single-sidedly we pursue what was called "Knowledge I," the knowledge of Darwinian survival.

Aided by new printing and communication technologies, a new superpower of indoctrination has arisen. This has been more often abused to spread "propaganda" than to educate. Even in democracies a destructive feedback loop formed between the public's raw instinctual curiosity in violence and bad taste, and the commercially-driven response of more violence and bad taste on the screen. This has destroyed a superb tool of education and a potentially promising art form. It has been reduced to the lowest common denominator when it could have served as a model of the highest possible aspirations of humanity.

Erotic love used to have the power to stimulate to the greatest personal achievements and provide education of the emotional sphere, of interpersonal respect, and of compassion, strongly aiding the process of individuation, and thus promoting happiness. The same tragic feedback loop has mislead the masses to confuse sex with love, and it has condemned millions to the life of emotional barbarians.

Add poverty to this and you have a prescription for disaster. In the United States, the number of children living in mother-only households has quadrupled since 1950, from 6 percent to 24 percent. In some neighborhoods, 95 percent of all children grow up under such circumstances. There are "fathers" who already for three generations have not had a male role model, and do not know the meaning of fathering. It is each for himself, survive if you can, have sex if you can, by any means, including crime and rape. Even among the well-to-do, divorce has become casual.

And the legislators, who possibly could improve much of this through wise legislation, are trapped in ancient ideologies. Adam Smith died more than two hundred years ago, and contrary to common belief, his thinking went much deeper than to simply push the blessings of an absolutely free market. Besides correctly understanding the forces unleashed by the free market, he also was correct in being skeptical as to whether ultimately that system would be good for humanity. Speaking of the laborer, whose activity due to the profitability of the division of labor is reduced to a "few simple operations," he says, "He . . . generally becomes as stupid and ignorant as it is possible for a human creature to become." This happens by necessity, "unless government takes some pain to improve it" (Smith 1776, pp. 781-82). Adam Smith also notes that the manufacturers and merchants "form an order of men whose interest is never exactly the same with that of the public, who generally have an interest to deceive and even to oppress the public, and who accordingly have, upon many occasions, both deceived and oppressed it" (Smith 1776, p. 267).

Today both of these points have not just come true; they have also been amplified to an extent Adam Smith could not have foreseen. Robert Heilbroner writes:

> A large-scale industrial economy is a very different kind of "commercial society" from the small scale economy of Smith's time. Smith's earlier form of capitalism can still be observed within the structure of modern industrial capitalism, tucked away in the listings of the Yellow Pages of the phone book, but it is no longer an accurate description of the organized power, the massive institutions, and the often ponderous workings of its modern form. (Heilbroner 1986, p. 152)

The small manufacturers and merchants of the eighteenth century have become giant global corporations with great and mostly unchecked power to deceive and to oppress. And the employed masses are not only not being educated much by a government's effort; also, they are increasingly stupefied by commercial media of communication who have an interest to keep them stupid.

They also are rapidly growing poorer. That is because of increased "productivity"; less and less labor is needed today to produce the same amount of goods. This process already had begun when agriculture became "business." Due to mechanization and artificial fertilizers, today less then 3 percent of the United States population are involved in agriculture, as compared with 12 percent as recently as 1950. The surplus workers had to move to the factories. For a few decades the manufacturing sector in developed nations was strong and healthy. It supported its labor relatively well. But technological progress caught up with the laborers soon. Automation, better productivity, and the export of work to poorer nations made domestic manufacture drop sharply during the 1980s. Now the globalization of industry has tapped into a huge oversupply of cheap labor.

Consequently, the real wages in developed nations like the United States have dropped dramatically while labor and the environment in poor countries are badly exploited. The income gap between laborers and owners has widened to such an extent that again the danger of social instability has come into view. Worldwide, the problem of not enough work is worsening. Often it is said that somehow computers will cause a "third wave" of prosperity. But computers, economically speaking, are nothing but labor-saving tools. How can they create more work?

Despite such simple facts and Adam Smith's skepticism, belief in the universally beneficent powers of the "invisible hand" of the free market, today has become an unexamined ideology of the leading establishment. From the insights on human nature as developed in this book, what can be added to the discussion in order to end this ideological sleep?

For the reader who has followed me to this point, this should not be difficult. Obviously this system of unhampered market forces is a genuine part and extension of what has been called the "ego system." More than that, it is a collective ego system that consists of many neocortices acting in unison, mutually reinforcing their beliefs and goals as is befitting a circular storytelling society. Even though all the rational activity that leads to economic success is driven by personal brain stem survival will of the participants (such as the ranking rivalries that become economic competition), it is a system freed from the human values of individuation and the quest for spiritual legitimacy. Human happiness, as we have seen, depends on progress in that quest, not on progress in accumulating capital and goods.

At the centers of today's commercial establishments we find less of what formerly was called a "managerial class." Since the advent of high technology and supercompetition we find "professionals," specialists of all kinds with advanced academic degrees, lawyers, financial consultants, systems managers, computer scientists, and engineers. They all have in common this: they work very long hours, are very good at their specialty, are competitive to a degree that virtually eliminates all private life—and they make good money. They think they are happy because they are addicted to their work and have money to spend. Besides sex for relaxation, they know little else of human self-experience. The money often is used to establish a luxurious house that rarely is a home, because both parties pursue the same addiction. They send the neglected children to private schools and camps, and, following the imperative of primate ranking order, exhibit to other people an image of great achievement. They have no problem to participate in the deceptions, oppressions, and environmental destruction on which their commercial success may depend. Such things are, for them, part of the progress of "business," which is good for mankind to pursue.

Such is the life for, through, and by the logic of the ego system alone. One other development reveals even more what is happening here. The new "lifestyle" began to flourish exactly at the time of computerization of the commercial enterprise. Now stocks are bought and sold by automatic computer programs. Billions of dollars are moved between continents by pressing a button.[50] Individuals talk through terminals to each other, in writing, while they are advised by their computers what to say or what to do. Any direct person-to-person encounter is inefficient and difficult in this world. Even the telephone is not any more answered by a person. This demonstrates how the whole system centers on rational neocortical transactions and their flawless continuation within computer logic. The ego system becomes enlarged again—by computers.

---

[50] This had not been foreseen by David Ricardo, the founder of today's theory of world trade. According to H. E. Daly (1993) and (1991), this is causing many of the world's problems, including pollution and the export and devaluation of labor, and is putting the dogma of free trade between nations in question.

The latest invention of this new order of lost souls is the information superhighway. Now all the individual ego systems of the world can unite. The irony of it all is that there really is no need for any more information. We already are flooded with more of it than any real person can ever digest. The knowledge that counts in the progress of real persons can be found in good novels, significant works of philosophy and science, and in works of great art, music, and dance. They have to be experienced, relived in long and quiet meditation, after being carefully selected out of the available material. But now one can quasi-communicate without having to encounter a full person at the other end. Things can be said that would not be said face to face. Experience with the network already has established that this medium tends to be appropriated by all kinds of cranks, religious fanatics who want to spread their maladaptation, irresponsible showoffs who spread falsehood, seekers of loveless sex, and political extremists, all suffering from the same modern malaise of many names: hypertrophic neurological dualism, inner void, cultural illiteracy, lack of emotional intelligence (Goleman 1995), or simply the lack of Knowledge II.

When it was said that simple and monotonous manual labor was stupefying, the new commercial system is super-stupefying in a much more radical sense. The assembly line worker at least could turn off his mind, or possibly even daydream to some extent. But this involvement is so total that it destroys the person by cutting it off from the richness of an inner life that our nature otherwise could have provided us. Our neocortex becomes just a logic chip among the silicon chips of the worldwide commercial network that knows no human values.

In today's "developed" countries, whoever is unwilling or unable to participate in that dehumanizing system is doomed to poverty. Also, because of the global nature of the network, and of capital flow, all the world's poor tend to be reduced to the same low level of the poorest nations.

We have slipped into this situation not because of a destructive power inherent in science, as one popular opinion wants it, but through thoughtless, free market driven misuse of technology. Since the ego system thrives on our central explainer's attachment to worldly achievements, the superachievements permitted through the use of technology become super-stimuli, which cause superdevelopment of the ego system. We are overwhelmed by the lure of our own achievements. We think we cannot live without the conveniences of technological consumer goods, and like in the drug addict, our only purpose in life becomes to do anything that will get us those goods. We gladly sell ourselves into slavery and do not know that there is a normal life to enjoy, tragically short as this may be. Knowledge II would help finding our true needs and responsibilities, but it remains undeveloped.

The shining example should not be today's superachiever, but the Bushman who can satisfy all his needs with three hours of daily work. The rest of the day can be filled with what we are designed to do: live and be happy. With all our knowledge and technology we should be able to do better than the Bushman. But instead, in this century Homo sapiens has turned into *Homo economicus,* that is to say, into mere cogwheels of a self-sustaining machinery that runs on human sweat and blood. No drug addict can live a life in harmony with his biological framework and be happy. The presently bemoaned erosion of culturally evolved social order is a direct result of this widespread ego-system addiction.

It almost seems as if George Orwell's vision, described in his novel *1984,* has arrived. But it is not yet a totalitarian state, but it's the anonymous commercial system that has reconditioned most of us to become willing victims of the "marketplace." And the perpetrators, like in *1984,* even think to do humanity a favor. The Gross National Product (GNP) grows. But what should grow instead is the Gross National Contentment (Goldsmith 1994). The latter is rapidly declining toward social instability.

In the course of the industrial epoch, we have thoughtlessly reproduced. We have enslaved. We have destroyed "less developed" cultures. We have let our own artistic expression degrade from the refined joy of classical music to hoarse wailing of raw sexual heat. We have created crime-ridden cities. We have done severe damage to the world ecosystem. The mineral treasures we have squandered will never come back. Future generations will curse us. Our time will be called "the century of waste and hypocrisy."

Interestingly, as this epoch is ending, the economic dogma that has sustained it also seems to crumble. Herman E. Daly, one of the main proponents of "ecological economics," has said, ". . . my major concern about my profession today is that our disciplinary *preference for logically beautiful results over factually grounded policies has reached such fanatical proportions that we economists have become dangerous to the earth and its inhabitants"* (Daly 1993, italics added). His mainstream opponent, J. Bhagwati (in the same issue) accuses the environmentalists of pushing "values" and "appeal to instincts, rather than to one's intellect," thus admitting that his own system is purely rational and removed from human values, as we have recognized the ego system driven industrial epoch to be.

Like in all ending epochs, intellectual uncertainty reigns supreme in many quarters. At the "brain stem end" of the chasm, so to speak, are the pupils of the likes of Martin Heidegger who have spawned the ideology of "postmodernism." Here is a gut movement that has completely renounced human progress and the power of reason (Gross and Levitt 1994). According to such beliefs there are other ways of "knowing," as in shamanic ecstasy. The Western way of finding truth is renounced. Other "multicultural" ways are pushed, including "Afrocentrism," on the way distorting the archeological record.

In fact, the study of history itself is declared a meaningless enterprise. This extreme reaction can only demonstrate to what extent the neurological dualism has widened and that the tension has reached a breaking point: the mind of these ideologues is broken. We see here near-psychotic inflation (in the Jungian sense) of the instinctual sphere. Direct action, gut feelings, heroism, and witch hunting have set in, as their grand master Heidegger already had prescribed when he urged the students in Freiburg to join the Nazi movement. "The great Mystery," asks Allan Bloom (Bloom 1987, p. 55), "is the kinship of all this to American souls that were not prepared by education or historical experience for it . . . was there something that the American self-understanding had not sufficiently recognized or satisfied?"

Our evolutionary synthesis can answer Bloom's question. America is the world's leading technology driven civilization. The neurological dualism is wider here then anywhere else. That is why Heidegger's nihilism is loved at American Universities, that is also why Christian Fundamentalism only grows here, and why earlier Freudian analysis had found the most receptive audience here.

Yet, fortunately, during that period of hypertrophic technological development, some good things also came along with the explosion of real knowledge, which bode well for whatever form the future may take:

1. Objective knowledge grew exponentially. It reached the point where the human spirit begins to understand itself.
2. Benevolent systems of government have been invented.
3. Mass education at least was introduced.
4. A system of worldwide cooperation was initiated—the beginnings of a future world federation.
5. A science based high technology has been created. Even though this will have to be radically reigned in, elements of it can form important ingredients of a future culture.

True, all of these achievements will have to be modified for improved future use. But they are created, and will be available to future generations. In fact, the long range evolutionary perspective of this book permits us to recognize the present difficulties, hopefully, as nothing but the birth pains of a new epoch.

Catastrophe or Smooth Transition?

We are rapidly approaching a critical period in the evolution of our species, if not of all life on Earth. A survival test of the human spirit is coming up which is of a kind that has not been known in all of the past 3.5 billion years of emergent Darwinian evolution in this solar system. Whether we like it or not, we are about to become initiated into adulthood. This is a test of our latest acquisition: the self-adapting cortical system and its cultural extension (Popper's World Three). Will this system be able to muster enough power of objective reason in order to check its own excesses and at the same time not fall prey to the most ancient phylogenetic imperatives: the instinct to reproduce at all cost, the instinct of tribal aggression, and the related instinct to compete at all cost? In other words, will Knowledge II also reach the necessary level of accomplishment in a sufficiently large population, so that a turnabout can happen?

This human predicament appears in its clearest distillation among the people who live in the developed world of technological civilization. It is clearest of all in the United States. But others, who live in the "zone of turmoil" (Heilbroner 1995), under Islamic fundamentalism, tyranny, or where the social system simply is broken, even they are not outside of the shadow cast by technology. Fundamentalists of all denominations thrive on protest against encroaching technological civilization. It is a wrong kind of protest, but it is against a real threat. This makes them part of the system. The poor in broken countries without a fundamentalist rule cannot escape their fate either, because instead of pragmatically focusing on solving their problems, they are drawn to the fata morgana of unattainable technological bliss. Therefore, the outlined problem is a problem of the entire world, and its solution would have worldwide implications.

Even if we cannot predict the future, we can talk of possible alternatives, of extremes, between which one could expect to find the actual outcome. This is not idle talk. It is necessary to sharpen our vision of the future even as little as the matter will permit. We are being confronted with urgent moral decisions. The future of the world and of our children will be determined by what action we take today.

One extreme scenario materializes if business should continue as usual. No awakening occurs. No significant reigning in of market-driven exploitation of resources and of the mental health of citizens is organized, and population continues to grow as it is today. The poor countries will be the first to disintegrate. More Somalias, Bosnias, Rwandas, and Haitis are bound to overwhelm the political will and the ability of intact countries to intervene. After mass starvation and epidemics, they will revert to small groups of bandits. Whoever survives, after generations, may see a new primitive social order reemerge.

But the industrialized countries, too, are not exempt from social chaos when the industrial base begins to collapse and food becomes scarce and expensive. At first we can expect social stratification to continue to grow. Later in the underclass, unemployment, starvation, new diseases, and anarchy will ensue. Actually a preview of this already plays in many inner cities of the world. Possibly, fascism-like totalitarian regimes will proliferate because the skinheads are ready and waiting everywhere. In the United States armed "militias," "survivalists," and white supremacists are eagerly awaiting an occasion to transform their paranoid fantasies into bloody action.

I admit this is a worst case scenario, but it certainly is not an impossible one. As we have seen earlier in this book, biologically we are just intelligent apes. It is the evolved cultural superstructure that uniquely defines us as human beings. When that collapses we have to start at the beginning again.

The most fortunate scenario is at the opposite end of the spectrum of possible futures. Externally it is characterized by what has become the catchword of "steady state" or "sustainable" economy (Daly 1991). Unfortunately, unlike the previous one, this scenario cannot be expected to arrive by itself. In order to become reality it will need intervention based on superior human insight and moral resolve. In other words, it needs a change in paradigm, a revolution of attitude of the kind that I have claimed is already in the works today.

But, even with all the human resolve, is such a development possible? I think it is. Hope can begin with the fact that there are alternate energy sources. For instance, if only 1 percent of the Arabian desert were covered with solar cells, this would produce as much power as the local current oil production, and this source would last forever. The United States has less sunshine, but is larger; 0.4 percent of the land surface would provide all the current electricity needs (Piel 1992). This is a lot of surface, but also there are wind and water, and energy waste can be reduced. Nuclear energy so far remains a problem. But here is an area where technological progress may still bring changes. France and Belgium have significantly improved their air quality by going nuclear. Hydrogen fusion may be on the horizon, as well as the direct generation of hydrogen by solar energy. Hydrogen may become the zero pollution fuel of the future. A change of energy sources will be costly, but is unavoidable.

Since energy is not an unsolvable problem, we should begin to replace oil and coal as energy sources immediately. They are much more valuable as chemical raw material. When oil is gone many products of the chemical industry will disappear. Plastics, will become much more expensive. When coal is gone the chemical industry will have to rely on plant material. This would severely reduce land surface available for food production. This is just one of the knowledgeable moral resolves to be made.

Another development invites guarded optimism: some limited progress in birth control seems to have been made recently. The latest population surveys in many developing countries show a somewhat encouraging trend. Apparently education in contraceptive techniques is beginning to have an effect.

Slow progress also can be seen in reducing material throughput and pollution, mainly by recycling. In Germany much unnecessary merchandise packaging has been eliminated by legally encouraging customers to simply leave the fancy boxes in the store. Manufacturers got the message. The European Union is considering legislation that would require car manufacturers to take back and recycle old automobiles. This already has prompted new modular designs to ease such operations and save material.

In all developed countries there is growing public awareness of the need to conserve and recycle. Some manufacturers have discovered that saving energy can make their operations more efficient with only little investment. In many cases internal recycling of chemicals, solvents, even water, also were found to be better solutions.

Slowly but steadily the environment is becoming cleaner, at least in developed countries. An international ban on CFCs has been signed. CO_2 emission reductions are being considered. The use of pesticides is being reduced, and some are outlawed. Natural predators of the pests are introduced instead. Organic farming methods are growing in use. Some animal species have returned from near extinction. Even though these positive signs are more than balanced by impending disasters, they inspire hope.

On the negative side, the depletion of minerals, oil, and water was already mentioned. Worldwide, 20 percent of the population is at least malnourished (World Resources Report 1992-93). The United States, the breadbasket of the world, is expected to become an importer of wheat in the foreseeable future. Worldwide, 1.2 billion hectares of agricultural land (the area of China and India combined) has been significantly degraded, some of it beyond recovery, other parts requiring major investment which is not available. In sixty-nine developing countries, the per capita food production has declined. On the average, the maximum sustainable amount of fish is being caught now in the seas. In some areas, as in the southeast Pacific, almost twice the sustainable amount is being harvested, ". . . it may take years of rehabilitation just to maintain current production levels" (ibid.), and a negative impact of ozone depletion is expected.

How bad our coming plagues will be, nobody knows. Possibly, today's underdeveloped regions will carry the brunt of it. Maybe the politically chaotic societies will mutually exterminate each other, thus, in a quasi Darwinian selection process, helping the more advanced ones to survive. But maybe we all will be lucky, and the human spirit will turn the corner just in time by a small margin. The question of catastrophe or smooth transition will have to remain unanswered.

Are there alternatives that lie outside of the domain between the two extremes? Can we undo history and start again at the beginning? No, the only way is forward, to the perfection of an objective understanding of the problem and its solution. We cannot throw out technology and go back to being hunter-gatherers, nor can we all live by Walden Pond. Can we become early agriculturists? I don't think we want to. This was a hard lot. But we can go forward to a life that is at the same time closer to nature as it is enriched with the knowledge and wisdom of two thousand years, and a sustainable technology that serves our true needs, not our imagined wants. And we can design a sociopolitical system that encourages spiritual development and outlaws as mass hypnotism, the indoctrination to consume ever more, and thus condemn the masses to remain in the atavistic state of mind forever.

Following the realistic attitude of Western civilization, however, such spiritual development cannot be toward the self-withdrawal of Indian mysticism. It will be open to the world and at the same time fully embrace the human potential for cultivation of the inner domain (Knowledge II). I see no biological obstacle whatsoever to such revolution. The old feedback loop that created the hypertrophic ego system is not preordained nor is it a biological necessity. It is just a cultural habit that once was very successful. It served to solidify our central explainer's effective model of reality. But today this project is not just finished; it is overdone. At the same time we have remained so backwards in developing Knowledge II, that we even fail to apply the great objective insights available today in order to educate us about ourselves. Jacob Bronowski in his widely circulated book, *The Ascent of Man*, and television series has had the vision to declare:

"And yet, fifty years from now, if an understanding of man's origins, his evolution, his history, his progress is not the common place of the schoolbooks, we shall not exist." (Bronowski 1973, pp. 436-37)

He was right. And of the fifty years, twenty-five already have passed. Time is running out.[51]

Conclusions on Political Theory

"Drawing blueprints for future societies is a favorite pastime of intellectuals and dreamers, and it is often dismissed as a waste of time. Detailed blueprints no doubt are a waste of time judged by the likelihood that future people will precisely follow their specific impositions. But a general outline or image of a desirable future is an absolute logical necessity for any kind of policy that is not a mere repetition of past practices" (Daly 1991)

I have compared the present situation in modern technological civilizations with the Orwellian vision. But fortunately, it is not yet political totalitarianism from which we suffer. The "big brother" is more subtle, and so far our political system has not directly come under his power. But I think it is in danger. The United States Congress so far has been unable to shake the influence of corporate money. Also, there are nations today where a government by and for the industrial giants is almost a reality: Japan and South Korea. On the planetary scale there is similar trouble on the horizon. Earth is becoming an integrated commercial sphere without any supervision. The only powers that fill this vacuum are global industrial giants pushing their own interests.

Ultimately it is not just the conditions in modern technological civilization that call for a re-engineering of the political system, the new insights on human nature demand it. From the very foundation of accepted democratic political theory there has remained one unexamined doctrine about human nature: the individual was taken as an absolute, self-determined entity, internally only answerable to God, from which this personal absoluteness was derived. This originated in the Christian medieval universe: man as a divine microcosm, a mirror image of the equally divine universe at large, endowed with the Holy Spirit and thus absolute. We have seen in chapter 8 how the idea of the inherent Holy Spirit had legitimized new pursuits of the ego power.

Later, Descartes' *"Cogito ergo sum"* ("I think, therefore I am") became the ultimate philosophical statement of ego-system autonomy. The Christian root of that doctrine became irrelevant. By the time this absoluteness of personal autonomy was pushed by Montesquieu, Condorcet, Voltaire, and the "Encyclopedists," in order to call for absolute individual liberty, it had lost all

---

[51] I feel the need to confess to my readers that Bronowski and his absolutley genuine "whole person" positive approach, twenty years ago, inspired me more than anyone else to overcome a certain cultural pessimism that had been conveyed on me by contemporary philosophy and wartime experience of social chaos. That positive influence is also greatly responsible for this present writing.

Christian undertones. Rousseau even had a completely different view that called for censorship and propaganda in order to better society. One common ground, however, was rebellion against church, authority, and unexamined tradition, all marks of ego system addiction.

Subsequently, the synthesis of Franklin, Jefferson, Adams, and Madison reverted to the "one empowered individual under God" concept. They reinforced it by declaring it to be a private religious value, and consequently an unquestioned and protected personal element of the political system.

But the preceding pages have uncovered quite a different human nature. Humans are members of a collection of "emotional puzzle assemblers," interacting, telling stories to each other, learning from each other. Thus, individual ego system attitudes form in a "Game of Life" fashion from the influence of neighbors, and in turn influencing them. The only fixed references are the phylogenetically acquired brain stem wishes and the realities of physical nature. These are the only constraints under which cultural traditions, accepted "values," and beliefs develop. But between these constraints no predetermined contents of consciousness are to be found. Hegel and Marx were right in this respect. Human beliefs and values can be molded to a degree, as the long success of the Russian communist rule has demonstrated. But its ultimate failure also has proven what this book has derived from the evolution of animal behavior and from functional brain architecture: what cannot be molded are the instinctual determinants of the brain stem. The Marxist empire died because it had eliminated competition from its social system.

We have seen that, ultimately, human well-being depends on the closing of the neurological dualism. If there are gifted individuals leading the way, then traditions should be expected to evolve to help this process rather than to hinder it. Under fortunate circumstances, before the impact of technology, such evolution indeed was seen to happen. It was driven by strong individuals in leadership positions who by themselves were able to achieve personal integration in order to become role models and wise lawgivers. Ruling elites, even churches, and enlightened monarchs appeared who were able to project their acquired wisdom upon the subjects by establishing laws, supporting cultural institutions, and promoting other exemplary individuals to leading positions. In this way the admired high cultures in eighteenth century Europe had come into existence. But this was an uncertain enterprise that more often failed than succeeded, and that often led to political repression. Unlike today, in such systems there was no belief in this absolute autonomy of the individual. The absoluteness was delegated to the monarch alone.

But all this is gone now. The splendor of good monarchies has disappeared together with the misery of bad ones. We now subscribe to the here contested belief that everyone is an absolutely free and autonomous agent, who knows best what is good for him or her, and government should restrict that freedom as little as common interest will permit, preferably even less. We also granted the privileges of such free persons to corporations as a whole, naming them "judicial persons," and thus giving them full protection of their freedom of expression.

Successful as this state of affairs has been in the past pioneering society in America, or in overthrowing despotism, or during vigorous industrialization, it now has become atavistic. The social problems that grew on the basis of this faulty assumption can be understood with the often used concept of the feedback loop. There are several such loops. The most visible one in America today came into existence through commercial mass media that have been given unconditional freedom of expression. It goes like this: once a significant fraction of the population suffers from a lack of inner integration, it is attracted to expressions of violence and loveless sex in suspenseful plots without artistic (educational) value.

Since the media establishment makes money by attracting customers, it offers more of the same. That further increases the level of public barbarization, which then demands more bad taste, and so on forever. Exactly the same mechanism has enshrined adolescent musical noises to become the accepted standard. Even in the magazine- and book-publishing establishment this loop was at work. There have been times when bestsellers were exemplary literary achievements. Now sensationalist junk leads the way.

Most American public schools have not resisted that downward spiral, because the ideology of the autonomous agent does not permit universal standards of achievement, nor does it permit the teaching of values. Thus, like in all positive feedback loops, personal disintegration has grown exponentially until we now have irrational "tribal militias" that worship assault weapons and bombs, paying homage to our primate nature of intertribal aggression. Yet nobody notices that his or her freedom to self-integration has been violated.[52] On the contrary, more of the same is demanded. Thus, our freewheeling, "naive" democracy in time is bound to reduce everyone to the lowest common denominator of genetic primate nature, ultimately destroying itself, if no corrective action is taken.

Instead, what should happen is that everyone is lifted toward the level of our most accomplished creative and exemplary personalities. This, of course, cannot be prescribed, nor directly legislated or achieved through censorship. But the same feedback loop is capable of reversing the trend. All we need is to insist that the media and the schools be run by truly artistic and exemplary personalities, who, like everyone else, are given all the personal freedom of expression they want. When they become role models, a small amount of emotional education (Knowledge II) can be injected into the population. When accepted, this will ask for more of the same. So the feedback loop now begins to amplify human self-integration, and again nobody notices being educated or deprived of any freedom, because personal autonomy is an illusion. There is no measuring stick other than, on the long run, an increase of social stability and happiness.

---

[52] As far as I can tell, the concept of freedom to self-integration has not been coined before. It depends on understanding of the human predicament in terms of neurological dualism.

All real social change includes corrections to the political system. Therefore I will now briefly look at the origins of current political theory, and search for improvements that can be made on grounds of a better understanding of human nature. The successful theory on which modern democracies are built, like this book, also began with an assessment of human nature. Thomas Hobbes (1588-1679) had noted that life in the state of nature is "nasty, brutish, and short." His "state of nature" can be observed in present-day street gangs, who are deprived of transmitted culture, including political culture. I am sure Hobbes had ample occasion to observe such conditions in Cromwell's England and to hear reports of the deprivation caused by the Thirty Years' War on the continent. Therefore, he concluded, a sovereign was needed to protect one's life. Out of self-interest it was necessary to obey the sovereign. This was a humble, but realistic beginning.

Soon John Locke (1632-1704) recognized that the wish to acquire property was a natural incentive to work and self-improve, and a positive force to improve man's lot. Therefore he extended the needed protection to "life, liberty and property." My reader will readily link the desire for property to the desire for territory and rank. Both are at the roots of our motivation. The energy they create can be redirected in many ways. Basically, this protection tries to insure a freely developing ranking order that already was found to be a positive force in animal societies.

Charles de Montesquieu (1689-1755) again returned to the drive for power and rank by warning that the sovereign would rule absolute and abolish liberty, even property, if not held in check by a separation of powers; so that one alone could not impose laws without being jealously checked by the others. This, again, was checking domination with ranking rivalries. Ever since, indeed, primate toughness runs rampant among the politicians themselves, but most of the time the citizens of the commonwealth are reasonably protected.

It testifies to the genius of Hobbes, Locke, Montesquieu, and other philosophers of the eighteenth century, and of the founding fathers, that long before Darwin and ethology, they had a picture of human nature that was realistic enough to build a long-lasting system of government on it.

Yet enlightenment philosophy and the political systems based on it are now more than two hundred years old. Only minor adaptations have been added later. Since that time our world has changed so radically that if John Locke would be transplanted to the present, he would not recognize this as his old planet Earth. If he lived today and saw humanity in "developed" societies uprooted, enslaved, and brainwashed in the name of commercial progress, and in "undeveloped" societies uprooted as well and starving, he undoubtedly would demand two more articles of protection to be installed by enlightened government: the protection from techno-fascism[53] and the protection of freedom to self-integration, or one could say, freedom from psychological indoctrination.

The question arises whether our traditional democratic systems can live up to the challenges posed by modern technological society. Without reforming the

---

[53] This term was introduced by Erich Fromm, the last wise old man of the psychoanalytic movement. It means the de facto domination of political decision making and public doctrine by the self-interest of technology-based corporations, large or small.

present state of affairs, I think the answer has to be no. Instead of adapting the political system to changing needs, in the United States we still quarrel about an eighteenth century ideological dinosaur: federalism versus populism. As if national corporations can be regulated by local decisions. As if objective truth can be decided by vote of the local school board.

The original vision was that elected representatives would make decisions, based on their own wisdom and conscience, in order to increase the long-term average common well-being of all. This rested on two premises: (1) that once elected, the representatives would be free to exercise their judgment, and (2) that for them there was a way to know what the common good was. But meanwhile both premises have collapsed. Instant news about every intended decision causes instant and amplified criticism from interest groups, especially groups that represent the big brother establishment, or from people who have no occasion nor wish to fully understand the matter. The representative is not free to follow his or her own conscience, for want to be reelected. Secondly it simply is not clear anymore what the common goods are, because human existence in today's world has become so complicated.

The collapse of the first premise has made us become something like a direct democracy, which already had been seen to fail in ancient Athens. The collapse of the second premise calls for representatives with exemplary knowledge and wisdom, like the philosopher-kings in Plato's Republic.

But instead there is this creeping problem already envisioned by Alexis de Tocqueville in 1835. The representatives are not "natural aristocrats" (Tocqueville's term), that is, not leaders who in themselves have achieved a level of personal development (realistic knowledge, individuation, wisdom), that is necessary in order to cause good judgment. But what he could not have known was that mass media of instant communication would become invented which would magnify the problems out of any imaginable proportion and further discourage personal growth of the representatives as well as everyone else's. He also could not have predicted the influence of large impersonal corporate systems on the representatives, nor that in the complex modern world simple "good judgment" is not enough. It must be true objective understanding.

Despite the difficulties, Tocqueville was convinced of the inevitability of democracy. Interestingly, in some way he linked this to God's providence: "Wherever we look, we perceive the same revolution going on throughout the Christian world" (Tocqueville 1835). Much has been speculated about the roots of his "inevitability theorem." It appears to me that the roots go back to his superb intuition for the underlying currents in the Christian cultures: for what I have called the "strongest spiritual legitimacy of all times." Only in these cultures could "natural aristocrats" arise (as opposed to members of a heritable caste), who, due to their well-integrated personality, could be legitimate leaders. But Tocqueville feared that in democracies such leaders could not easily arise. How correct he was can be seen in Washington today. The process has relapsed into power play between individuals and traditional, but unfounded ideologies. In the absence of a clear and contemporary common goal, without objective self-understanding, and without understanding of the sociopsychological mechanisms, one could hardly expect otherwise.

So there are four different problems to be solved:

1. According to Tocqueville, our democracy is flawed because it is not very good in producing capable leaders.
2. The challenge of a creeping technological totalitarianism has to be met.
3. There is lack of a contemporary common goal.
4. Insightful legislation is hampered by a lack of objective knowledge.

I will address the first point first. Tocqueville's "natural aristocrats" are personalities that need the best possible circumstances to arise. He was thinking of the security, leisure, and educational opportunity that existed in old aristo-cratic families. But even there, of course, success was not insured. It took a lifetime, special gifts, a sense of esthetics, learning of reason, and the experience of a healthy religion in order to really progress in what could be called "practical wisdom."

I believe that all this honorable effort was rather inefficient, because no clear understanding existed of what was being achieved. Becoming a natural leader was an accidental byproduct of an educational process of no small magnitude. The new objective self-understanding, however, cannot only better focus such process of education, but it also makes it accessible to persons without exceptional gifts. The knowledge to be transmitted is straightforward and factual. As technological engineering is built on objective natural science, so will social engineering be built on the objective understanding of human nature and predicament.

Therefore, once objective humanism has become common knowledge, good leaders will arise more easily. Also, a common goal will exist again. This will remain "the pursuit of maximum happiness" for all, but we will know better what this is, and what it is not. The creeping techno-fascism can be checked through democratic legislation, once the situation is understood. All will depend on being able to inject into a social system, that presently is so much dominated by doctrine and emotion, the tendency to learn and respect the facts of objective human truth. How can we achieve this and make it a permanent habit?

All this cannot be done by fiat. The time has to be ripe before visionary leaders can precipitate the necessary changes. In the meantime I shall try to propose my own limited vision.

Given the shifting nature of our circular storytelling society, a kind of incorruptible standard of the insights of objective humanism has to be created, and of whatever social engineering knowledge has been built on it. But this cannot be a rigid standard, like for instance, the platinum-iridium meter that is kept in Paris. Life's essence still is change. Any rigid norm would soon become outgrown and would have a totalitarian smell to it. The standard has to be a living standard, in short a panel of constantly striving and excelling individuals who are as independent as the Supreme Court is today. They accumulate among themselves the objective knowledge of social and environmental affairs, and have an institutional power to be heard. They also initiate and supervise new collective efforts to expand that knowledge.

Today, panels of learned people are not in high esteem because in human affairs there are many opinions and not much truth. But I am talking about a time when there will be enough objective knowledge so that social systems can be engineered the way bridges are built, so that rivers can be crossed safely. Both kinds of engineering need high moral standards, but the social engineers have the highest responsibility of all.

At the core there are two truths to be upheld by such an institution: Human nature demands the protection and nurture by a proper cultural environment, and the spaceship Earth needs proper protective maintenance.

I will now summarize some conclusions. This will partly resemble a "humanist manifesto," forming a basis for future improvements of the social system, and partly it shall contain some more detailed thoughts for such future social engineering.

1. Despite all wishful beliefs to the contrary, man is alone. We are the result of natural tendencies of the universe. There is no other "cosmic" intelligence at work besides our own. We have the absolute responsibility for everything we do to ourselves and to the planet. No one will rescue us and our children from any natural events, or from the consequences of our own folly.

2. Our nature is such that we find in ourselves wants from a mode of existence that we share with other animals, and which often are opposing personal and common well-being. Among them is the tendency to get involved in competition, aggression, tribal chauvinism, and violence. We also have the natural ability for love, compassion, and a desire to know. The latter should be systematically cultivated so that the former can be held in check.

3. Human progress is cultural progress. Advanced cultural tradition is not transmitted between generations without effort. It must be built on a well-designed system of education. Such education must cultivate reason and the application of reason, the development of good taste for art forms that foster positive human development (Knowledge II), and of practical skills for improving the well-being of present and future generations (Knowledge I).

4. "Values" are not arbitrary, there must be an effective "hygiene" of them. Knowledge and proper cultural tradition can foster such hygiene. It builds personal contentment and mental health. Such progress depends on the cultivation of wants which are conducive to a maximum well-being for all.

5. Like human physical health, the development of the human personality has to be protected. In addition to the human rights already recognized in advanced democratic societies, there must be a human right of freedom to self-integration. There are types of indoctrination that destroy this freedom,

such as attempts to inoculate wants that serve commercial gain, or gain of political power, while destroying positive cultural values. Freedom of speech cannot be claimed by impersonal commercial enterprises that engage in such activities, particularly through the mass media. Especially the superpower of suggestion, as exercised by radio and television, and unrestrained commercial access to it, are incompatible. The solution is not censure, but decommercialization and a good personnel policy.

6. Man's well-being depends on living in a healthy natural environment. Among the taught cultural values must be objective knowledge about the preservation and rebuilding of the natural environment.

7. We have to respect people who are unable to shed dogmatic religion. But the history and true nature of religion should be taught in schools. It should be made equally clear that cosmic mysteries remain, and will remain forever. Artistic expression of the unique and responsible calling of humanity should be an important part of future education. This may lead to a new spirituality based on understanding man's true predicament.

8. Since cultural values are not arbitrary, government has to include an institution that consists of a group of highly respected individuals—top universal minds, scientists and artists, who are independent. They should have the power to veto any laws that may be detrimental to a positive development of humanity. They are sworn to, and are guardians of objective knowledge about human nature, the social system, and the world ecosystem.

9. Advanced human society cannot succeed on old accidental ideologies. It requires real technical knowledge and wisdom. Legislators, therefore, should have to satisfy strict educational requirements, and proven personal achievements in the objective humanities before they can become eligible to seek elected office. Even today, in order to protect society, every important professional has be licensed.

10. National states that compete with other states for territory or status are a remnant from ape society. Nevertheless we have to recognize that nation states are natural units of social organization. To foster their positive global interaction, establishment of an Earth federation should be the most urgent and immediate political goal. The most important duties of such a federation would be to foster universal education, to see that global commercial activities are not destructive of social structures, and to set global environmental policy and population control policies.

Many suggestions for detailed social engineering already exist. For instance, Kenneth Boulding's system of transferable birth licenses could be taken seriously in future planning of population control laws (Boulding 1964). The other two institutions proposed by Herman Daly (1991), (the distributist institution and the depletion quota) can form a basis for further discussion.

Marxist socialism has failed, but untempered free-for-all capitalism also has shown its destructiveness. Also, the old ideological divide between right and left has lost its meaning. Today's complex social problems can only be approached with true objective knowledge. So, instead of right versus left, it should be more like knowledge versus ignorance.

I am aware that many of the above points are in strict opposition to the erroneous "individualistic" self-understanding, that prevails today. In America such cultural habit is stronger than in any other industrialized country. This had its origin in the pioneering frontier society of the past, where it had contributed much to civilizing success. Today this trait survives by cultural retention alone. It has become atavistic. European nations, due to their monarchic history, generally have been saved from this extreme overvaluation of the individual. For them, mild socialist interventionism was not a problem. It has borne fruits of which the inhabitants of the United States are still deprived.

Nature cannot be exploited forever. The marketplace, particularly if joined by technology driven commercialism, cannot be permitted to control our lives forever. It rapidly is turning into tyranny. Let us hope that the free pioneering spirit of the past can be reawakened and will accept the new frontier of the future: human freedom and self-preservation in the face of a new Leviathan. If this theme sounds familiar, this is because here is nothing but the logical continuation of the political project of the enlightenment period.

Epilogue

Let me finish with some personal speculations that are less designed to predict the future, but to convince that different, and happier, modes of social existence are conceivable.

In the future my conjecture sees a rural population of many small farmers who have left the cities to live a better life in intimacy with nature and their family. In fact, first signs of this exodus are already here. Thanks to computer networks, the rural population in the United States is growing again. So maybe, after some period of maturation, the network is yet to become a positive contribution. Karl Ohm, publisher of the Rural Enterprise quarterly sees a quiet revolution happening (O'Neill 1992). Educated professionals are seen to begin part-time farming and then are drawn into a new and more satisfying lifestyle.

Visionaries like Wes Jackson[54] of Salina, Kansas, experiment with new food crops which prevent soil erosion. Rapid advances in biology are leading to completely novel ways to produce food or to genetically improve crops. For instance, there is hope to genetically engineer cassava to have the protein content of meat. Alfalfa that has more protein and requires less fertilizer already is available. But even without such advances, more intensive farming gives better yield and soil protection. Organic farmers and the Amish of Pennsylvania demonstrate this. When the radical times of change come, it should not be impossible to interest and to empower the urban poor to go and do intensive farming.

I am looking to the not so distant future, after the unavoidable changes are behind us. Nature by then may have become more precious than ever before

---

[54] Jackson is a recipient of the McArthur Foundation's "Genius Award" for his efforts.

because so much of it had been destroyed during the wild age of adolescence, so many species were exterminated, and the climate changed. Those farmers, however, do not resemble farmers of past centuries. They manage to work and improve their land with limited effort. Biotechnology and simplified machinery have much improved their lot. They do little more than subsistence farming because the great population centers have disappeared. Possibly much of the land is being reverted to a natural state. A good fraction of the food is produced in solar heated, energy self-sufficient installations that operate on the principle of closed loop biological microsystems.

A significant amount of free time is available and used in the pursuit of knowledge, wisdom, and happiness. They are more like academic gardeners. Villages do exist and small towns, and travelling theaters, philosophers, and artists. Social cohesion has much improved as compared with today's lonely suburban crowd. But, in addition, they participate in a matured global electronic network that provides them with much of the individual intellectual and artistic challenge and communication, and the information necessary for their scientific and sustainable farming. Fortunately electronic communication and data storage is not material intensive. It can become a sustainable, low-energy technology.

One of the improvements came through the evolution of a two-tiered network, an edited and an unedited one. Trusted editors make a living by shielding their customers from having to sort out sense from nonsense and beauty from ugliness. The unedited tier has shrunk into insignificance. Compared with today's mass of published junk, not many books are actually printed on paper, but the valuable ones, that foster true inner development, have become beautifully bound and cherished possessions.

Dogmatic religions are gone. Instead a new spirituality has arisen that beholds as the highest good the beauty of nature, including our own inner natural self. It sets the highest demands on education and a harmonious inner development of that self.

Under an enlightened world federation, the future citizens of our planet keep population small and stable. Again, the matured computer network has proven to be a valuable asset to a smoothly functioning world democratic system. Immediate access to relevant facts has made decisions easier and voters more informed. Simulations of the future can be accessed that play the consequences of any contemplated decision. Voting is done through this network.

Education is built on the needs of each individual child. Objective insight into the neuro-psychological and motivational structure of the learning instinct has made learning fun and much more effective. Active personal learning explorations are aided by immediate access to all of the world's libraries and to presentations by internationally recognized teachers, but good books certainly have not disappeared. The schoolteacher's role is to guide and to challenge. He or she has acquired the highest social status, because the transmission of a realistic and civil culture, finally, has been recognized to be the most important factor in human happiness and long-term survival. Civilized life in a high culture is the proper emergent level of human existence.

The technology is characterized by durability and advanced simplicity. It is not only sustainable, but also designed to reverse previous environmental damage. Even though there are automobiles of a simple, practical, and durable construction, it has become fashionable to have horses for riding as well as for pulling carriages. The public transportation system has been revitalized. Life has become slower and more pleasant. These people know their own place and the needs of their own animal nature. They have the knowledge that enables them to live the best possible life for all.

After the "feminist" confusions of the late twentieth century, women finally have found their collective cultural identity as separate and different from the male dominated aggressive competitive ego system civilization of today. This, in fact, was the turning point that made cultural rejuvenation possible. Women will be centers of the family enterprise on their patch of preserved nature. They will set the tone and rebuild sense and sensitivity in interpersonal relations. Schoolchildren will have difficulties to understand the confused "values" of the late twentieth century, where women thought of themselves as "liberated," but in fact had been indoctrinated to compete as men in a horde of nastily aggressive males.

Children will grow up happy, because they will grow the way Homo sapiens forever had been growing to become the evolutionary success he is: within a close and active family.

Dreams about Arcadian life in the future, of course, may remain dreams. What counts, however, is that a human potential exists for such a state of affairs, and that hope does not have to be abandoned. This, I think, has been made quite believable through the real history of evolution of the human spirit, as I have tried to trace it on the preceding pages.

The emotional puzzle assembler finally has finished with the puzzle of human existence. The quest for objective knowledge is completed. We are ready to enter into a new level of being. There might be limits set by the world ecosystem. But the limits for cultural evolution of the human spirit certainly are out of sight so far. There probably are internal limits set by the quantity and complexity of cultural contents that can be assimilated by any one individual. But those limits certainly have not been reached anywhere. In fact, nowhere yet do we find a conscientious effort in education that is based on the reality of the human situation, on the full understanding of the learning instinct, and of what should be transmitted in order to make life most enjoyable for all.

Among the core educational areas still to be developed are: evolution of the human spirit (realistic self-knowledge), the spaceship Earth (the world ecosystem), mental health and creative life design (objective insight applied to polity, marital life, and especially to the hygiene of values) and sustainable technologies. As Mortimer Adler, the great popularizer of philosophy, has said,

"Happiness is not to do what we want to do, but what we ought to do." What that is can only be determined from our evolved nature, and it has to be conveyed through education anew to every generation. I think that will be the new frontier of the future. Here is an eternal inexhaustible challenge.

In that vein, the more I let unroll before my inner eye the incredible emergent evolution, as it so far has happened on this planet, the more I become convinced that here is a process that has some inherent power of necessity. I have no idea why that is so, nor where this comes from. But it is an observable entity whose origin is shrouded in mystery. Yet I believe that it will not stop until the whole planet has become an expression of this one positive human spirit. At that time there will prevail harmony between humanity's goals and the needs of other life.

Because other life is less conscious, or altogether unconscious, the human spirit will be viewed like the flower that all life on planet Earth unwittingly has brought forth through a common effort that has lasted 3.5 billion years. Or such will be the images expressed by poets of the future. Our life then will be experienced as deeply active in the inner domains, compassionate, and darkly beautiful. We will feel to be serene bearers of the spirit of this blue planet Earth. There will be mytho-poets, but their poetry will not create castles in the air. It will adorn and praise the miraculous reality of nature and of human existence, and will set aesthetic and ethic goals for an ever higher refinement of the civilized human condition.

Vignette 1: Cell Biology

By now it is common knowledge that DNA carried the code for construction of an organism. But if there is a single invention of evolution that has made advanced life possible, it is not DNA but the cell membrane. Cells can live for a while and even multiply without the DNA of the nucleus (Holtzman and Novikoff 1984, p. 92). Red blood cells, for instance, have no nucleus. But without the bag around it, there is no cell. Viral parasites essentially consist of DNA only. In order to multiply they have to enter a host cell and take advantage of its internal chemistry, which proves the point. The organelles within the cell again are individually enclosed by membranes. And those membranes do much more than confine. They are intimately involved in the busy chemistry that goes on.

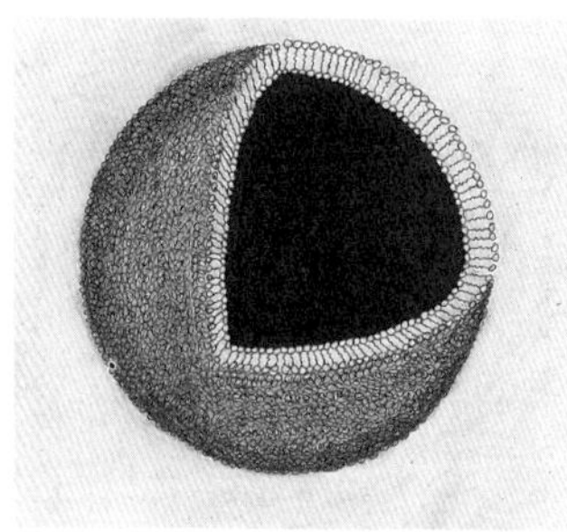

Fig. V1.1
The bimolecular lipid membrane has the tendency to spontaneously form a closed spherical shell (vesicle), since this is a state of lowest energy

The membrane is a bimolecular layer of long molecules, called "lipids" (fatty acids). One of the two ends of the lipid molecule is attracted by water (hydrophilic) the other is repelled (hydrophobic). Thanks to this property, if free lipid molecules are put into water, they spontaneously assemble to bilayers. This is similar to crystallization. First, pairs are formed by the hydrophobic ends clinging together in order to escape the water, then pairs assemble sideways to further avoid contact with water at the inner link. Figure V1.1 shows the arrangement. Because of internal stress there is a tendency of the membrane to bend into a spherical shell. In this way even the last open edges become hidden from water, when the sphere closes. This is the lowest possible energy state. All this is the result of electromagnetic forces similar to the tension by which a drop of water clings to a wettable surface, or conversely, is repelled by a nonwettable one. We recognize one more example of a surprising result of simple rules played. The cell bag is a self-organizing, complex dynamic system that consists of lipids under random thermal agitation.

Imbedded in the bilayer are occasional large protein molecules (fig. V1.2), which may form doors or tunnels from the inside of the cell (the cytoplasm) to the outside world. They have various specific functions. Some bind to specific protein molecules on other cells (cell adhesion molecules). Others selectively transport specific molecules across the membrane, thus making cell metabolism possible. Often the transport is gated. It turns "on" or "off," depending on the current need for input or output of the little chemical factory inside. The transport sometimes is unidirectional, in this way ions (electrically charged molecules), predominantly potassium, sodium and calcium, are used to maintain a voltage difference across the membrane. This is necessary to keep the factory going.

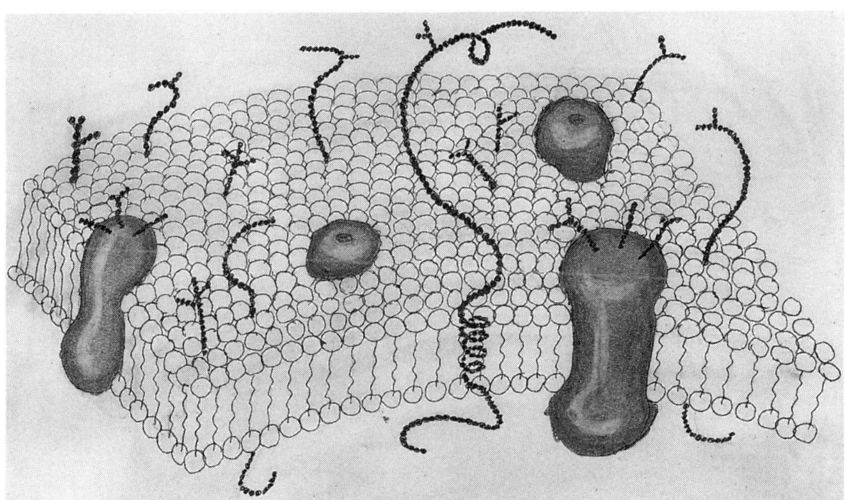

Fig. V1.2
Cell membrane with various imbedded and attached molecules

Another mechanism of transport across the membrane is endocytosis. Bulky clusters of molecules can be taken in by the cell membrane which bulges inward, ultimately enclosing (fig. V1.3) the material in a bag (vacuole or vesicle) that separates from the outer membrane and then opens to the inside. The process can be reversed to bring something from the inside out. We will find this to play an important part in the transmission of signals between neurons. As long as the cell is growing in size after division, more new vesicles join the membrane from inside, thus enlarging it.

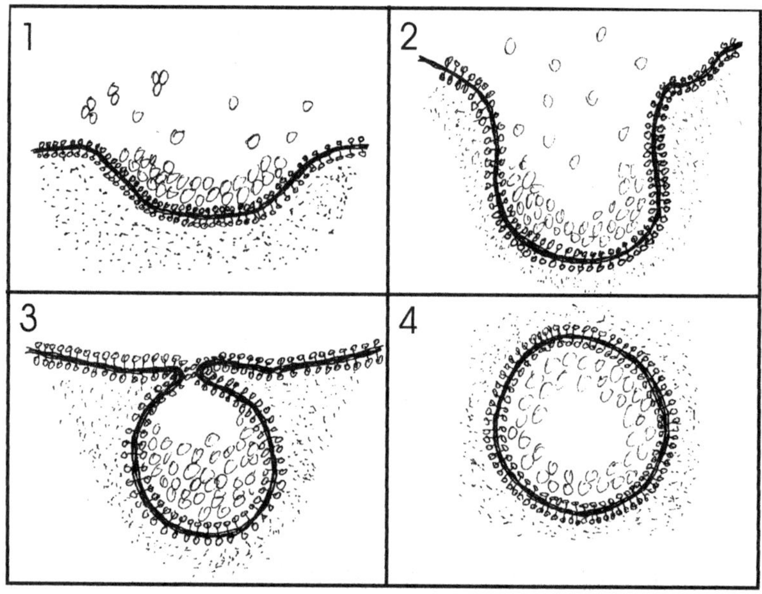

Fig. V.1.3
Stages of vesicle formation.
Cytoplasm is on the lower side of the membrane.
(1) shallow indentation, (2) (3) succesive stages of bulging, (4) detached vesicle

Inside the cell is the nucleus containing the genetic material and various "organelles." The main organelles are mitochondria, lysosomes, centrosomes, peroxysomes, cavities called "endoplasmic reticula," and the Golgi apparatus. The nucleus is also counted among the organelles. In the organelles the idea of a chemically active semipermeable membrane is repeated. Most of the above mentioned organelles are bags of various contorted shapes, including tubes, made from the same material as the outer membrane. Some move around in the cell. All perform chemical tasks. Everything is in a dynamic equilibrium; many organelles replace their own molecules (turnover) within days, if not hours. The general picture is that of vibrant activity.[55] Like on a factory floor, material is transported between the organelles (via vesicles).

Interesting is the hypothesis that in the course of evolution, some organelles originated from symbiosis with different and simpler single cell organisms (Margulis 1981, p. 419). One living inside the other had advantages for both

---

[55] In the last three decades, new techniques (particularly in electron microscopy), improved light microscopy, and autoradiography have yielded a wealth of information. There is no single task that any given organelle performs. Among the known functions are the following: the endoplasmic reticulum produces enzymes called "oxidases," and it controls calcium ion levels within the cell and synthesizes the lipids to form new vesicles to be used in enlarging the cell membrane. It also interacts with the work of the Golgi apparatus. Sometimes the two organelles are joined, and sometimes they exchange vesicles. The Golgi apparatus produces many of the proteins lodged in the outer cell membrane and is involved in the cell's secretion to the outside world of a number of enzymes and carbohydrates. The materials are transported via vesicles through the cell wall.

(endosymbiosis). Ultimately the genetic material was joined, but the bag within the bag remained, and with it the functional specialization. Mitochondria even retain their own DNA. They give the best evidence for this hypothesis (fig. V1.4). Besides their own DNA, they also contain all the machinery necessary for self-replication—but this machinery is markedly different from any of the two cell architectures that survive today (eukaryotes and prokaryotes). It is possible that here is a third kind of primordial cell, that only survived as a symbiont in the larger eukaryotes.

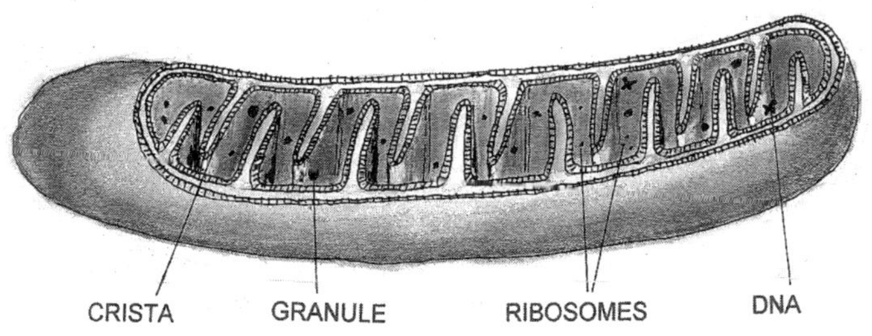

Fig. V1.4
Mitochondrion, part of membrane removed

Mitochondria have been found to divide independently of the host cell (Holtzman and Novikoff 1984, p. 194.). They can appear in great numbers (in the thousands) within one cell. They are the main sites of ATP (adenosine triphosphate) production, and they make other enzymes which all drive the body's oxygen metabolism, also known as "Krebs cycle." ATP is the fuel for virtually all cell functions, such as movement, cell division, secretion, and the chemical reactions inside of the cell. In view of this role in oxygen metabolism, it is possible to speculate that when Earth's atmosphere began to be oxygen rich, pro-mitochondria were the first among cells to specialize in the use of oxygen. This would have made them welcome guests inside of older, successful, and more complex cells, which, however, had difficulties to adapt to the new milieu by themselves. An interesting feature of the mitochondria's independent reproduction is that their DNA is not included in the male spermatozoon. Thus, mitochondrial DNA is exclusively passed on along the female lineage.

It is the nucleus, however, whose function is most instructive of life's basic design. It contains the DNA (deoxyribonucleic acid) strands. The molecular structure of DNA resembles a long rope ladder that is twisted, forming the famous "double helix." Each rung of this ladder is made up of a junction between two opposing bases. There are only four such bases: adenine (A), guanine (G), thymine (T) and cytosine (C). Adenine likes to bind with thymine, and guanine with cytosine. So there are only two kinds of rungs: AT and GC. But each can be reversed: TA and CG.

Figure V1.5 shows the flattened-out structure of the ladder. In the sequence of the four kinds of rungs along the long ladder resides the code of construction of an individual animal or plant. Since there are four different kinds of rungs, it can be said that this code is written in a base four number system. The ladder is long indeed. The human DNA contains about 109 (one thousand million) of such base pairs. It is likely that there are twenty-three separate pieces of continuous strand. During cell division each strand coils up to form a characteristic shape and becomes visible as a chromosome. One of the chromosomes is different between male and female. When it is Y-shaped, it determines male sex; otherwise, it has the shape of an X.

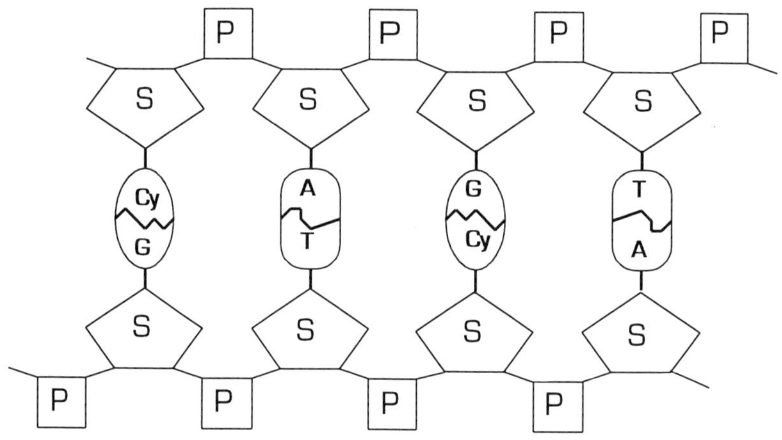

Fig. V1.5
Flattened structure of DNA

Cell division and DNA replication have been described often enough. I will concentrate on the role of DNA in the cell's chemical activity. Functionally, DNA is nothing more than a template for the construction of specific proteins. Human DNA specifies thirty thousand to one hundred thousand different proteins, each of them making up a "gene" (Lewin 1983, ch. 18). Many of those proteins we share with other mammals, others with lower animals. Each average protein is specified by not more than five thousand to eight thousand base pairs, probably less. Simple arithmetic shows that thirty thousand to one hundred thousand genes do not use up all of the 10^9 available base pairs. In fact, it is believed that only 10 percent (10^8) of the pairs encode any used, or even usable, information. Some flowering plants and amphibians contain up to fifty times more base pairs than humans do. This probably is due to metamorphoses (tadpole to frog) and the complexities of flowering followed by production of durable seeds. But this also might be the mark of an early and inefficient blueprint. We share our number (within 10 percent) with all other mammals.

But how is the template used? For this the rest of the cell is needed. Certain specific enzyme proteins (polymerases) already have been present in the egg cell. More of them and all other proteins are produced in the following way. A polymerase molecule binds to one of the specific regulatory sites (addresses) along the DNA strand. Here, they cause the base pairs to temporarily separate, whereupon one side of the strand binds to free floating bases present in the nuclear bag, thus forming its complement (negative replica) (fig. V1.6). This single strand separates, and the DNA pairs come together again. The replica is called "RNA" (ribonucleic acid) and is different from DNA in a minor but

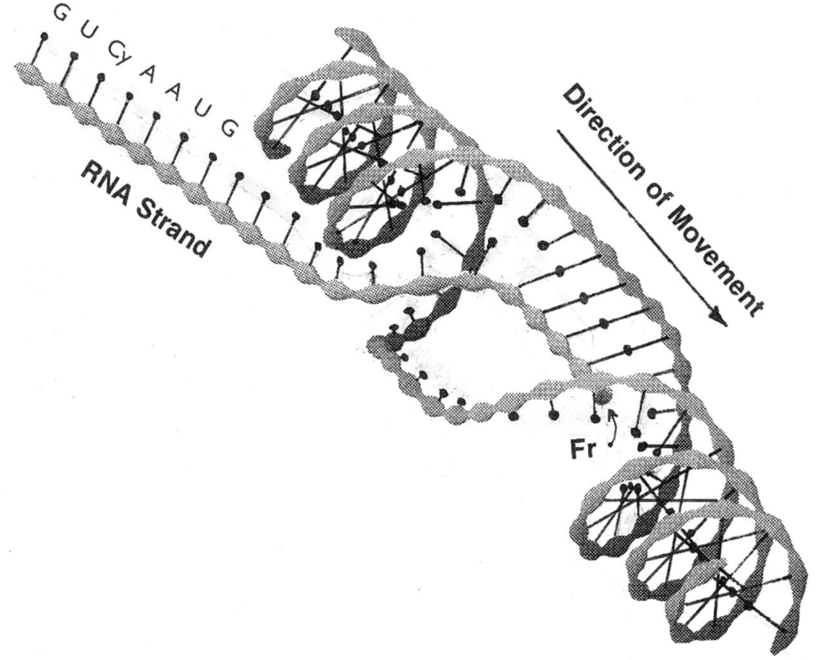

Fig. V1.6
RNA Replication
A molecule of RNA polymerase slides to the right, generating a
negative of one of the DNA strands. Fr is the free end to which free
floating bases, sugars, and phosphates are attracted.

important way: one of the bases, thymine (T), is replaced by a similar molecule, uracil (U), that has the same tendency to bind adenine (A).

Strands of RNA are shorter than DNA. The ones that carry the code for a specific protein are called "messenger RNA" or "mRNA." Besides this, two other kinds of short strands have been replicated from DNA, and are present in the nucleus: transfer RNA (or tRNA) and ribosomal RNA (or rRNA). The rRNA associates with some already existing proteins to form the ribosomes, small particles which leave the nuclear bag. Of the tRNA, there are twenty specific kinds, one for each of the twenty free floating amino acids, the building blocks of proteins. Each tRNA chain assumes a typical curled-up form, resembling a cloverleaf. It can bind to its specific amino acid at one end, and carries a triplet of bases at the other end. A triplet, such as AGG, AUA, or AAC is the code for the carried amino acid. Then tRNA and mRNA leave the nuclear bag. Outside, in the cytoplasm, each tRNA picks up its free floating specific amino acid.

Now one of the most dramatic molecular events takes place: A ribosome attaches itself to the mRNA strand on one end and begins to slide along it (fig. V1.7). As it moves it "reads" the protein code, three bases at a time, and attracts to the site the one tRNA with its attached amino acid, that is called for by the code. As the ribosome slides along the mRNA, out comes a chain of amino acids in the proper sequence. This is the specific protein. The tRNA is free to be reused. The protein strand folds to form its specific shape, as dictated by the amino acid sequence. Actually there is a train of many ribosomes, sliding simultaneously

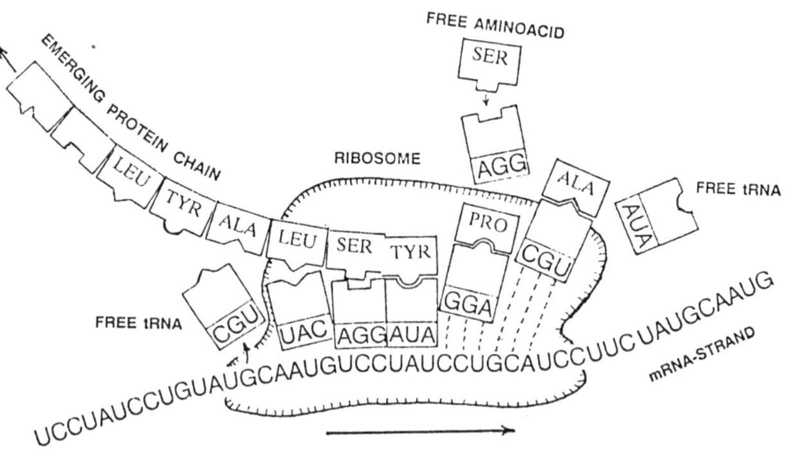

Fig. V1.7

Synthesis of specific protein

The free tRNA with a specific 3-base codon on one end and its amino acid on the other, bind temporarily to the specific sited on the mRNA strand, are then released one by one, leaving the coded aminoacid sequence behind. This is the protein.

along the mRNA strand, each sprouting a protein chain. When a given sequence is finished, the ribosome detaches itself to find another loose end of a mRNA chain. Thus, one DNA ladder makes many mRNA strands, and each of them makes many protein molecules simultaneously in order to satisfy the current demand for the specific protein.

As wonderful as this process appears, again it is but the result of a game of selectively "sticky" molecules under random thermal agitation. Two molecules stick together if their shapes complement each other, like a key and keyhole. The whole machinery only works over a narrow range of temperatures. Everything is slowed down as the temperature drops and then comes to a halt when water freezes. If the temperature is too high the increased random agitation overcomes the selective stickiness. The cell dies.

If the synthesized protein is an enzyme, it will cause the reactions necessary to respond to the stimulus that started the process. It can also be a structural protein. Of the thirty thousand to one hundred thousand different proteins our DNA can synthesize, each is the only and unique expression of one particular gene. Every genetic regulatory function is a consequence of reactions of these specific proteins.

The stimulus to synthesize could have come from a chain of reactions purely within the cell. Thus, the cell is in constant activity within itself, replicating, growing, moving, and secreting as well as responding to chemical signals from outside.

Many researchers believe (Darnell 1985, p. 68 ff) that in the course of evolution of life, RNA came before DNA. RNA is more easily produced by spontaneous reaction. One could argue that life began when an accidental strand of RNA synthesized the first useful protein. RNA can transcribe to a DNA strand, which can complete itself to the standard double helix. DNA has more stability and thus became the carrier of the code standard which is the valuable record of previously gained knowledge of how to succeed best.

We have to note that already on the cellular level the overall process can only follow along lines prescribed by the possibilities of chemistry. In more technical terms, it is constrained by valences, structural architecture, reaction potentials, temperature, and available reactants. Thus, the cell is not a product of the genetic code alone. Rather the cell and its genetic code are one complex chemical unit responding in a "knowledgeable" way to stimuli from its environment. The cell is a machine that selects and reads pages from the DNA code, then follows instructions. A bottle full of DNA will never come to life unless the DNA is put into the proper cell environment.

Indeed, in the course of evolution each cell has acquired very sophisticated knowledge. This is not only stored in the codes along the DNA strand, but also in the organic compounds present in the cytoplasm and the mitochondria. The cell's chemical knowledge outside of the nuclear DNA is passed on unaltered via the female lineage.

If the cell is a single-cell organism, this knowledge enables it to survive in a relatively variable environment. If the cell is part of a multicellular animal it can only live in a stable environment, but it can respond more radically to external stimuli during embryonic development and adapt itself to become specialized as, for instance, a muscle, bone, or liver cell. Or it may become a neuron. In the process of becoming specialized it will begin to secrete messenger molecules, which in turn will influence neighboring cells in their development so that the overall system can grow and function in harmony and efficiency among all parts, and be an overall success in the competition with other systems.

Vignette 2: Natural Selection

The blueprint is in the genetic code. How in the course of Darwinian evolution, is knowledge acquired by the genome? A simple diagram (fig. V2.1) shall illustrate this process. It is not different from any other acquisition of a heritable characteristic.

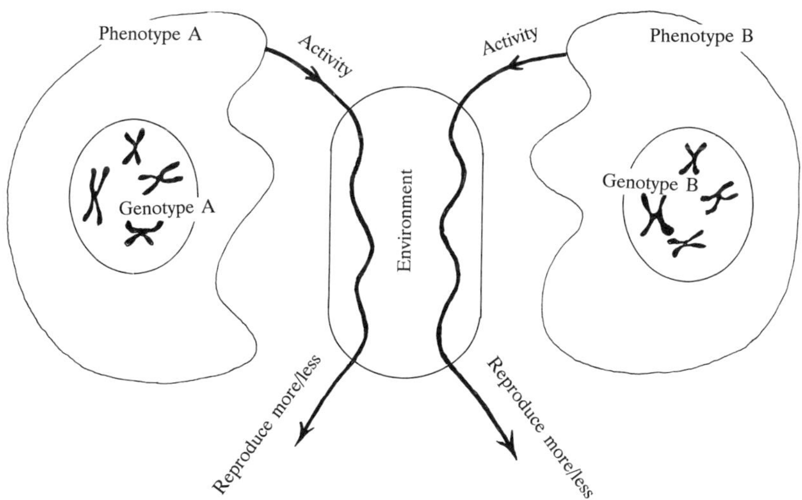

Fig. V2.1
Feedback from environment.
Schema of natural selection.

Two organisms are interacting with the same environment that elicits genetically programmed activities. Now, if the two genomes are not identical, either due to mutation or due to variation inherent in sexual reproduction, the activities might differ. For instance, B, might search for food algae in water at a lesser depth than A. If the algae tend to float to the surface, B, will grow faster and reproduce more. So the next generation will contain more individuals of the variant B than of A.[56] For mathematically inclined readers, this process can be expressed in form of simple equations.

We define p_k as the size of population in the current (k^{th}) generation, carrying the original strain, and v_k as the population of the slightly better reproducing variant. Now, as the equations state, the next, $(k+1)^{th}$ generation will be larger (or smaller) by a factor called "reproduction rate," designated by a, and b.

A reproduction rate of exactly one means that there will be exactly the same number of individuals in the next generation. Or the population may increase or decrease if the reproduction rate is larger or smaller than one.

$$p_{k+1} = a\, p_k$$
$$v_{k+1} = b\, v_k$$

If there is no shortage of food and space, according to these equations, both populations will grow exponentially. With a reproduction rate of 1.4, the population will almost double every second generation. After ten generations it will have increased twenty-nine-fold.

But if the total population (p + v) grows, and the available food resources remain constant, reproduction rates have to drop. This needs to be expressed more quantitatively. In this illustration the dependence of a and b on the total population, $p_k + v_k$, is assumed as in figure V2.2. It is not important to give the exact mathematical form of a and b. The initial value for a is taken as 1.4. For b, it is taken slightly higher at 1.6. Both drop as the total population grows, but the variant can sustain a larger population of 2,800 (as compared to 2,400 of the original strain) on the same limited resource, because it has a better method of food gathering.

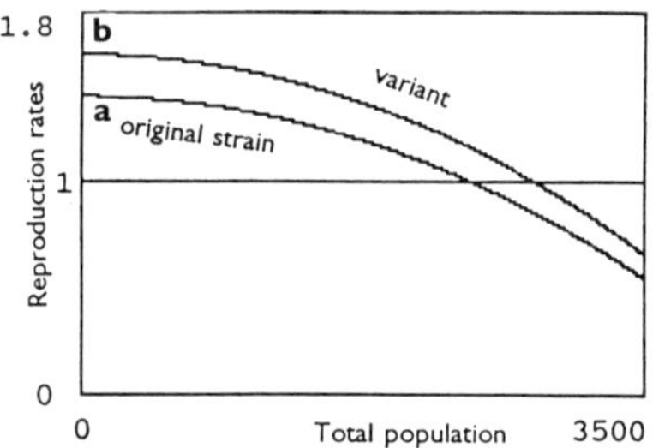

Fig. V2.2
Decline of reproduction rates due to overcrowding

When the above equations are played out on the computer, the result shown in figure V2.3 appears. After fifty generations the better adapted variant has "crowded out" the original strain. This despite the fact that initially the variant only had a population of ten, whereas the original strain started out with one thousand.

---

[56] Figure V2.1 shows that selection works on the phenotype. This seems to be a most reasonable assumption. Yet there is a controversy among theoretical biologists. Some say that genes should be uses as units of selection, while others consider "extended phenotypes" (kindship groups) to be units. I believe that the matters discussed here are not touched by this issue.

One could say that the genome now has acquired the knowledge that algae can be found higher up in the water.

The population S-curve of the new variant is typical for the growth of populations under limited resources. Ultimately an equilibrium is reached between the food available and the number of feeding organisms. This equilibrium is precarious because the ecological system that produces the food also enters into the equation. It may break down from overgrazing, resulting in a catastrophic collapse of the population. But even in the stable state of maximum sustainable population, life is merciless and generally miserable, if animals could say so. Malnutrition dominates and disease is rampant. Humans living under such circumstances certainly have heartbreaking stories to tell.

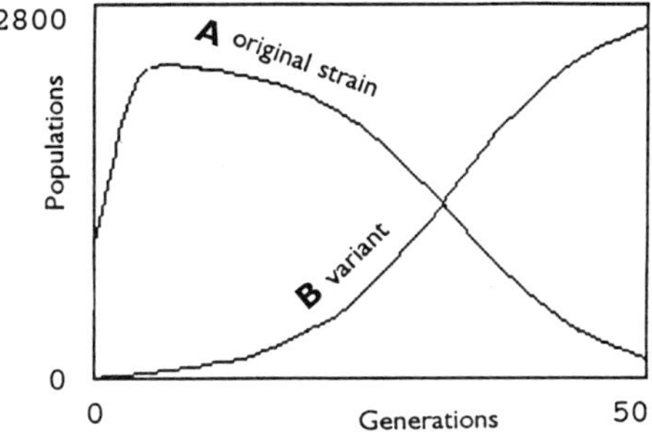

Fig. V 2.3
Crowding out of the less adapted variant.

In this selected example, after fifty generations, the less adapted variant, A, has virtually disappeared. A recently observed adaptation proceeded even faster. Around cities in Britain the moth *Biston betularia* has changed its camouflage color to black as a result of environmental pollution with soot. It is reported that after recent progress in cleanup, the moth has already begun to readapt to a lighter color (Bishop and Cook 1980).

Vignette 3: Artificial Neural Networks

Figure 5.1 on page 102 showed the excitation characteristic of a generic neuron. Already there it was mentioned that neurons can be modeled with electronic circuits. Such a circuit would essentially be an amplifier with a number of resistors at the input, as in figure V3.1. Each resistor's value represents the strength (or weight) of a synapse. Some inputs are positive, or excitatory, others are negative, or inhibitory.

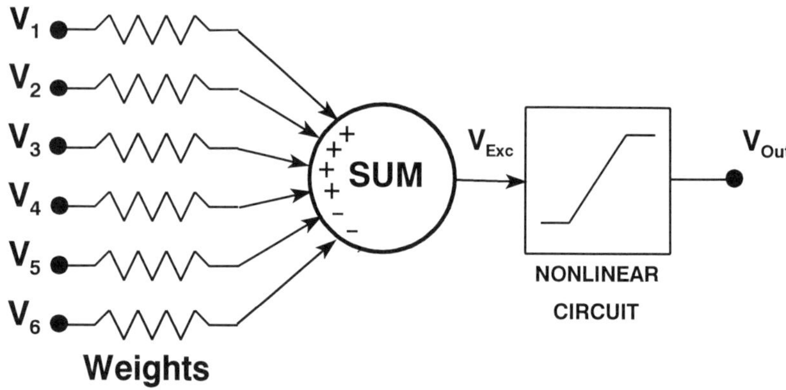

Fig. V3.1
Schematic representation of an artificial neuron.

The amplifier accepts the sum of all the currents as input. The overall excitation then will be:

$$V_{exc} = w_1V_1 + w_2V_2 + w_3V_3 - w_4V_4 - w_5V_5 - w_6V_6$$

Here, only six synapses of possibly a very large number have been considered with their positive or negative weights, w_k. After summation the total excitatory strength V_{exc} can be subjected to a nonlinear circuit with a sigmoid characteristic as in figure 5.1. Even though neural networks consisting of such electronic circuits have been built and are commercially available to perform various control and "cognitive" tasks in robots, it mostly is easier to simulate them on the computer. Such artificial neurons can be connected to form networks, for instance, of the type of figure 5.2. But such schematic representations become unwieldy for larger networks. Therefore an equivalent, if slightly more abstract form, is introduced in figure V3.2. The synapses have been moved away from the cell body and assembled in a square array.

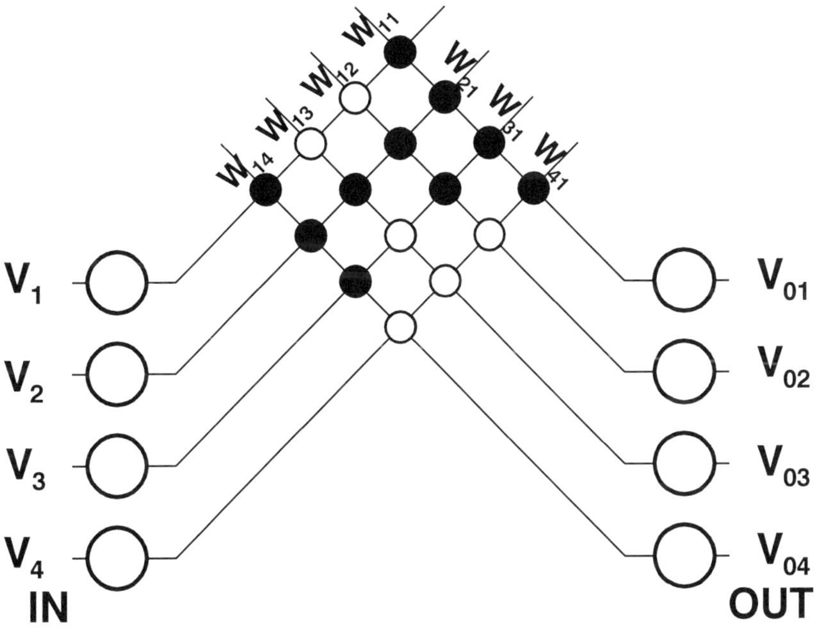

Fig. V3.2
Simple feed-forward classifier.

Chapter 5 demonstrated that such networks can be used as classifiers of images or patterns. But there is more to it. The mathematically inclined reader will recognize that if all neurons work near the linear middle section of the sigmoid, this network approximates a linear transformation of the input array of excitations (V_1, V_2, V_3, V_4 . . .) into an output array (Vo_1, Vo_2, Vo_3, Vo_4 . . .). Such arrays generally are called "vectors." But they also can be thought to represent an image, each V_k being the brightness value of a pixel. Now, if the input vector represents an image containing a letter, A, for instance, using 120 pixels, as in figure 5.3, this can be transformed into a letter, B, at the output, also having 120 pixels. A network to do this will have 120 inputs, and 120 times 120, that is 14,400 weighted connections, wik, and 120 outputs. In the linear approximation, simple algebra shows that the transformation of A into B can be accomplished with many different combinations of the 14,400 weights. Thus, many different transformations, like B into D or V into Z can be produced with one and the same fixed network. The precise result is that a square array of N times N weights can transform exactly N vectors with N components (pixels) each into any other desired set of N vectors with N pixels each (Kohonen 1984).

A three-layer network, or classifier of classes, would take the form of figure V3.3. Here, a detailed explanation of its working is more difficult. But roughly what happens is this: in the hidden (middle) layer there appear patterns that are neither a single lighted pixel for each letter nor a transformation of one letter into another, but rather patterns that are characteristic of, for instance, the presented letter being a lower case, a capital one, or from the Gothic alphabet. The output layer then classifies this pattern, so that now one output only lights up when any Gothic letter is presented.

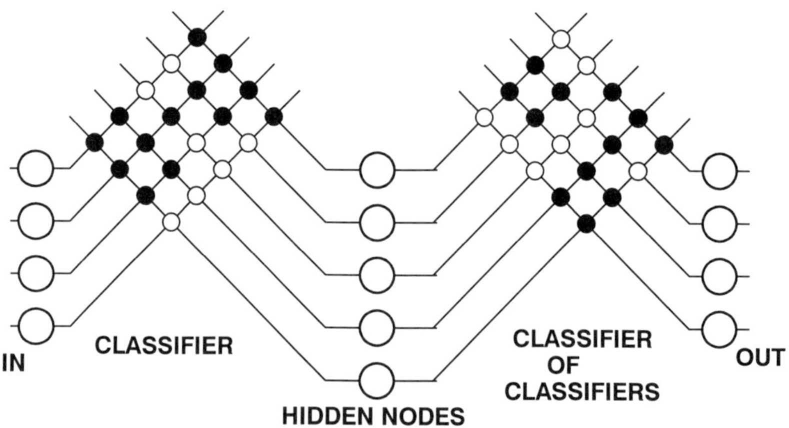

Fig. V3.3
Layered classifier.

Natural neural networks mostly are trans-connected in the sense of figure V3.4. There probably is an evolutionary reason for that architecture. The simplest organisms transmitted excitation directly from sensory cells to motor neurons. Later, as complexity of responses grew, "interneurons" were inserted, forming the hidden layers. In fact, trans-connected networks seem to be able to do more with less neurons.

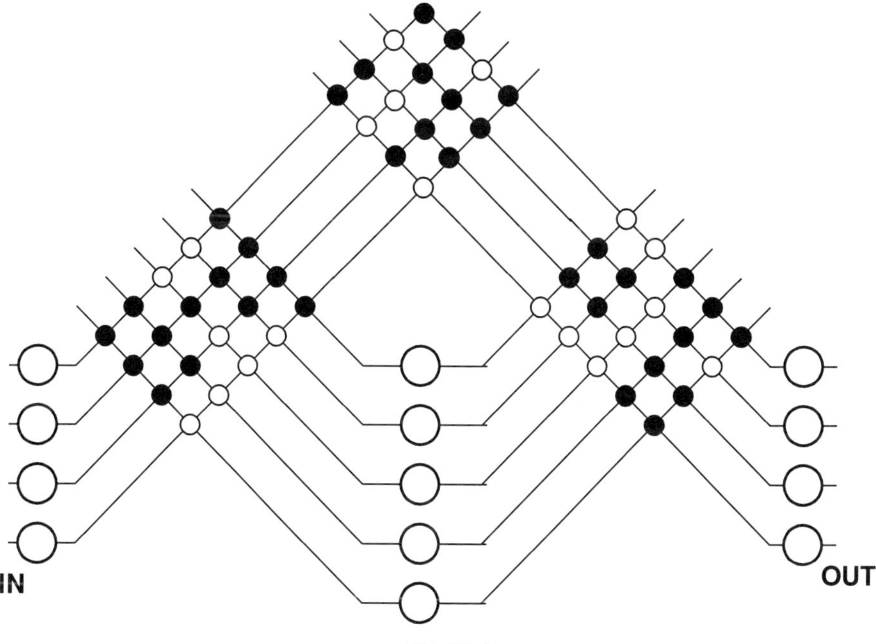

Fig. V3.4
A fully transconnected layered network.

Figure 5.3 gives the impression of "binary" images; pixels are either "on" or "off." Sometimes networks, indeed, are used in this way. But since the voltages (or excitations) are gliding "analog" variables, gray scales can be processed as well, or color intensities. In other applications the gradual excitation can be used to move "actuators" in robots, resembling muscles in real animals being stressed with variable strength. Such use of neural networks solves a vexing problem in the operation of robotic devices, such as robot arms: The tip of a finger of a mechanical arm plus hand, has to be moved to a prescribed point in space, which may be given by the three coordinates x, y, and z, plus an angle for the direction of approach. From the shoulder joint to the fingertip there are about six different joints to be properly flexed to solve this problem. And some of the joints also can be rotated. So there is a large number of flexors (motors) involved in this movement, each has to achieve a prescribed angle.

Of course, this is a mathematically well-defined problem, yet a powerful computer would be needed to move the arm rapidly. An analog neural network, though, can move the arm without delay. Three inputs may consist of the three target coordinates. One output is connected to each flexor. If there are enough hidden layers, the network can be trained to move the fingertip to any desired position within reach of the arm. Such networks have been demonstrated as actual hardware (Eberhardt et al. 1990).

When neural networks are used as classifiers, it was said that the pattern to be classified can have variations from the norm. This can be carried to the extreme, in particular if the number of available synapses is larger than necessary (or the number of patterns smaller than the maximum N patterns with N pixels each), then it becomes possible to devise a scheme for adjusting the weights in such a manner that an incomplete or strongly garbled input image will, nevertheless, produce an accurate output image. This is called "associative recall" (Kohonen 1984). Very impressive is the case when the output image is chosen to be identical to the uncorrupted input image (auto-associative recall). Figure V3.5 shows two pairs of images, garbled or incomplete on the top, which have been used as input, and the reconstructed output images on the bottom. It is as if the network remembers: "Oh yes, right, I know that face." This particular network could remember one hundred different photographs. It had 3,024 pixels, was fully connected and had a linear (not a sigmoid) characteristic.

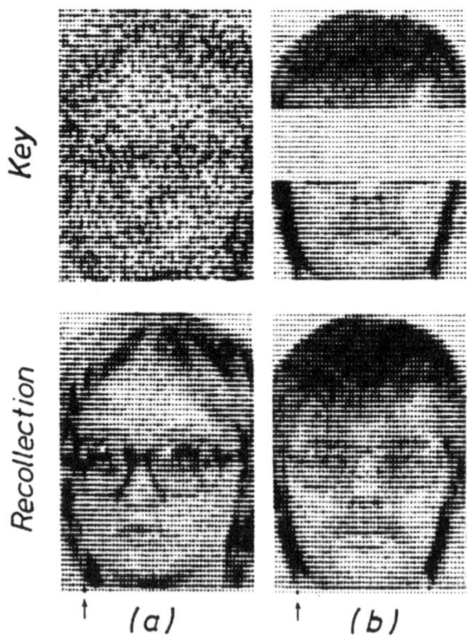

Fig. V3.5
Two images restored via associative memory
[From: Kohouen (1984), by permission of Springer Verlag]

Another early and striking illustration of the power of neural nets was the system called "NETtalk" (Sejnowski and Rosenberg 1987). It is a three-layer feed forward network of only 309 neurons and 18,629 connections, which has been simulated on a computer. At the present time not too many artificial neural networks have actually been built in hardware. Computer simulations are quite adequate even though they may be slower. This network can pronounce English text. Figure V3.6 shows the basic setup. At the bottom are seven groups of twenty-nine neurons. A text sequence of seven letters, including spaces, commas and periods is presented simultaneously to the seven groups. Each group recognizes one letter at a time. The middle letter is pronounced and then the text is moved by one letter to the left. It is necessary to present letters adjacent to the one to be read because a phoneme either depends on the context or consists of more than one letter.

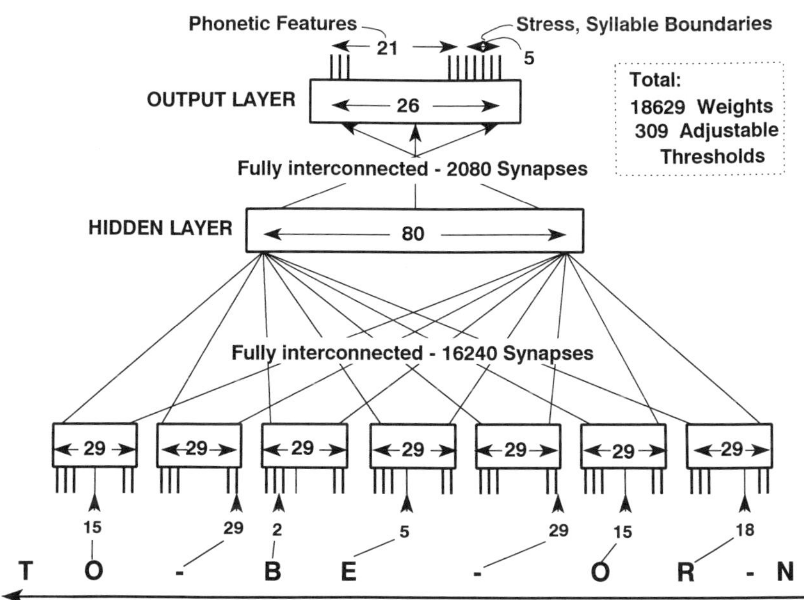

Fig. V3.6
The circuits of NETtalk,
an artificial neural network that can pronounce English from written text.
[After: Sejnowsky and Rosenberg (1987)]

Actually, in the reported version of NETtalk, no pattern recognition of letters is attempted. This has been done elsewhere. In this simulation only one of the twenty-nine neurons in a group is excited for a given letter, space, comma, or period. The hidden layer consists of eighty neurons. The output layer has twenty-six neurons, each activating a certain phonemic feature in a speech synthesizer, such as position in the mouth (labial, dental), whether it is nasal or fricative, pitch of a vowel, punctuation, and stresses. After training, the network could produce quite understandable speech from unfamiliar written text.

We have seen an estimate of the capacity of linear classifiers. But the interesting question remains how many words can a nonlinear network like this learn to pronounce? Or more generally, how many different patterns can a network learn to distinguish? It turns out that this is a very difficult question. The answer will depend on how the question is exactly formulated. How much different is different? Is there an implicit order in the sounds to be recognized that may indicate a hidden simplicity underlying the apparent complexity? We are told that NETtalk was trained on a text containing one thousand different words. After fifty readings and subsequent corrections it had "perfect scores" in 55 percent of the phonemes. In 95 percent of all phonemes the pronunciation was sufficiently distinct to be recognized as different from any other. If presented with new text the scores dropped to 35 percent and 78 percent. But the speech was still recognizable. From these figures one can draw two kinds of conclusions:

1. On the English language: Since new text was pronounced with diminished success, the one thousand words did not contain a complete representation of the rules of English pronunciation. This is not unexpected.

2. On the potential of neural nets: Since the familiar one-thousand-word text was pronounced recognizably but not perfectly, a network of 309 neurons is not of sufficient size in order to learn to good perfection the pronunciation rules appearing in a sample of one thousand words of English language. This is not unexpected either.

Still, performance is very impressive: A network of 309 neurons is sufficient to learn marginal but understandable pronunciation of the entire English language. I think this example alone well demonstrates the power of neural networks. It provides a reference scale for the cognitive power of the 10^{11} or so neurons in our brain.

In conclusion, let me ask two philosophical questions:

1. Why does a language have only a limited number of rules of pronunciation? Answer: Because it has been invented by the neural networks around the human speech center. They are huge as compared with NETtalk, but still limited.

2. Why is it so difficult to establish an accurate set of rules of pronunciation that does not contain exceptions? Or: Why is it difficult to establish formal rules of grammar without having to deal with exceptions? Answer: Because neural networks do not deal in "rules." They adapt.

The rules of English pronunciation reside in the distribution of weights in NETtalk. If one knows the rules, one could in principle compute the necessary weights. Such a procedure would be programming, which in this case would be extremely difficult. But neural nets do much better with learning. Generally we learn a language without bothering to understand the rules of grammar or pronunciation. These rules are only abstracted after the fact by academic specialists. They use abstract maps of the world that are quite different from the weaved life of networks that live language.

NETtalk is not a self-adaptive system. It has to be taught correct pronunciation. Would it be possible to construct self-adapting networks that read and pronounce written text? The answer to this is illuminating: Self-adapting schemes like the ones described by Grossberg (Grossberg 1987) and Carpenter and Grossberg (1987) can be used to translate groups of letters into sounds (as produced by an attached electronic sound synthesizer). After self-organization, if the system has enough capacity, it would produce sounds that are uniquely related to sequences of letters, but it will not be any known spoken language. Still one would expect it to be a language that can be learned by a human being. This, obviously, would be "NETpidgin."

A final word about the speed of operation of neural nets: It has been frequently pointed out that even the biggest and fastest computers would have difficulties to perform the complex automatic tasks involved in visual perception alone. Images consisting of hundred thousands of pixels have to be sorted out, rotated, enlarged or reduced and recognized as three-dimensional objects of a certain real size, then compared with others in order to be classified. Von Neumann sequential computers work pixel by pixel. That is why it takes so long to process images on such computers. Neural nets, on the other hand, process all pixels at once. Thus, individual neurons can be very slow on the scale of a computer's logic circuits. Nevertheless the processing is accomplished in a few hundredths of a second because all neurons work simultaneously.

Literature Cited

Alkon, D. L. 1988. *Memory traces in the brain.* Cambridge: Cambridge University Press. 40

Allison, T., H. Van Twyver, and W. R. Goff. 1972. Electrophysiological studies of the echidna,Tachyglossus aculeatus. In Waking and sleep, Archives of Italian Biology, 110: 145, 204.. 161

Ammerman, A. J., and L. L. Cavalli-Sforza. 1984. *The neolithic transition and the genetics of the populations in Europe.* Princeton: Princeton University Press. 210

Armstrong, K. 1994. *A history of God.* New York: Alfred A. Knopf. . . . 246

Baker, J. F., S. F. Petersen, W. T. Newsome, and J. M. Allman. 1981. Visual response properties of neurons in four extrastriate visual areas of the owl monkey. *Journal of Neurophysiology* 45: 397. 137

Balmer, R. 1993. *Mine eyes have seen the glory: A journey into the evangelical subculture in America.* New York: Oxford University Press. 246,247

Banks, E. M. 1962. A time and motion study of prefighting behavior in mice. *Journal of General Psychology* 101: 165-183.. 53

Beach, F. A. 1948. Hormones and Behavior. New York : Cooper Square. 47

Bertalanffy, L. V. 1949. *Das biologische weltbild.* Bern, Switzerland: A. Francke A. G. 16

Bertalanffy, L. V. 1968. *General system theory.* New York: George Braziller. 16

Bishop, J. A. and L. M. Cook. 1980. Industrial melanism and the urban environment. Advances in Ecological Research 11: 373-404. 295

Bleek, D. F. 1924. *The mantis and his friends.* London: Blackwell Ltd. . . 213

Bleek, W. H. I. and L. C. Lloyd. 1911. *Specimens of Bushman folklore.* Reprint, Capetown: C. Struik Ltd. 213

Fox, M. W. 1980. *The soul of the wolf.* Boston: Little, Brown and Co. . . 55

Fox, R. L. 1992. *The unauthorized version, truth and fiction in the Bible.* New York: A. L. Knopf. 228, 240

Frings, H., and M. Frings. 1959. Reactions of American and French species of Corvus and Larus to recorded communication signals tested reciprocally. *Ecology* 39: 26-131. 63

Fujita, N. S. 1986. *A crack in the jar, What ancient Jewish documents tell us about the New Testament.* New York: Paulist Press. 240

Fuster, J. M. 1985. The prefrontal cortex, mediator of cross-temporal contingencies. *Human Neurobiology* 4: 169-179. 143

Gardner, H. 1985. *The mind's new science.* New York: Basic Books Inc.. . 101

Gazzaniga, M. S., J. D. Holtzman, and C. S. Smylie. 1987. Speech without conscious awareness. *Neurology* 35: 682-685. 147

Gazzaniga, M. S., C. S. Smylie, K. Baynes, W. Hirst, and C. McCleary. 1984. Profiles of right hemisphere language and speech following brain bisection. *Brain and Language* 22: 206-220.. 147

Gazzaniga, M. S. 1983. Right-hemisphere language, a twenty-year perspective. *The American Psychologist* 83: 525-537. 147

Gazzaniga, M. S. 1988. Brain modularity: towards a philosophy of conscious experience. In Consciousness in contemporary science. Oxford: Clarendon Press. 147, 148

Gazzaniga, M. S., ed. 1995. *The cognitive neurosciences.* Cambridge, Mass.: MIT Press. 100

Gell-Mann, M. 1994. *The quark and the jaguar.* New York: W. H. Freeman and Co. 14

Ghiglieri, M. P. 1975. *East of the mountains of the moon: Chimpanzee society in the African rainforest.* New York: The Free Press, Macmillan Inc. 71

Goldsmith, J. 1994. *The trap.* New York: Carroll and Graf Publishers, Inc. 266

Goleman, D. 1995. *Emotional intelligence.* New York: Bantam Books. 101, 265

Goodall, J. 1986. *The chimpanzees of Gombe.* Cambridge, Mass. and London: Harvard University Press. 71, 72, 76-79, 87, 88

Gould, S J., and R. C. Lewontin. 1979. The Spandrels of San Marco and the Panglossian paradigm: A critique of the adaptionist programme. *Proceedings of the Royal Society of London (Biology)* 205: 581-98. 14

Gould, S. J. 1981. *The mismeasure of man*. New York and London: W. W. Norton & Company. 186

Gould, S. J. 1989. *Wonderful life*. New York and London: W. W. Norton and Co. 18, 32

Gross, P. R., and N. Levitt. 1994. *Higher superstition—The academic left and its quarrels with science*. Baltimore: Johns Hopkins University Press. 266

Grossberg, S. 1987 "Competitive Learning: From Interactive Activation to Adaptive Resonance." *Cognitive Science*. Vol. 2 P. 23-63. 302

Hadamard, J. 1949. *The psychology of invention in the mathematical field*. Princeton, NJ: Princeton University Press. 161

Haken, H., and M. Stadler, eds. 1990. *Synergetics of cognition*. Berlin, Heidelberg, and NewYork: Springer Verlag. 12

Haken, H. 1978. *Synergetics, an introduction*. Berlin, Heidelberg, and New York: Springer Verlag. 12

Haken, H. 1991. *Synergetics computers and cognition*. Berlin, Heidelberg, and New York: Springer Verlag. 12

Harlow, H. F. 1960. Primary affectional patterns in primates. *American Journal of Orthopsychiatry* Vol. 30. 48

Hawkes, J. 1973. *The great civilizations*. London: Hutchinson of London. 212, 213, 215

Heath, R. G. 1963. Electrical self-stimulation of the brain of man. *American Journal of Psychiatry* 120: 571-577. 142

Hebb, D. O. 1949. *The organization of behavior: A neuropsychological theory*. New York: Wiley. 39, 104

Heilbroner, R. 1986. *The essential Adam Smith*. New York and London: W. W Norton & Company. 263, 268

Heilbroner, R. 1995. *Visions of the future*. New York: Oxford University Press. 268

Henry, J. P., and P. M. Stephens. 1977. *Stress, health and social environment*. New York: Springer Verlag. 170

Hilgard, E. R. 1977. *Divided consciousness: Multiple controls in human thought and action*. New York: John Wiley. 148

Hillis, W. D. 1988. Intelligence as an emergent behavior; or, The songs of Eden. In *The artificial intelligence debate.* Edited by S. R. Graubard. Cambridge, Mass.: MIT Press. 4

Ho, Ping-ti. 1975. *The cradle of the East.* Chicago: University of Chicago Press.. .... 222, 223

Hokanson, J. E., and S. Shetler. 1961. The effect of overt aggression on physiological tension level. *Journal of Abnormal Social Psychology* 63: 446-448. 57

Holtzman, E., and A. B. Novikoff. 1984. *Cells and organelles.* Philadelphia: Saunders College Publishing.. .... 284, 287

Hubel, D. 1959. Single unit activity in striate cortex of unrestrained cats. *Journal of Psychology* 147: 226-240.. 133

Hubel, D. H., T. N. Wiesel, and M. P. Stryker. 1978. Anatomical demonstration of orientation columns in macaque monkeys. *Journal of Comparative Neurology* 177: 361-379. 133

Hubel, D. H., and T. N. Wiesel. 1979. Brain mechanisms of vision. *Scientific American* 241: 150-162.. .... 132, 137

Igarashi, S., and T. Kamiya. 1972. *Atlas of the vertebrate brain.* Baltimore, London, and Tokyo: University Park Press. 113

Jacobi, J. [1942] 1973. *The psychology of C. G. Jung.* New Haven and London: Yale University Press.. .... 170, 173

Jaines, J. 1976. *The origin of consciousness in the breakdown of the bicameral mind.* Boston: Houghton Mifflin Company. 170, 232

Jennings, J. D. 1979. *The prehistory of Polynesia.* Cambridge, Mass.: Harvard University Press. 252

Jones, E. G., and T. P. S. Powell. 1970. An anatomical study of converging sensory pathways within the cerebral cortex of the monkey. Brain 93: 793-820.. 126

Jung, C. G. 1933. *Modern man in search of a soul.* New York: Harcourt, Brace and World.. 172

Jung, C. G. 1965. *Memories, dreams and reflections.* New York: Random House, Vintage Press. 208, 209, 218

Kandel, E. R. 1976. *Cellular bases of behavior.* San Francisco: W. H. Freeman and Co.. 38

Kandel, E. R. 1979. Small systems of neurons. *Scientific American,* September. 38

Kauffman, S. A. 1991. Antichaos and adaptation. *Scientific American,* August. 14

Kauffman, S. A. 1993. *The Origin of Order.* Princeton, NJ: Oxford University Press.. . . . . 14, 28

Kimber, G., and R. S. Athwal. 1972. A reassessment of the course of evolution of wheat. Proceedings of the National Academy of Science 69: 447-450.. 210

King, J. A., and N. L. Gurney. 1954. Effect of early social experience on adult aggressive behavior in C 57 BL/10 mice. *Journal of Comparative Physiological Psychology* 47: 326-336. 53

King, L. 1966. *Weeds of the world.* New York: Interscience.. 222

Koestler, A. 1964. *The act of creation.* New York: Macmillan. 161

Köhler, P. A., and H. F. Zacher, eds. 1982. *The evolution of social insurance 1881-1981.* London: Frances Pinter, and New York: St. Martin's Press.. 258

Köhler, W. 1925. *The mentality of apes.* London: Routledge and Kegan Paul. 83

Kohonen, T. 1984. *Self-organization and associative memory.* New York: Springer Verlag.. . . . . 134, 298, 300

Konishi, M. 1964. *The effect of deafening on song development in two species of juncos.* Condor 66: 85-102.. 45

Krieg, W. J. S. 1953. *Functional neuroanatomy.* New York and Toronto: Blackiston Company, Inc.. 109

Lagerspetz, K., and S. Talo. 1967. Maturation of aggressive behavior of young mice. Reports of the Institute for Psychology. University of Turku 28: 1-9. 53

Lagerspetz, K. 1964. Studies on aggressive behavior of mice. Suomalaisen Tiedeakatemian Toimituksia. Annual Academy of Science, Fennice. B131 Helsinki. 53

Langton, C. G., C. Taylor, J. D. Farmer, S. Rasmussen, eds. 1992. *Artificial life II,* Redwood City, Calif.: Addison-Wesley.. . . . . 11, 27

Langton, C. G., ed. 1989. Artificial life: the proceedings of an interdisciplinary workshop on the synthesis and simulation of living systems. Reading, Mass.: Addison-Wesley. 11

Langton, C. G., ed. 1994. *Artificial life III,* Redwood City, Calif.: Addison-Wesley. 11, 27

Lawrence, R. D. 1986. *In praise of wolves*. New York: Bellantine Books. . . . 55

LeDoux, J. E., and W. Hirst, eds. 1986. The neurobiology of emotion. In Mind and brain. Cambridge: Cambridge University Press. 141

Lee, R. B., and I. DeVore. 1968. *Man the hunter*. Chicago: Aldine Publishing Company.. 195

Lee, R. B. 1979. *The !kung san: Men, women and work in a foraging society*. Cambridge: Cambridge University Press.. 192

Lewin, B. 1983. *Genes*. New York: J. Wiley and Sons. 288

Lhermitte, F., J. Deroulsne, and J. L. Signoret. 1972. Analyse neuropsy-chologique du syndrome frontal. *Revue Neurologiqque* 127: 415-440.. 145

Lorenz, K., and N. Tinbergen. 1938. Taxis und instinkthandlung in der eiroll-bewegung der graugans. Zeitschr. für Tierpsychologie. 2: 1-29. 43

Lorenz, K. [1940] 1982. In Learning, development and culture: Essays in evolutionary epistemology. New York: Wiley.. 95

Lorenz, K. 1962. *The function of colour in coral reef fishes*. London: Royal Institute of Great Britain, 39: 282-296. 52

Lorenz, K. 1965. *Evolution and modification of behavior*. Chicago: University of Chicago Press. 45

Lorenz, K. 1966. *On aggression*. New York and London: Harcourt, Brace and Jovanovich. 45, 55, 57, 59, 61, 77, 187

Lorenz, K. 1973. *Behind the mirror*. New York and London: Harcourt, Brace and Jovanovich.. . . . . 46, 95

Lovelock, J. 1979. *Gaia: A new look at life on earth*. New York: Oxford University Press. 252

Lovelock, J. 1988. *The ages of Gaia*. New York and London: W. W. Norton and Co.. 252

Lovelock, J. E. 1972. Gaia as seen through the atmosphere. *Atmospheric Environment* 6: 579-580.. 251

Ludwig, E., and J. Klingler. 1956. *Atlas cerebri humani, The inner structure of the brain*. Boston: Little & Brown.. 109

Lumsden, C., and E. O. Wilson. 1981. *Genes, mind and culture*. Cambridge, Mass.: Harvard University Press. 187

Maclean, P. 1973. *A triune concept of the brain and behavior*. Edited by J. Boag and D. T. Campbell. Toronto: University of Toronto Press. . . . 170

Magoun, H. W. 1963. *The waking brain*. Springfield, Ill.: C. C. Thomas.. . . 124

Nauta, W. J. H., and M. Feirtag. 1986. *Fundamental neuroanatomy.* New York: W. H. Freeman and Co. 107

Nauta, W. J. H. 1971. The problem of the frontal lobe: A reinterpretation. *Journal of Psychiatric Research* 8: 167-181. . . . . 141, 143

Nicolai, J. 1959. Familientradition in der gesangstradition des Gimpels. *Journal of Orinthology* 100: 39-46. 45

Nicolis, G., and I. Prigogine. 1977. *Self-organization in nonequilibrium systems.* New York: Wiley. 12

Nieuwenhuys, R., J. Voogd, and C. van Huijzen. 1978. The human central nervous system. A synopsis and atlas. New York and Berlin: Springer. . . 107

Nisbet, E. G. 1991. *Leaving Eden, to protect and manage the earth.* Cambridge: Cambridge University Press. 252

Odum, E. P. 1983. *Basic ecology.* Philadelphia: Saunders College Publ. Co. 252

Ojemann, G. A. 1990. Organization of language cortex derived from investigations during neurosurgery. Seminars in Neuroscience 2: 297-305. . . . 102

Olds, J. E., and P. Milner. 1954. Positive reinforcement produced by electrical stimulation of the septal area and other regions of the rat brain. *Journal of Comparative Physiology and Psychology* 47: 419-427. 142

Ophuls, W. 1992. Ecology and the politics of scarcity revisited. New York: W. H. Freeman and Co. 261

Pais, A. 1982. *Subtle is the Lord, The science and life of Albert Einstein.* New York: Oxford University Press. 245

Pais, A. 1991. *Niels Bohr's times, in physics, philosophy and polity.* New York: Oxford University Press. 245

Partridge, E. 1983. *Origins: A short etymological dictionary of modern English.* New York: Greenwich House. 205

Patterson, F., and E. Linden. 1981. *The education of Koko.* New York: Holt, Rinehart and Winston. 89

Penfield, W., and P. Perot. 1963. The brain's record of auditory and visual experience. *Brain* 86: 595-696. . . . . 120, 146

Penfield, W., and T. Rasmussen. 1952. *The Cerebral Cortex of Man.* New York: Macmillan Co. 119

Penfield, W., and L. Roberts. 1959. *Speech and brain mechanisms.* Princeton, NJ: Princeton University Press. 119

Penfield, W. 1955. The role of the temporal cortex in certain psychical phenomena. *Journal of Mental Science* 101: 451-465. 145

Penfield, W. 1975. *The mystery of the mind.* Princeton, NJ: Princeton University Press. 120

Penrose, R. 1989. *The emperor's new mind.* New York: Oxford University Press.. . . . . 20, 98

Penrose, R. 1994 *Shadows of the Mind: A search for the Missing Science of Consciousness.* Oxford University Press. New York. 98

Petrarca, F. 1975. *Rerum familiarum libri I-VIII.* Translated by A. S. Bernardo. Albany, NY: State University of New York Press.. 242

Petsche, H., P. Richter, A. von Stein, S. C. Etlinger, and O. Filz. 1993. *EEG coherence and musical thinking, music perception.* 11: 117-151. Berkely, CA: University of California Press.. 202

Piaget, J. 1927. La premiere annee de l'enfant. British *Journal of Psychology* 18: 97-120.. . . . . 153, 155

Piel, G. 1992. *Only one world, our own to make and to keep.* New York: W. H. Freeman and Co.. . . . . 254, 256, 260, 269

Pilbeam, D. 1972. *The ascent of man.* New York: Macmillan.. 191

Pinker, S. 1994, *The language instinct.* New York: William Morrow.. . . . 85

Poincare, H. 1921. *Chapter 3 in The foundations of science.* New York: Science Press.. 161

Poliak, S. 1957. *The vertebrate visual system.* Chicago: University of Chicago Press.. 129

Ponting, C. 1991. *A green history of the world.* New York: St. Martin's Press.. . . . . 252, 254

Popper, K. 1972. *Objective knowledge.* New York: Oxford University Press. 16, 101

Popper, K. R., and J. C. Eccles. 1977. *The self and its brain.* Berlin and New York: Springer International.. 5

Premack, D., and A. J. Premack. 1983. *The mind of an ape.* New York: W. W. Norton and Co.. 88

Raine, A. 1993. *The psychopathology of crime.* San Diego: Academic Press. 143

Rasmussen, K. 1930. Intellectual Culture of the Hudson Bay Eskimos. Report of the Fifth Thule Expedition. Vol. 7, Glydenalske, Copenhagen. Cited in J. Campbell, The way of animal powers. San Francisco: Harper & Row. 208

Renfrew, C. 1988. *Archeology and language: The puzzle of Indo-European origins.* Cambridge: Cambridge University Press.. 211

Renfrew, C. 1989. The origins of Indo-European languages. *Scientific American,* October, 106-114. . . . . 210, 215, 216, 221

Rensch, B., and J. Doehl. 1967. Spontanes Öffnen verschiedener Kistenverschlüsse durch einen Schimpansen. Zeitschr. für Tierpsychologie. 24: 476-489. 88

Rietman, E. 1993. *Creating artificial life: Self-organization.* New York: Windcrest/McGraw-Hill.. 11

Runes, D. D. 1984. *Dictionary of philosophy.* Totowa, N.J.: Rowman and Allanheld. 19, 85

Sacks, O. 1992. The last hippie. New York Review of Books, March 26, 53-62. 218

Sapolsky, R. M., and J. C. Ray. 1989. Styles of dominance and their endocrine correlates among wild olive baboons (Papio Anubis). American Journal of Primatology 18: 1-13. 69

Sapolsky, R. M. 1987. Stress, social status, and reproductive physiology in free living baboons. In Psychobiology of reproductive behavior. D. Crews, ed. Englewood Cliffs, NJ Prentice Hall. 69

Sapolsky, R. M. 1990. Stress in the wild. *Scientific American,* 262: No. 195. 69

Sauer, F. 1954. Die Entwicklung der Lautäusserungen vom Ei ab schalldicht gehaltener Dorngrasmücken. Zeitschr. für Tierpsychologie. 11: 1-93. . . 44

Savage-Rumbaugh, S. 1986a. *Ape language.* New York: Columbia University Press.. 90

Savage-Rumbaugh, S. 1986b. Spontaneous symbol acquisition and communicative use by pygmy chimpanzees (Pan paniscus). *Journal of Experimental Psychology.* 115: 211-235. 92

Savage-Rumbaugh, S. 1988. A new look at ape language: Comprehension of vocal speech and syntax. In Nebraska Symposium on Motivation. Edited by D. Leger. Lincoln, Neb.: University of Nebraska Press. 93

Schank, R. C. 1990. *Tell me a story: A new look at real and artificial memory.* New York: Charles Scribner's Sons. 165

Schjelderup-Ebbe, T. 1922. Beiträge zur Sozialpsychologie des Haushuhns. Zeitschr. für Tierpsychologie. 88: 225-252.. 53

Schoeman, P. J. 1957. Hunters in the desert land. Cape Town: Howard Timmins. 213

Schönborn, F. von. 1993. Eugen Drewermann–Rebell oder prophet? Solothurn und Düsseldorf: Walter Verlag.................... XV

Searle, J. R. 1992. *The rediscovery of the mind*. Cambridge, Mass.: MIT Press.. 101, 145

Sejnowski, T., and C. Rosenberg. 1987. Parallel networks that learn to pronounce English text. Complex systems 1: 145-168........ 300, 301

Smith, A. [1776] 1976. *An inquiry into the nature and the causes of the wealth of nations*. R. H. Campbell and A. S. Skinner, eds. Oxford: Clarendon Press..................................... 263

Steiniger, F. 1950. Zur Soziologie und sonstigen Biologie der Wanderratte. Zeitschr. für Tierpsychologie. 7: 356-379...................... 55

Stevens, A. 1982. *Archetypes, a natural history of the self*. New York: William Morrow and Co. 170, 173

Stevens, A. 1993. *The two-million-year-old self*. College Station: Texas A. & M. University Press.................................. 170, 173

Stevens, A. 1996. *Private myths*. Cambridge, Mass.: Harvard University Press....................................... 170, 173, 248

Strum, S. C. 1987. *Almost human*. New York: Random House........ 66

Syrup-Neuloi. 1953. 70 Jahre Sozialversicherungsrecht. Bundesarbeitsblatt, a publication of the government of the German Federal Republic..... 258

Szentagothai, J. 1978. The neuron network of the cerebral cortex: A functional interpretation. Proceedings of Royal Society of London, Ser. B 201: 219-248.. 111

Tawney, R. H. 1926. *Religion and the rise of capitalism*. New York: Harcourt, Brace and Co. 257

Temerlin, M. K. 1972. Lucy: *Growing up human*. Palo Alto, Calif.: Science and Behavior Books. Inc.................................. 82, 88

Thompson, T. I. 1963. Visual reinforcement in Siamese fighting fish. *Science* 141: 55-57.. 53

Thompson, T. I. 1964. Visual reinforcement in fighting cocks. Journal of Experimental Animal Behavior. 7: 45-49...................... 53

Thompson, T. I. 1969. In. Aggressive behavior. Edited by S. Garattini and E. B. Amsterdam: Excerpta Medica Foundation, 15-31. 53

Tocqueville, A. de [1835] 1945. *Democracy in America, vol. 1,*. Edited by P. Bradley. New York: Vintage Books. 275

Tolstaya, T. 1991. In cannibalistic times. In New York review of books. Vol. 38., No. 7. 248

Treisman, A. M., and G. Gelade. 1980. A feature–Integration theory of attention. *Cognitive Psychology* 12: 97-136. 139

Turing, A. M. [1950] 1963. Computing machinery and intelligence. In Computers and Thought. E. A. Feigenbaum and J. Feldman, eds. New York: McGraw-Hill. 100

Uexküll, J. von. 1934. Streifzüge durch die Unwelten von Tieren and Menschen. Berlin: Springer Verlag. 47, 122

Ungerleider, L. G., and C. A. Chistensen. 1977. Pulvinar lesions in monkeys produce abnormal eye movements during visual discrimination training. Brain Research 136: 189-196. 132

van der Post, L. [1958] 1986. *The lost world of the Kalahari*. New York and London: Harcourt, Brace and Jovanovich. 191, 192, 194

Vollmer, G. 1983. Mesocosm and Objective Knowledge–Our Problems Solved by Evolutionary Epistemology. In: F.M. Wuketits (ed.): Concepts and approaches in evolutionary epistemology. Dordrecht: Reidel. . . . 95

Vollmer, G. 1985. Was können wir wissen? Band 1, Die Natur der Erkenntnis. Band 2 (1986), Die Erkenntnis der Natur, S. Hirzel Verlag, Stuttgart. 19, 95, 150

Waddington, C. H. 1960. *The ethical animal*. London: George Allen and Unwin, Ltd. 183

Wadsworth, B. J. 1979. Piaget's theory of cognitive development. New York: Longman Inc. 155

Walther, F. R. 1958. Zum Kampf und Paarungsverhalten einiger Antilopen. Zeitschr. für Tierpsychologie. 15: 340-380. 56

Washburn, S. L., and I. DeVore. 1961. The social life of baboons. *Scientific American* 204/6: 62-71. 54

Watson, A. 1966. Social status and population regulation in the red grouse (Lagopus lagopus scoticus). Royal Society Population Study Group. Proc. 2, Royal Society of London, 22-30. . . . . .52, 53

Watts, P. M. 1982. Nicolaus Cusanus, *A fifteenth century vision of man*. Leiden, Netherlands: E. J. Brill. 243

Weiskrantz, L., E. K. Warrington, M. D. Sanders, and J. Marshall. 1974. Visual capacity in the hemianopic field following a restricted occipital ablation. *Brain* 97: 709-728. 132